Plant Extracts
Applications in the Food Industry

Plant Extracts

Applications in the Food Industry

Edited by

Shabir Ahmad Mir

Department of Food Science and Technology, Government College for Women, Srinagar, India

Annamalai Manickavasagan

School of Engineering, College of Engineering and Physical Sciences, University of Guelph, ON, Canada

Manzoor Ahmad Shah

Department of Food Science and Technology, Government Degree College for Women, Anantnag, India

Academic Press is an imprint of Elsevier
125 London Wall, London EC2Y 5AS, United Kingdom
525 B Street, Suite 1650, San Diego, CA 92101, United States
50 Hampshire Street, 5th Floor, Cambridge, MA 02139, United States
The Boulevard, Langford Lane, Kidlington, Oxford OX5 1GB, United Kingdom

Notices

Knowledge and best practice in this field are constantly changing. As new research and experience broaden our understanding, changes in research methods, professional practices, or medical treatment may become necessary.

Practitioners and researchers must always rely on their own experience and knowledge in evaluating and using any information, methods, compounds, or experiments described herein. In using such information or methods they should be mindful of their own safety and the safety of others, including parties for whom they have a professional responsibility.

To the fullest extent of the law, neither the Publisher nor the authors, contributors, or editors, assume any liability for any injury and/or damage to persons or property as a matter of products liability, negligence or otherwise, or from any use or operation of any methods, products, instructions, or ideas contained in the material herein.

British Library Cataloguing-in-Publication Data
A catalogue record for this book is available from the British Library

Library of Congress Cataloging-in-Publication Data
A catalog record for this book is available from the Library of Congress

ISBN: 978-0-12-822475-5

For Information on all Academic Press publications
visit our website at https://www.elsevier.com/books-and-journals

Publisher: Charlotte Cockle
Acquisitions Editor: Nina Bandeira
Editorial Project Manager: Allison Hill
Production Project Manager: Vijayaraj Purushothaman
Cover Designer: Matthew Limbert

Typeset by MPS Limited, Chennai, India

Contents

List of contributors

Ravindra Kumar Agarwal
Department of Food Technology, Centre for Health and Applied Sciences, Ganpat University, Mehsana, India

Usha Antony
Department of Biotechnology, Centre for Food Technology, Anna University, Chennai, India; College of Fish Nutrition and Food Technology, Dr. J. Jayalalithaa Fisheries University, Chennai, India

Fazilah Ariffin
Food Biopolymer Research Group, Food Technology Division, School of Industrial Technology, Universiti Sains Malaysia, Malaysia

V. Geetha Balasubramaniam
Department of Biotechnology, Centre for Food Technology, Anna University, Chennai, India

Naseer Ahmad Bhat
Department of Food Science & Technology, University of Kashmir, Srinagar, India

Monika Choudhary
Department of Bioscience and Biotechnology, Banasthali Vidyapith, Jaipur, India

Kanica Chauhan
Department of Forest Products and Utilization, College of Horticulture and Forestry, Jhalawar, India

Lavanya Devraj
Department of Food Engineering, National Institute of Food Technology Entrepreneurship and Management (formerly Indian Institute of Food Processing Technology), Thanjavur, India

Saqib Farooq
Department of Food Technology, Islamic University of Science and Technology, Awantipora, India

Anju Jayachandran
Department of Fruit Science, College of Agriculture, Thrissur, India

Sunanda Joshi
Department of Bioscience and Biotechnology, Banasthali Vidyapith, Jaipur, India

A.A. Karim
Food Biopolymer Research Group, Food Technology Division, School of Industrial Technology, Universiti Sains Malaysia, Malaysia

Toiba Majeed
Department of Food Science & Technology, University of Kashmir, Srinagar, India

Annamalai Manickavasagan
School of Engineering, College of Engineering and Physical Sciences, University of Guelph, Guelph, ON, Canada

Nirmal Kumar Meena
Agriculture University, Kota, India; Department of Fruit Science, College of Horticulture and Forestry, Jhalawar, India

Vijay Singh Meena
NBPGR, Regional Station, Jodhpur, India

Manohar Meghwal
Department of Fruit Science, College of Agriculture, Thrissur, India

Shabir Ahmad Mir
Department of Food Science & Technology, Govt. College for Women, Srinagar, India

Subhrajyoti Mishra
Junagarh Agriculture University, Junagarh, India

Nikitha Modupalli
Department of Food Engineering, National Institute of Food Technology Entrepreneurship and Management (formerly Indian Institute of Food Processing Technology), Thanjavur, India

Abdorreza Mohammadi Nafchi
Food Biopolymer Research Group, Food Technology Division, School of Industrial Technology, Universiti Sains Malaysia, Malaysia; Food Biopolymer Research Group, Food Science and Technology Department, Damghan Branch, Islamic Azad University, Damghan, Iran

Venkatachalapathy Natarajan
Department of Food Engineering, National Institute of Food Technology Entrepreneurship and Management (formerly Indian Institute of Food Processing Technology), Thanjavur, India

Pradeep Singh Negi
Department of Fruit and Vegetable Technology, CSIR-Central Food Technological Research Institute, Mysore, India

Jyoti Nishad
Department of Food Technology, SRCASW, University of Delhi, New Delhi, India

Nazila Oladzadabbasabadi
Food Biopolymer Research Group, Food Technology Division, School of Industrial Technology, Universiti Sains Malaysia, Malaysia

Havalli Bommegowda Rashmi
Department of Fruit and Vegetable Technology, CSIR-Central Food Technological Research Institute, Mysore, India

Chagam Koteswara Reddy
Department of Biochemistry and Bioinformatics, Institute of Science, GITAM (Deemed to be University), Visakhapatnam, India; Department of Food Technology, Centre for Health and Applied Sciences, Ganpat University, Mehsana, India

Manzoor Ahmad Shah
Department of Food Science & Technology, Govt. Degree College for Women, Anantnag, India

Gauri Singhal
Department of Biotechnology, School of Medical and Allied Sciences, Sanskriti University, Mathura, India

Nidhi Srivastava
Department of Biotechnology, National Institute of Pharmaceutical Education and Research (NIPER), Lucknow, India

Sudha Rani Ramakrishnan
Department of Biotechnology, Centre for Food Technology, Anna University, Chennai, India

M. Suriya
Centre for Food Technology, Anna University, Chennai, India

M. Verma
SK Rajasthan Agriculture University, Bikaner, India

Vartika Verma
Department of Bioscience and Biotechnology, Banasthali Vidyapith, Jaipur, India

CHAPTER

Sources of plant extracts

1

Shabir Ahmad Mir[1], Manzoor Ahmad Shah[2] and Annamalai Manickavasagan[3]

[1]*Department of Food Science & Technology, Govt. College for Women, Srinagar, India* [2]*Department of Food Science & Technology, Govt. Degree College for Women, Anantnag, India* [3]*School of Engineering, College of Engineering and Physical Sciences, University of Guelph, Guelph, ON, Canada*

1.1 Introduction

There is a great demand for natural ingredients in the food industry. Plant materials are good sources of natural ingredients with different functionalities, and these can be used for various applications. The advantages of plant materials are their abundance in nature and wide geographic distribution. Plant extracts are concentrated bioactive compounds with potential antimicrobial, antioxidant, and several other beneficial properties. Plant extracts vary in their properties because of structural diversity and complexity based on their source.

Researchers and industries are working on conventional and emerging extraction techniques for obtaining the extracts from plant material with highest yield, stability, and efficacy. Various types of plants are valuable sources of these extracts such as fruits, vegetables, plantation crops, cereals, spices, and herbs. Plant extracts are obtained from different parts of plants such as leaves (Hauser et al., 2016; Nouri & Nafchi, 2014; Rodríguez, Sibaja, Espitia, & Otoni, 2020), fruits (Ferysiuk, Wójciak, Materska, Chilczuk, & Pabich, 2020; Wang, Marcone, Barbut, & Lim, 2012), seeds (Guo, Huang, Chen, Hou, & Huang, 2020; Reddy et al., 2013; Tan, Lim, Tay, Lee, & Thian, 2015), peels (Aliyari, Bakhshi Kazaj, Barzegar, & Ahmadi Gavlighi, 2020; Zhang, Li, & Jiang, 2020), roots (Hedayati et al., 2021; Jaśkiewicz, Budryn, Nowak, & Efenberger-Szmechtyk, 2020; Norajit, Kim, & Ryu, 2010), and flowers (Khan et al., 2019; Samsudin, Soto-Valdez, & Auras, 2014) (Fig. 1.1).

In the recent years, there has been an increasing demand for natural additives in the food industry. Market trend has been changed due to the healthy lifestyle and consumer awareness about the additives. Plant extracts are simple and safe and have been effectively utilized in the food industry for different applications. They are used as antioxidants, antimicrobial agents, flavoring agents, coloring agents, enzymes, nutrient enhancers, and packaging additives.

Plant Extracts: ***Applications in the Food Industry*****. DOI: https://doi.org/10.1016/B978-0-12-822475-5.00011-9**

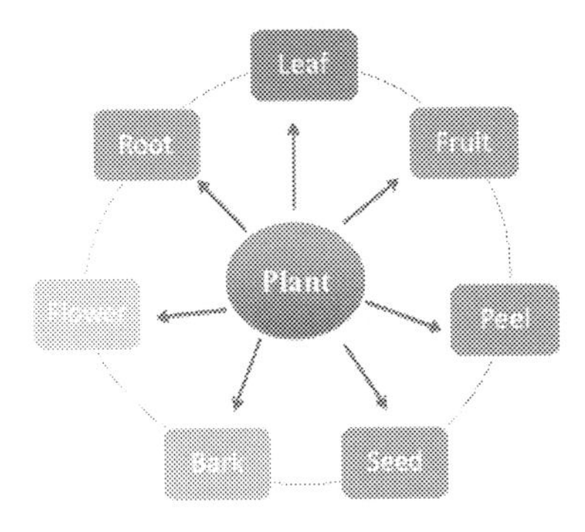

FIGURE 1.1

Plant parts used for the preparation of extracts.

1.2 Plant parts used for the preparation of extracts

1.2.1 Leaf

Leaves are important parts of plant responsible for the photosynthesis and consequent production of oxygen. According to various literature, leaves are rich sources of health beneficial bioactive compounds. Leaves contain carotenoids and anthocyanins responsible for the yellow to orange to red color of leaves during autumn in addition to chlorophyll, which is responsible for green color. These color compounds of leaves are responsible for the rich antioxidant capacity of leaves. Additionally, some leave extracts posses antimicrobial activity due to the presence of flavonoids (Botsoglou, Govaris, Ambrosiadis, Fletouris, & Botsoglou, 2014; Han & Song, 2021a). Leaf extract possess more bioactive compounds as compared to stem, seed, and bark of a tree. Leaf extracts have gained popularity for use in different types of food applications. Leaf extracts have been used as preservative, coloring and antimicrobial agent.

Various studies have indicated the use of leaf extracts for different types of food products (Table 1.1). Lotus leaf is a potential source of phytochemicals and is used in ground pork (Choe et al., 2011). Lotus leaves are rich in polyphenols, flavonoids, and alkaloids (Wang et al., 2021). Amaranthus leaf is rich in betalains and other active compounds. This leaf extract is used in packaging films for shelf life enhancement of chicken and fish meat during storage (Kanatt, 2020). Mango leaf extract is rich in bioactive components like gallic acid, mangiferin, glucosides, and other phenolic compounds (Rambabu, Bharath, Banat, Show, & Cocoletzi, 2019). This extract is incorporated into chitosan-based packaging films for food applications.

Mint leaf extract was used for lamb meat preservation by Kanatt, Chander, and Sharma (2008). Mint leaf extracts are excellent source of phenolic and flavonoid compounds. Mint leaves help to prevent the lipid oxidation and improve the taste and flavor of meat. The leaf extracts of myrtle, rosemary, lemon balm, and nettle were used in beef patties to reduce lipid oxidation and enhance the shelf life (Akarpat, Turhan, & Ustun, 2008). Olive leaf extract used in pork patties delayed the protein oxidation and lipid oxidation of conjugated dienes, hydroperoxides, and malondialdehyde (Botsoglou et al., 2014). Leaf extract obtained from cork oak (*Quercus suber* L.) was used as a natural antioxidant in cooked meat (Lavado, Ladero, & Cava, 2021). *Moringa oleifera* leaf extracts

Table 1.1 Application of leaf extracts in the food industry.

Plant	*Genus*	*Species*	Uses	Application industries	Remarks	References
Green tea	*Camellia*	*Sinesis*	Antioxidant	Packaging films	Increased mechanical, water vapor barrier and antioxidant properties	Siripatrawan and Harte (2010)
			Antioxidant antimicrobial	Beef patties	Delayed microbial spoilage, redness loss and lipid oxidation	Banon et al. (2007)
			Antioxidant	Goat meat	Reduces lipid oxidation during storage	Rababah et al. (2011)
			Antioxidant	Pork meat	Inhibit lipid oxidation	Wójciak et al. (2011)
Rosemary	*Salvia*	*Rosmarinus*	Antioxidant	Packaging film for peanuts	Inhibits lipid oxidative rancidity in peanuts	Wambura, Yang, and Mwakatage (2011)
			Antioxidant	Pork sausage	Shelf life enhancement of sausage	Sebranek, Sewalt, Robbins, and Houser (2005)
			Antioxidant	Beef patties	Reduced lipid oxidation	Akarpat et al. (2008)
			Antioxidant	Pork meat	Inhibit lipid oxidation	Wójciak et al. (2011)
Nettle	*Urtica*	*Dioica*	Antimicrobial, Antioxidant	Ground beef	Inhibition of *Pseudomonas* and psychrotrophic bacteria in the stored ground beef Inhibition of ipid oxidation	Alp and Aksu (2010)
			Antioxidant	Fermented sausage	reduced TBARS value and biogenic amine formation	Karabacak and Bozkurt (2008)
Olive	*Olea*	*Europea*	Antioxidant	Pork patties	Reduced lipid oxidation in cooked pork patties	Hayes, Stepanyan, O'Grady, Allen, and Kerry (2010)
			Antioxidant	Pork sausages	Reduced lipid oxidation in raw and cooked pork sausages	Hayes, Stepanyan, Allen, O'grady, and Kerry (2011)
			Antioxidant	Pork patties	Extract significantly inhibit protein and lipid oxidation in patties	Botsoglou et al. (2014)
			Antioxidant	Packaging film	Films exhibit good barrier and mechanical properties	da Rosa, Vanga, Gariepy, and Raghavan (2020)
			Antimicrobial	Lamb meat	Potential antimicrobial activity during the storage of lamb meat	Martiny, Raghavan, Moraes, Rosa, and Dotto (2020)

(Continued)

Table 1.1 Application of leaf extracts in the food industry. ***Continued***

Plant	*Genus*	*Species*	Uses	Application industries	Remarks	References
Mint	*Mentha*	*Spicata*	Antioxidant Antimicrobial	Meat	Potential antimicrobial and antioxidant agent improving the shelf life of meat	Kanatt et al. (2008)
			Antioxidant	Pork meat	Minimize lipid oxidation in pork products	Biswas, Chatli, and Sahoo (2012)
Drumstick	*Moringa*	*Oleifera*	Antioxidant	Goat meat patties	Protect goat meat patties against oxidative rancidity	Das et al. (2012)
			Antioxidant	Pork patties	Inhibit lipid oxidation in ground pork patties	Muthukumar et al. (2014)
			Antioxidant	Raw beef	Significantly reduced the TBARS value of modified atmosphere packaged raw beef	Shah et al. (2015)
Betel	*Piper*	*Betel*	Antimicrobial	Packaging films	Inhibitory activity against the growth of Gram positive and Gram negative bacteria, Barrier properties of packaging film such as ultraviolet protection, water vapor and oxygen permeation increased	Nouri and Nafchi (2014)
Murta	*Ugni*	*Molinae*	Antimicrobial Antioxidant	Packaging films	Enhanced antimicrobial and antioxidant properties of packaging films	Hauser et al. (2016)
Ginkgo	*Gingko*	*Biloba*	Antioxidant	Packaging films	Excellent antioxidant properties of packaging films, improved barrier properties against UV light and moisture	Li, Miao, Wu, Chen, and Zhang (2014)
Curry	*Murraya*	*Koenigii*	Antioxidant	Pork meat	Minimize lipid oxidation in pork products	Biswas et al. (2012)
Lotus	*Nelumbo*	*Nucifera*	Antioxidant	Porcine and bovine ground meat	Inhibiting meat oxidation	Huang et al. (2011)
Myrtle	*Myrtus*	-	Antioxidant	Beef patties	Reduced lipid oxidation	Akarpat et al. (2008)
Satureja	*Satureja*	*Thymbra*	Antioxidant	Packaging for potato chips	Protects fried potato chips against oxidation	Choulitoudi, Velliopoulou, Tsimogiannis, and Oreopoulou (2020)
Amaranth	*Amaranthus*	*Tricolor*	Antioxidant	Chicken and fish	Inhibit bacterial growth and oxidative rancidity	Kanatt (2020)
Mango	*Mangifera*	*Indica*	Antioxidant Antimicrobial	Packaging films	Extract based films showed good mechanical, UV-light barrier, antioxidant and antimicrobial properties	Bastante et al. (2021)
Blueberry	*Vaccinium*	*Corymbosum*	Antioxidant	Packaging films	Enhanced antioxidant activity of packaging films	Han and Song (2021b)
Cork oak	*Quercus*	*Suber*	Antioxidant	Meat	Inhibit lipid oxidation in cooked meat	Lavado et al. (2021)

were applied in meat for shelf life enhancement by Das, Rajkumar, Verma, and Swarup (2012), Muthukumar, Naveena, Vaithiyanathan, Sen, and Sureshkumar (2014) and Shah, Bosco, and Mir (2015). These leaf extracts are rich source of phenolic compounds, which reduce lipid and protein oxidation in meat and meat products. The barley leaf extract is effective for retarding lipid oxidation in ground pork (Choe et al., 2011).

Tea extract is one of the most popular leaf extracts used for food applications due to the abundance of active compounds. The predominant bioactive components in tea extract are flavanols, flavonoids, phenolic acids, etc. These extracts are widely applied in meat and other food products to delay the lipid oxidation and to enhance their shelf-life (Banon, Díaz, Rodríguez, Garrido, & Price, 2007; Rababah et al., 2011; Wójciak, Dolatowski, & Okoń, 2011). Roasted coffee extracts are rich sources of antioxidants and effective for lowering the lipid oxidation in beef (Lin, Toto, & Were, 2015).

Herb extracts are excellent source of active compounds, which improve the quality of food products. Extract of rosemary, sage, and marjoram were used to minimize the lipid oxidation, enhance the color and sensory properties of irradiated ground beef (Mohamed, Mansour, & Farag, 2011). Extracts of peony, sappanwood, rehmania, and angelica with antioxidant properties were used in the meat industry (Han & Rhee, 2005). These herbal extracts effectively reduced lipid oxidation and improved the quality of ground meat. Epazote herb extracts were used as antioxidants in ground beef by Villalobos-Delgado, González-Mondragón, Ramírez-Andrade, Salazar-Govea, and Santiago-Castro (2020). The epazote extract effectively prevented the lipid oxidation and enhanced the sensory attributes of beef.

Plant extracts have been used as an important ingredient in active packaging films. Bastante et al. (2021) impregnated the mango leaf extract into packaging films and showed the antimicrobial activity against gram positive and gram negative bacteria. Han and Song (2021b) developed the active packaging material by incorporating the blueberry leaf extract into polysaccharide based films. Sago starch film incorporated with betel leaf extract showed inhibitory activity against the bacteria and also improved mechanical properties of developed packaging films (Nouri & Nafchi, 2014). Coffee and cocoa extracts were used in polylactide films for food packaging applications (Moraczewski et al., 2019). Tea extracts are used as antimicrobial and antioxidant agents in active packaging films (Siripatrawan & Harte, 2010).

1.2.2 Fruits and vegetables

Fruits are diverse in nature and grown in different geographical regions such as temperate, tropical, sub-tropical, and arid zones of the world. Fruits are rich sources of different types of functional compounds. Extracts are obtained from different sources and varieties of fruits. These extracts are rich sources of phenolic compounds and include phenolic acids, stilbenes, lignans, flavonoids, and tannins or proanthocyanidins. The most abundant phenolic compounds in the fruits are phenolic acids such as derivatives of benzoic and cinnamic acids and flavonoids. These special compounds of fruit extracts have considerable applications in food industry. Extracts are obtained from different fruits such as grape, blueberry, raspberry, pomegranate, murta, chestnut, etc. have been prepared by various extraction technologies and are used in different types of food products such as meat, cheese, poultry and fish (Table 1.2).

Table 1.2 Application of fruit extracts in the food industry.

Plant	*Genus*	*Species*	Uses	Application industries	Remarks	References
Black mulberry	*Morus*	*Nigra*	Antioxidant Antimicrobial	Beef patties	Lipid oxidation and microbial growth decreased during storage, improving redness of patties	Turan and Şimşek (2021)
Chestnut	*Castanea*	*Sativa*	Antioxidant Antimicrobial	Packaging films for pasta	Prevents microbial spoilage of pasta	Kõrge, Bajić et al. (2020)
Chestnut	*Castanea*	*Sativa*	Antimicrobial	Cheese	Potential antimicrobial properties	Kõrge, Šeme, Bajić, Likozar, & Novak, (2020)
Jabuticaba	*Myrciaria*	*Jaboticaba*	Antioxidant Color enhancer	Bakery products	Enhances the nutritional value and color of product	Albuquerque et al. (2020)
Hawthorn	*Crataegus*	*Pinnatifida*	Antioxidant	Beef and chicken breast meat	Inhibition of heterocyclic aromatic amine formation at higher temperature	Tengilimoglu-Metin et al. (2017)
Murta	*Ugni*	*Molinae*	Antioxidant Antimicrobial	Packaging films	Improved mechanical, antioxidant and antimicrobial properties	de Dicastillo et al. (2016)
Red raspberry	*Rubus*	*Strigosus*	Packaging aid	Packaging films	Enhanced tensile strength and percent elongation at break	Wang et al. (2012)
Blueberry	*Vaccinium*	*Vitis-idaea*	Antioxidant	Lard	Delayed the oxidation of packaged lard	Zhang et al. (2010)
Carob	*Ceratonia*	*Siliqua*	Antioxidant	Cooked pork	Reduce fat oxidation	Bastida et al. (2009)

Howthorn fruit extracts were used in beef and chicken breast meat by Tengilimoglu-Metin, Hamzalioglu, Gokmen, and Kizil (2017). Howthorn extracts reduced the formation of carcinogenic heterocyclic aromatic amine in meat. Acorn fruit extract exhibited the potential antioxidant activity in raw chicken meat during refrigerated storage (Özünlü, Ergezer, & Gökçe, 2018). This extract inhibited the lipid oxidation and also enhanced the flavor of the chicken meat. Black mulberry fruit extracts were used in beef patties stored under aerobic and vacuum packaged conditions (Turan & Şimşek, 2021). The black mulberry extracts were rich in phenolics and anthocyanins and showed remarkable effect on antioxidant, antimicrobial and color properties of patties. Mulberry extract decreased the lipid oxidation and microbial spoilage and prevented the formation of metmyoglobin and improved the redness in patties. Red pitaya fruit extract acted as an antioxidant and natural colorant in pork patties (Bellucci, Munekata, Pateiro, Lorenzo, & da Silva Barretto, 2021). Bastida et al. (2009) used carob fruit extract in cooked pork.

Blueberry fruit extract is used in active packaging films (Zhang et al., 2010). Raspberry fruit extract is used in soy protein based packaging films (Wang et al., 2012). The murta fruit extract is incorporated in methyl cellulose based films (de Dicastillo, Bustos, Guarda, & Galotto, 2016). Kõrge, Bajić, Likozar, and Novak (2020) used chestnut fruit extract in chitosan films for shelf life enhancement of pasta.

Vegetables are also used for the production of plant extracts. Vegetables are rich in phenolic compounds, flavonoid glycosides, carotenoids, tannins, vitamins C and E. Furthermore, attention has been given to the extracts from waste material of vegetables, which are also good sources of bioactive compounds. Vegetable-based extracts have widely used in the food industry such as meat products, fish products, and poultry products, etc. Chamnamul (*Pimpinella brachycarpa*), fatsia (*Aralia elata*), chinese chives (*Allium tuberosum*), broccoli (*Brassia oleracea*) exhibit signifiant antioxidant and antimicrobial properties. These extracts were used as prevertaive for the shelf life enhancement of raw beef patties (Kim, Cho, & Han, 2013). The extract of broccoli are rich in phenolic antioxidants which have excellent radical scavanning capacity. The extract significantly reduced the lipid peroxidation and improved the shelf life of goat meat nuggets (Banerjee et al., 2012). Potato extracts used in fresh and precooked beef patties improved their shelf life stability and sensory characteristics (Colle et al., 2019). Red pepper extract was used in cooked pork by Wójciak et al. (2011). Sweet pepper extract was used in refrigerated pork by Ferysiuk et al. (2020).

1.2.3 Peel and skin

Peel is considered as a waste of the food processing and represents a good percentage and an excellent source of photochemical compounds. Peel is superior in antioxidant potential as compared to pulp. Peel extracts have potential application as antimicrobial and antioxidant agents in various food products (Table 1.3). Kinnow and pomegranate peel extracts were applied to goat meat patties (Devatkal, Narsaiah, & Borah, 2010). Pomegranate peel extract was used in ground goat meat and nuggets (Devatkal, Thorat, & Manjunatha, 2014). Kanatt, Rao, Chawla, and Sharma (2012) used pomegranate peel extract into chitosan based films. Potato peel contains phenolic compounds and glycoalkaloids which showed the potential applications in food industry (Sampaio et al., 2020). Pitaya peel extracts were used to improve the oxidative stability of refrigerated ground pork patties (Cunha et al., 2018). Chitosan packaging films were incorporated with banana peel extract by

Table 1.3 Application of peel extracts in the food industry.

Plant	*Genus*	*Species*	Uses	Application industries	Remarks	References
Pomegranate	*Punica*	*Granatum*	Antibacterial Antioxidant	Raw pork	Reduce numbers of pathogenic bacteria, color degradation and lipid oxidation in raw pork	Shan, Cai, Brooks, and Corke (2009)
			Antioxidant	Ground goat meat and nuggets	Significantly reduced TBARS value	Devatkal et al. (2014)
			Packaging aid	Packaging films	mechanical properties, antioxidant properties and antibacterial properties considerably increased	Moghadam, Salami, Mohammadian, Khodadadi, and Emam-Djomeh (2020)
			Antimicrobial	Packaging films	Antimicrobial activity of extract were more pronounced against gram-positive as compared to gram-negative bacteria	Emam-Djomeh, Moghaddam, and Yasini Ardakani (2015)
			Antimicrobial Antioxidant	Sausages	Reduces lipid oxidation and microbial growth	Aliyari et al. (2020)
Blood orange	*Citrus*	*Sinensis*	Antimicrobial Antioxidant	Cheese	Enhancing the microbial stability of cheese during storage, improving nutritional value of cheese	Jridi et al. (2020)
Kiwifruit	*Actinidia*	*Chinensis*	Antioxidant	Chicken meat	Retarded lipid oxidation in chicken thigh meat	Han and Song (2021a)
Banana	*Musa*	*Paradisiaca*	Antioxidant	Packaging films	Films exhibit excellent antioxidant activity	Zhang et al. (2020)
Pitaya	*Hylocereus*	*Costaricensis*	Antioxidant	Pork patties	Enhanced oxidative stability	Cunha et al. (2018), Cunha et al. (2021)

Zhang et al. (2020). The developed films showed excellent antioxidant activity and enhanced the shelf life of apple fruit during storage.

Yu, Ahmedna, and Goktepe (2010) utilized the peanut skin extract for raw and cooked ground beef for shelf life enhancement. They reported that peanut skin extracts inhibit lipid oxidation and enhance the shelf life of beef. The addition of rice bran extracts in meat products have been reported to increase the nutritional value, retard the oxidative degradation, and increase the health promoting properties (Kim, 2000). Hao and Beta (2012) incorporated the barley hull and flaxseed hull extracts into Chinese steamed bread. These extracts significantly increased the phytochemical profile and antioxidant capacity of developed bread. Rye and wheat bran extracts containing phenolic compounds were tested in beef hamburgers (Šulniūtė, Jaime, Rovira, & Venskutonis, 2016). The bran extracts showed promising results by inhibiting the production of oxidation products, hexanal and malondialdehyde during storage. Furthermore, these extracts also enhanced the nutritional profile of beef hamburgers.

1.2.4 Seeds

Seed is one of the inedible parts of fruits and vegetables, which is considered as a waste in the food industry. Nowadays, seed extracts have been widely used in food products as a source of natural ingredients in the food industry, as these are rich sources of diverse bioactive compounds (Table 1.4). Grape seeds have been used for wide applications in the food industry. Grape seeds extracts are rich in phenolic compounds such as catechin, epicatechin, proanthocyanidin, gallate, flanonols, etc. (Chen et al., 2020). Grape seed extract was used for shelf life enhancement of meat (Reddy et al., 2013), sausage (Jayawardana et al., 2011) and Suisse cheese (Deolindo et al., 2019) due to its potential antioxidant properties. Tajik, Farhangfar, Moradi, and Razavi Rohani (2014) investigated the effect of grape seed extract on lipid oxidation of buffalo meat patties during storage. Grape seed extract were used in restructured mutton for shelf life enhancement stored at refrigerated temperatures by Reddy et al. (2013). This extract was also evaluated in roasted chicken stored under modified atmospheric packaging (Guo et al., 2020). Grape seed extract contains potential antioxidant properties which inhibit the lipid oxidation and was used in raw and cooked ground muscle during different storage conditions (Brannan & Mah, 2007). This extract was also used in frankfurters (Özvural & Vural, 2012). The grape fruit seed extract was incorporated in carrageenan-based packaging films by Kanmani and Rhim (2014) and Tan et al. (2015). Sivarooban, Hettiarachchy, and Johnson (2008) used the grape seed extract in soy protein based films.

Baobad seed extracts were used to assess the qualities of beef patties (Al-Juhaimi et al., 2020). Baobad extract showed good antioxidant potential and antibacterial properties, which help to improve the storage stability of beef patties. Guarana seed extract is a power source of phenolic antioxidant used in food industry. Guarana extract preserves the color and also protects the patties against lipid and protein oxidation (Pateiro et al., 2018). Adzuki bean extract has been utilized for cured and uncured cooked pork sausages (Jayawardana et al., 2011). The extract effectively reduced the lipid oxidation and results are par with the artificial antioxidants.

Spice seeds posses excellent antioxidant activity because of the presence of flavonoids, carotenoids, terpenoids, polyphenolics, coumarins, and curcumins. Zhang, Kong, Xiong, and Sun (2009) used extracts of pepper, aniseed, fennel, cardamom etc. in modified and vacuum packaged pork. The extracts of these spices was capable of inhibiting the spoilage causing microorganisms such as

Table 1.4 Application of seed extracts in food industry.

Plant	*genus*	*species*	uses	Application industries	Remarks	References
Grape	*Vitis*	*Vinifera*	Antimicrobial	Active packaging films	Extract increased the thickness, puncture and tensile strength of packaging films Showed inhibitory activity against microbes	Sivarooban et al. (2008)
			Antioxidant	Active packaging films	Extract improved the barrier to UV light and moisture of packaging films, Films posses excellent antioxidant activity	Li et al. (2014)
			Antioxidant	Refrigerated cooked beef and pork patties	Reduces oxidative rancidity	Rojas and Brewer (2007)
			Antioxidant	Frozen, vacuum packaged beef and pork patties	Reduces oxidative rancidity	Rojas and Brewer (2008)
			Antioxidant	Beef patties	Reduces oxidative rancidity	Colindres and Susan Brewer (2011)
			Antioxidant Flavor enhancer	Beef sausage	Reduces oxidative rancidity	Kulkarni, DeSantos, Kattamuri, Rossi, and Brewer (2011)
			Antioxidant Color enhancer	Raw and cooked goat meat	Reduces lipid oxidation	Rababah et al. (2011)
			Antioxidant	Frankfurters	Reduces oxidative rancidity	Özvural and Vural (2012)
			Antioxidant Antimicrobial	Restructured mutton slices	Reduces oxidative rancidity	Reddy et al. (2013)
			Antioxidant	Raw and cooked ground muscle	Reduces oxidative rancidity	Brannan and Mah (2007)
			Antioxidant	Mayonnaise	Reduces oxidative rancidity	Altunkaya et al. (2013)
			Antioxidant	Prevents shrimp discoloration	Inhibition melanosis in shrimp	Gokoglu and Yerlikaya (2008)
			Antimicrobial, flavor and color enhancer	Fresh produce	Modified color and flavor	Hollis et al. (2020)

Grapefruit	*Citrus*	*Paradisi*	Antimicrobial	Packaging films	Improved antimicrobial activity of packaging film	Lim, Jang, and Song (2010)
			Antimicrobial	Packaging films	Bacteriocidal and bacteriostatic capacity against different food borne pathogens	Wang, Lim, Tong, and San Thian (2019)
			Antimicrobial	Packaging films	Strong antimicrobial activity against various Gram-positive and Gram-negative food borne pathogens	Kanmani and Rhim (2014)
			Antimicrobial	Packaging films	Antimicrobial activity of packaging film. Extract increased the elongation at break of films	Tan et al. (2015)
			Antimicrobial Antioxidant	Salmon packaging	Increased shelf life of packaged salmon	Song, Shin, and Song (2012)
			Antimicrobial	Strawberry packaging	Extract increased the shelf life of packaged strawberries	Jang, Shin, and Song (2011)
Cumin	Nigella	Sativa	Antioxidant	Beef patties	Reduces oxidative rancidity	Rahman et al. (2021)
Guarana	*Paullinia*	*Cupana*	Antioxidant Antimicrobial Enhances color	Pork patties	Enhances shelf life, Protect patties against protein and lipid oxidation	Pateiro et al. (2018)
Baobab	Adansonia	Digitata	Antimicrobial Antioxidant	Beef patties	Improves lipid and microbial stability	Al-Juhaimi et al. (2020)

Listeria monocytogenes, Escherichia coli, and *Pseudomonas fluorescens* in pork. Black cumin extract used in beef patties shows excellent antioxidant activity and enhances the shelf life during refrigerated storage (Rahman, Alam, Monir, & Ahmed, 2021).

1.2.5 Flowers

Many species of flowers are considered as more than a delicacy or a garnish due to their nutritional value and the presence of special compounds. Flowers have attained significant attention during the recent years in food industry. Floral extracts contain abundant active compounds exhibiting rich bioactivity and promising health beneficial properties. Flower extracts are endowed with antioxidant, antimicrobial, color, and flavoring properties. Flower extracts have been incorporated in many products such as meat, beverages, packaging films, sausages etc. (Table 1.5)

Chrysanthemum morifolium flower extract was incorporated in goat meat patties cooked at different temperature (Khan et al., 2019). The flower extracts are effective for decreasing the heterocyclic amines in deep fried patties. Karabacak and Bozkurt (2008) utilized the *Urtica dioica* and *Hibiscus sabdariffa* L. flowers extracts in Turkish fermented sausage. Litchi flower extracts were found to be rich source of antioxidant used in pork meatballs significantly reduced the lipid and protein oxidation in frozen cooked meat products (Ding, Wang, Yang, Chang, & Chen, 2015). The incorporation of marigold flower extract into packaging film was found to be effective for enhancing the oxidative stability of fatty foods (Samsudin et al., 2014). Wang, Wang, Tong, and Zhou (2017) explored the use of honeysuckle flower extract in chitosan films for active food packaging. Extract-based films showed the potential application in food industry. Pullulan packaging films were incorporated with meadow sweet flower extracts. These extracts have excellent antimicrobial properties, which improve the shelf life and market value of apples (Gniewosz et al., 2014).

1.2.6 Barks

Tree bark is a major byproduct produced from the pulp and paper industries as well as during the processing of wood. Bark contains higher amounts of phenolic compounds as compared to wood, which plays an important role in the defense mechanisms. Limited studied are available on the exploitation of tree bark extract for food applications (Table 1.6). Pine bark extract is a source of flavanols and flavonoids such as catechin and taxifolin and have been used as antioxidant in meat industry (Mármol, Quero, Jiménez-Moreno, Rodríguez-Yoldi, & Ancín-Azpilicueta, 2019). Han, Yu, and Wang (2018) observed that pine bark extract is a good source of oligomeric procyanidins and has excellent antioxidant activity. These authors used this extract into packaging films which showed the antimicrobial and improved barrier properties. Edible films were prepared from hydroxypropyl methylcellulose and K-carrageen and incorporated with cork bark extract (Zhang, Sun, Li, & Wang, 2021). The developed packaging films incorporated with bark extract were endowed with potential antimicrobial and antioxidant properties. Tayel and El-tras (2012) used cinnamon bark extract in ground beef. Antimicrobial and antioxidant properties of *Citrus paradise* fruit bark extracts were investigated in turkey sausage. The results demonstrate that bark extracts reduced the lipid oxidation and microbial storage of turkey sausage during storage (Sayari et al., 2015).

Table 1.5 Application of flower extracts in food industry.

Plant	*Genus*	*Species*	Uses	Application industries	Remarks	References
Litchi	*Litchi*	*Chinensis*	Antioxidant	Pork meatballs	Reduced lipid peroxidation and protein degradation in meat balls	Ding et al. (2015)
Marigold	*Tagetes*	*Erecta*	Antioxidant	Fatty foods	Reduces oxidative rancidity	Samsudin et al. (2014)
Meadowsweet	*Filipendulae*	*Ulmariae*	Antimicrobial	Apple fruit	Protects fruit from bacterial and mold growth	Gniewosz et al. (2014)
Bitter orange	*Citrus*	*Aurantium*	Antimicrobial	Rice pudding	Potential preservative against food spoilage microorganisms	Degirmenci and Erkurt (2020)
Florist's daisy	*Chrysanthemum*	*Morifolium*	Preservative	Goat meat patties	Reduces formation of heterocyclic amines in pan and deep fat frying	Khan et al. (2019)
Broccoli	*Brassica*	*Oleracea*	Antioxidant Nutrient enhancer	Goat meat nuggets	Reduced TBARS value, improves phenolic content	Banerjee et al. (2012)
Hibiscus	*Hibiscus*	*Sabdariffa*	Antioxidant	Fermented sausage	Reduced TBARS value and biogenic amine formation	Karabacak and Bozkurt (2008)

Table 1.6 Application of bark extracts in food industry.

Plant	*Genus*	*Species*	Uses	Application industries	Remarks	References
Pine bark	*Pinus*	*Densiflora*	Antioxidant	Meat	Inhibit protein and lipid oxidation in meat	Mármol et al. (2019)
			Antioxidant	Packaging films	Extract improved thermal stability, water and oxygen barrier properties of films	Han et al. (2018)
Grapefruit	*Citrus*	*Paradisi*	Antioxidant Antimicrobial	Sausage	Minimize lipid oxidation and improve microbial stability	Sayari et al. (2015)
Cork bark	*Quercus*	*Suber*	Packaging aid	Packaging films	Enhanced flexibility and light resistance of films, improved antioxidant and antimicrobial properties	Zhang et al. (2021)
Cinnamon	Cinnamomum	Verum	Antimicrobial Antioxidant	Raw pork	Reduce number of pathogenic bacteria and lipid oxidation in raw pork	Shan et al. (2009)
			Antioxidant	Chicken meat	Improves oxidative stability and redness of chicken meat balls	Chan et al. (2014)

1.2.7 Roots

Roots have been studied for their potential health benefits as they are rich source of phytochemical compounds. Root extracts have been also used in various food applications (Table 1.7). Chinese chive (*A. tuberosum*) root extract is a good source of polyphenols and antimicrobial compounds (Riaz et al., 2020). Chive root extract has been used in packaging films in the food industry. The incorporation of liquorice extract into biocomposite films prevents the food from lipid oxidation (Singh, Agrawal, Mendiratta, & Chauhan, 2021). Chicory root extract shows potential antimicrobial activity in biodegradable starch films (Jaśkiewicz et al., 2020). The utilization of lotus root extract in the pork patties led to the inhibition of lipid oxidation (Shin, Choe, Hwang, Kim, & Jo, 2019).

Table 1.7 Application of root extracts in food industry.

Plant	*Genus*	*Species*	Uses	Application industries	Remarks	References
Chinese chive	*Allium*	*Tuberosum*	Antimicrobial Antioxidant	Packaging films	Extract improved physical properties, decreased water vapor permeability, films exhibited good antioxidant and antimicrobial property	Riaz et al. (2020)
Beetroot	*Beta*	*Vulgaris*	Antioxidant	Packaging films	Protect polymer against photo-degradation, plasticizing property	Akhtar et al. (2012)
Carrot	*Daucus*	*Carota*	Antioxidant	Packaging films	Protect polymer against photo-degradation, plasticizing property	Akhtar et al. (2012)
Licorice	*Glycyrrhiza*	*Glabra*	Antioxidant antimicrobial	Packaging films for chhana balls	Protects against microbial spoilage and lipid oxidaiton	Singh et al. (2021)
Hooker chives	*Allium*	*Hookeri*	Preservative	Pork patties	Retards lipid oxidation	Cho et al. (2015)
Lotus	*Nelumbo*	*Nucifera*	Antioxidant	Pork patties	Reduces oxidative rancidity	Shin et al. (2019)
Kudzu	*Pueraria*	*Radix*	Nutrient enhancer	Beef patties	Increased nutritive value of patties	Kumari, Raines, Martin, and Rodriguez (2015)
Chicory	*Cichorium*	*Intybus*	Antimicrobial	Packaging films	Broad-spectrum inhibitory effect on microbial growth	Jaśkiewicz et al. (2020)

Beetroot is known as a health promoting food due to the presence of special compounds such as phenolics, carotenoids, vitamins, ascorbic acids, and betalains that promote health. The beetroot extract is used as a coloring and antioxidant agent in different types of food products such as jams, jellies, breakfast cereals, ketchups, dairy products, candy, etc. (Chhikara, Kushwaha, Sharma, Gat, & Panghal, 2019). Akhtar et al. (2012) exploited beetroot and carrot extracts in the hydroxypropyl methylcellulose film.

1.3 Conclusion

In the food industry, various synthetic additives are used as antimicrobial, antioxidant, flavoring, and coloring agents. However, the use of synthetic additives is limited because of their negative health effects. Recently, plant extracts have gained increased popularity in the food industry and will reduce the application of artificial additives. Furthermore, the demand from consumer and market trend for natural additives leads to the identification and extraction of compounds from plant sources. In this regard, extracts from fruits and vegetables, spices and herbs and plantation crops are being used for different applications in the food industry as preservative, nutrient enhancer, as well as coloring and flavoring agents.

References

Akarpat, A., Turhan, S., & Ustun, N. S. (2008). Effects of hot-water extracts from myrtle, rosemary, nettle and lemon balm leaves on lipid oxidation and color of beef patties during frozen storage. *Journal of Food Processing and Preservation*, *32*(1), 117–132.

Akhtar, M. J., Jacquot, M., Jasniewski, J., Jacquot, C., Imran, M., Jamshidian, M., & Desobry, S. (2012). Antioxidant capacity and light-aging study of HPMC films functionalized with natural plant extract. *Carbohydrate Polymers*, *89*(4), 1150–1158.

Albuquerque, B. R., Pereira, C., Calhelha, R. C., Alves, M. J., Abreu, R. M., Barros, L., & Ferreira, I. C. (2020). Jabuticaba residues (Myrciaria jaboticaba (Vell.) Berg) are rich sources of valuable compounds with bioactive properties. *Food Chemistry*, *309*, 125735.

Aliyari, P., Bakhshi Kazaj, F., Barzegar, M., & Ahmadi Gavlighi, H. (2020). Production of functional sausage using pomegranate peel and pistachio green hull extracts as natural preservatives. *Journal of Agricultural Science and Technology*, *22*(1), 159–172.

Al-Juhaimi, F., Babtain, I. A., Ahmed, I. A. M., Alsawmahi, O. N., Ghafoor, K., Adiamo, O. Q., & Babiker, E. E. (2020). Assessment of oxidative stability and physicochemical, microbiological, and sensory properties of beef patties formulated with baobab seed (Adansonia digitata) extract. *Meat Science*, *162*, 108044.

Alp, E., & Aksu, M. İ. (2010). Effects of water extract of Urtica dioica L. and modified atmosphere packaging on the shelf life of ground beef. *Meat Science*, *86*(2), 468–473.

Altunkaya, A., Hedegaard, R. V., Harholt, J., Brimer, L., Gökmen, V., & Skibsted, L. H. (2013). Oxidative stability and chemical safety of mayonnaise enriched with grape seed extract. *Food & Function*, *4*(11), 1647–1653.

Banerjee, R., Verma, A. K., Das, A. K., Rajkumar, V., Shewalkar, A. A., & Narkhede, H. P. (2012). Antioxidant effects of broccoli powder extract in goat meat nuggets. *Meat Science*, *91*(2), 179–184.

Banon, S., Díaz, P., Rodríguez, M., Garrido, M. D., & Price, A. (2007). Ascorbate, green tea and grape seed extracts increase the shelf life of low sulphite beef patties. *Meat Science*, *77*(4), 626–633.

Bastante, C. C., Silva, N. H., Cardoso, L. C., Serrano, C. M., de la Ossa, E. J. M., Freire, C. S., & Vilela, C. (2021). Biobased films of nanocellulose and mango leaf extract for active food packaging: Supercritical impregnation vs solvent casting. *Food Hydrocolloids*, *117*, 106709.

Bastida, S., Sánchez-Muniz, F. J., Olivero, R., Pérez-Olleros, L., Ruiz-Roso, B., & Jiménez-Colmenero, F. (2009). Antioxidant activity of Carob fruit extracts in cooked pork meat systems during chilled and frozen storage. *Food Chemistry*, *116*(3), 748–754.

Bellucci, E. R. B., Munekata, P. E., Pateiro, M., Lorenzo, J. M., & da Silva Barretto, A. C. (2021). Red pitaya extract as natural antioxidant in pork patties with total replacement of animal fat. *Meat Science*, *171*, 108284.

Biswas, A. K., Chatli, M. K., & Sahoo, J. (2012). Antioxidant potential of curry (Murraya koenigii L.) and mint (Mentha spicata) leaf extracts and their effect on color and oxidative stability of raw ground pork meat during refrigeration storage. *Food Chemistry*, *133*(2), 467–472.

Botsoglou, E., Govaris, A., Ambrosiadis, I., Fletouris, D., & Botsoglou, N. (2014). Effect of olive leaf (Olea europea L.) extracts on protein and lipid oxidation of long-term frozen n-3 fatty acids-enriched pork patties. *Meat Science*, *98*(2), 150–157.

Brannan, R. G., & Mah, E. (2007). Grape seed extract inhibits lipid oxidation in muscle from different species during refrigerated and frozen storage and oxidation catalyzed by peroxynitrite and iron/ascorbate in a pyrogallol red model system. *Meat Science*, *77*(4), 540–546.

Chan, K. W., Khong, N. M., Iqbal, S., Ch'Ng, S. E., Younas, U., & Babji, A. S. (2014). Cinnamon bark deodorised aqueous extract as potential natural antioxidant in meat emulsion system: A comparative study with synthetic and natural food antioxidants. *Journal of Food Science and Technology*, *51*(11), 3269–3276.

Chen, Y., Wen, J., Deng, Z., Pan, X., Xie, X., & Peng, C. (2020). Effective utilization of food wastes: Bioactivity of grape seed extraction and its application in food industry. *Journal of Functional Foods*, *73*, 104113.

Chhikara, N., Kushwaha, K., Sharma, P., Gat, Y., & Panghal, A. (2019). Bioactive compounds of beetroot and utilization in food processing industry: A critical review. *Food Chemistry*, *272*, 192–200.

Cho, H. S., Park, W., Hong, G. E., Kim, J. H., Ju, M. G., & Lee, C. H. (2015). Antioxidant activity of Allium hookeri root extract and its effect on lipid stability of sulfur-fed pork patties. *Korean Journal for Food Science of Animal Resources*, *35*(1), 41.

Choe, J. H., Jang, A., Lee, E. S., Choi, J. H., Choi, Y. S., Han, D. J., & Kim, C. J. (2011). Oxidative and color stability of cooked ground pork containing lotus leaf (Nelumbo nucifera) and barley leaf (Hordeum vulgare) powder during refrigerated storage. *Meat Science*, *87*(1), 12–18.

Choulitoudi, E., Velliopoulou, A., Tsimogiannis, D., & Oreopoulou, V. (2020). Effect of active packaging with Satureja thymbra extracts on the oxidative stability of fried potato chips. *Food Packaging and Shelf Life*, *23*, 100455.

Colindres, P., & Susan Brewer, M. (2011). Oxidative stability of cooked, frozen, reheated beef patties: Effect of antioxidants. *Journal of the Science of Food and Agriculture*, *91*(5), 963–968.

Colle, M. C., Richard, R. P., Smith, D. M., Colle, M. J., Loucks, W. I., Gray, S. J., & Doumit, M. E. (2019). Dry potato extracts improve water holding capacity, shelf life, and sensory characteristics of fresh and precooked beef patties. *Meat Science*, *149*, 156–162.

Cunha, L. C., Monteiro, M. L. G., Costa-Lima, B. R., Guedes-Oliveira, J. M., Alves, V. H., Almeida, A. L., ... Conte-Junior, C. A. (2018). Effect of microencapsulated extract of pitaya (Hylocereus costaricensis) peel on color, texture and oxidative stability of refrigerated ground pork patties submitted to high pressure processing. *Innovative Food Science & Emerging Technologies*, *49*, 136–145.

Cunha, L. C. M., Monteiro, M. L. G., da Costa-Lima, B. R. C., Guedes-Oliveira, J. M., Rodrigues, B. L., Fortunato, A. R., & Conte-Junior, C. A. (2021). Effect of microencapsulated extract of pitaya (*Hylocereus costaricensis*) peel on oxidative quality parameters of refrigerated ground pork patties subjected to UV-C radiation. *Journal of Food Processing and Preservation*, *45*(3), e15272.

da Rosa, G. S., Vanga, S. K., Gariepy, Y., & Raghavan, V. (2020). Development of biodegradable films with improved antioxidant properties based on the addition of carrageenan containing olive leaf extract for food packaging applications. *Journal of Polymers and the Environment*, *28*(1), 123–130.

Das, A. K., Rajkumar, V., Verma, A. K., & Swarup, D. (2012). Moringa oleiferia leaves extract: A natural antioxidant for retarding lipid peroxidation in cooked goat meat patties. *International Journal of Food Science & Technology*, *47*(3), 585–591.

de Dicastillo, C. L., Bustos, F., Guarda, A., & Galotto, M. J. (2016). Cross-linked methyl cellulose films with murta fruit extract for antioxidant and antimicrobial active food packaging. *Food Hydrocolloids*, *60*, 335–344.

Degirmenci, H., & Erkurt, H. (2020). Chemical profile and antioxidant potency of Citrus aurantium L. flower extracts with antibacterial effect against foodborne pathogens in rice pudding. *LWT*, *126*, 109273.

Deolindo, C. T. P., Monteiro, P. I., Santos, J. S., Cruz, A. G., da Silva, M. C., & Granato, D. (2019). Phenolic-rich Petit Suisse cheese manufactured with organic Bordeaux grape juice, skin, and seed extract: Technological, sensory, and functional properties. *LWT*, *115*, 108493.

Devatkal, S. K., Narsaiah, K., & Borah, A. (2010). Anti-oxidant effect of extracts of kinnow rind, pomegranate rind and seed powders in cooked goat meat patties. *Meat Science*, *85*(1), 155–159.

Devatkal, S. K., Thorat, P., & Manjunatha, M. (2014). Effect of vacuum packaging and pomegranate peel extract on quality aspects of ground goat meat and nuggets. *Journal of Food Science and Technology*, *51*(10), 2685–2691.

Ding, Y., Wang, S. Y., Yang, D. J., Chang, M. H., & Chen, Y. C. (2015). Alleviative effects of litchi (Litchi chinensis Sonn.) flower on lipid peroxidation and protein degradation in emulsified pork meatballs. *Journal of Food and Drug Analysis*, *23*(3), 501–508.

Emam-Djomeh, Z., Moghaddam, A., & Yasini Ardakani, S. A. (2015). Antimicrobial activity of pomegranate (Punica granatum L.) peel extract, physical, mechanical, barrier and antimicrobial properties of pomegranate peel extract-incorporated sodium caseinate film and application in packaging for ground beef. *Packaging Technology and Science*, *28*(10), 869–881.

Ferysiuk, K., Wójciak, K. M., Materska, M., Chilczuk, B., & Pabich, M. (2020). Modification of lipid oxidation and antioxidant capacity in canned refrigerated pork with a nitrite content reduced by half and addition of sweet pepper extract. *LWT*, *118*, 108738.

Gniewosz, M., Synowiec, A., Kraśniewska, K., Przybył, J. L., Bączek, K., & Węglarz, Z. (2014). The antimicrobial activity of pullulan film incorporated with meadowsweet flower extracts (Filipendulae ulmariae flos) on postharvest quality of apples. *Food Control*, *37*, 351–361.

Gokoglu, N., & Yerlikaya, P. (2008). Inhibition effects of grape seed extracts on melanosis formation in shrimp (Parapenaeus longirostris). *International Journal of Food Science & Technology*, *43*(6), 1004–1008.

Guo, Y., Huang, J., Chen, Y., Hou, Q., & Huang, M. (2020). Effect of grape seed extract combined with modified atmosphere packaging on the quality of roast chicken. *Poultry Science*, *99*(3), 1598–1605.

Han, H. S., & Song, K. B. (2021a). Antioxidant properties of watermelon (Citrullus lanatus) rind pectin films containing kiwifruit (Actinidia chinensis) peel extract and their application as chicken thigh packaging. *Food Packaging and Shelf Life*, *28*, 100636.

Han, H. S., & Song, K. B. (2021b). Noni (Morinda citrifolia) fruit polysaccharide films containing blueberry (Vaccinium corymbosum) leaf extract as an antioxidant packaging material. *Food Hydrocolloids*, *112*, 106372.

Han, J., & Rhee, K. S. (2005). Antioxidant properties of selected Oriental non-culinary/nutraceutical herb extracts as evaluated in raw and cooked meat. *Meat Science*, *70*(1), 25–33.

Han, Y., Yu, M., & Wang, L. (2018). Bio-based films prepared with soybean by-products and pine (Pinus densiflora) bark extract. *Journal of Cleaner Production*, *187*, 1–8.

Hao, M., & Beta, T. (2012). Development of Chinese steamed bread enriched in bioactive compounds from barley hull and flaxseed hull extracts. *Food Chemistry*, *133*(4), 1320–1325.

Hauser, C., Peñaloza, A., Guarda, A., Galotto, M. J., Bruna, J. E., & Rodríguez, F. J. (2016). Development of an active packaging film based on a methylcellulose coating containing murta (Ugni molinae turcz) leaf extract. *Food and Bioprocess Technology*, *9*(2), 298–307.

Hayes, J. E., Stepanyan, V., Allen, P., O'grady, M. N., & Kerry, J. P. (2011). Evaluation of the effects of selected plant-derived nutraceuticals on the quality and shelf-life stability of raw and cooked pork sausages. *LWT-Food Science and Technology*, *44*(1), 164–172.

Hayes, J. E., Stepanyan, V., O'Grady, M. N., Allen, P., & Kerry, J. P. (2010). Evaluation of the effects of selected phytochemicals on quality indices and sensorial properties of raw and cooked pork stored in different packaging systems. *Meat Science*, *85*(2), 289–296.

Hedayati, S., Niakousari, M., Damyeh, M. S., Mazloomi, S. M., Babajafari, S., & Ansarifar, E. (2021). Selection of appropriate hydrocolloid for eggless cakes containing chubak root extract using multiple criteria decision-making approach. *LWT*, *141*, 110914.

Hollis, F. H., Denney, S., Halfacre, J., Jackson, T., Link, A., & Orr, D. (2020). Human perception of fresh produce treated with grape seed extract: A preliminary study. *Journal of Sensory Studies*, *35*(2), e12554.

Huang, B., He, J., Ban, X., Zeng, H., Yao, X., & Wang, Y. (2011). Antioxidant activity of bovine and porcine meat treated with extracts from edible lotus (Nelumbo nucifera) rhizome knot and leaf. *Meat Science*, *87* (1), 46–53.

Jang, S. A., Shin, Y. J., & Song, K. B. (2011). Effect of rapeseed protein–gelatin film containing grapefruit seed extract on 'Maehyang'strawberry quality. *International Journal of Food Science & Technology*, *46* (3), 620–625.

Jaśkiewicz, A., Budryn, G., Nowak, A., & Efenberger-Szmechtyk, M. (2020). Novel biodegradable starch film for food packaging with antimicrobial chicory root extract and phytic acid as a cross-linking agent. *Foods*, *9*(11), 1696.

Jayawardana, B. C., Hirano, T., Han, K. H., Ishii, H., Okada, T., Shibayama, S., & Shimada, K. I. (2011). Utilization of adzuki bean extract as a natural antioxidant in cured and uncured cooked pork sausages. *Meat Science*, *89*(2), 150–153.

Jridi, M., Abdelhedi, O., Salem, A., Kechaou, H., Nasri, M., & Menchari, Y. (2020). Physicochemical, antioxidant and antibacterial properties of fish gelatin-based edible films enriched with orange peel pectin: Wrapping application. *Food Hydrocolloids*, *103*, 105688.

Kanatt, S. R. (2020). Development of active/intelligent food packaging film containing Amaranthus leaf extract for shelf life extension of chicken/fish during chilled storage. *Food Packaging and Shelf Life*, *24*, 100506.

Kanatt, S. R., Chander, R., & Sharma, A. (2008). Chitosan and mint mixture: A new preservative for meat and meat products. *Food Chemistry*, *107*(2), 845–852.

Kanatt, S. R., Rao, M. S., Chawla, S. P., & Sharma, A. (2012). Active chitosan–polyvinyl alcohol films with natural extracts. *Food Hydrocolloids*, *29*(2), 290–297.

Kanmani, P., & Rhim, J. W. (2014). Antimicrobial and physical-mechanical properties of agar-based films incorporated with grapefruit seed extract. *Carbohydrate Polymers*, *102*, 708–716.

Karabacak, S., & Bozkurt, H. (2008). Effects of Urtica dioica and Hibiscus sabdariffa on the quality and safety of sucuk (Turkish dry-fermented sausage). *Meat Science*, *78*(3), 288–296.

Khan, I. A., Liu, D., Yao, M., Memon, A., Huang, J., & Huang, M. (2019). Inhibitory effect of Chrysanthemum morifolium flower extract on the formation of heterocyclic amines in goat meat patties cooked by various cooking methods and temperatures. *Meat Science*, *147*, 70–81.

Kim, J. S. (2000). Functional beef product containing rice bran extracts influence cholesterol oxidation and nutritional profile. *Food Industry and Nutrition*, *5*(3), 66–73.

Kim, S. J., Cho, A. R., & Han, J. (2013). Antioxidant and antimicrobial activities of leafy green vegetable extracts and their applications to meat product preservation. *Food Control*, *29*(1), 112–120.

Kõrge, K., Bajić, M., Likozar, B., & Novak, U. (2020). Active chitosan–chestnut extract films used for packaging and storage of fresh pasta. *International Journal of Food Science & Technology*, *55*(8), 3043–3052.

Kõrge, K., Šeme, H., Bajić, M., Likozar, B., & Novak, U. (2020). Reduction in spoilage microbiota and cyclopiazonic acid mycotoxin with chestnut extract enriched chitosan packaging: Stability of inoculated Gouda Cheese. *Foods*, *9*(11), 1645.

Kulkarni, S., DeSantos, F. A., Kattamuri, S., Rossi, S. J., & Brewer, M. S. (2011). Effect of grape seed extract on oxidative, color and sensory stability of a pre-cooked, frozen, re-heated beef sausage model system. *Meat Science*, *88*(1), 139–144.

Kumari, S., Raines, J. M., Martin, J. M., & Rodriguez, J. M. (2015). Thermal stability of kudzu root (Pueraria radix) isoflavones as additives to beef patties. *Journal of Food Science and Technology*, *52*(3), 1578–1585.

Lavado, G., Ladero, L., & Cava, R. (2021). Cork oak (*Quercus suber* L.) leaf extracts potential use as natural antioxidants in cooked meat. *Industrial Crops and Products*, *160*, 113086.

Li, J. H., Miao, J., Wu, J. L., Chen, S. F., & Zhang, Q. Q. (2014). Preparation and characterization of active gelatin-based films incorporated with natural antioxidants. *Food Hydrocolloids*, *37*, 166–173.

Lim, G. O., Jang, S. A., & Song, K. B. (2010). Physical and antimicrobial properties of Gelidium corneum/nano-clay composite film containing grapefruit seed extract or thymol. *Journal of Food Engineering*, *98*(4), 415–420.

Lin, C., Toto, C., & Were, L. (2015). Antioxidant effectiveness of ground roasted coffee in raw ground top round beef with added sodium chloride. *LWT-Food Science and Technology*, *60*(1), 29–35.

Mármol, I., Quero, J., Jiménez-Moreno, N., Rodríguez-Yoldi, M. J., & Ancín-Azpilicueta, C. (2019). A systematic review of the potential uses of pine bark in food industry and health care. *Trends in Food Science & Technology*, *88*, 558–566.

Martiny, T. R., Raghavan, V., Moraes, C. C. D., Rosa, G. S. D., & Dotto, G. L. (2020). Bio-based active packaging: Carrageenan film with olive leaf extract for lamb meat preservation. *Foods*, *9*(12), 1759.

Moghadam, M., Salami, M., Mohammadian, M., Khodadadi, M., & Emam-Djomeh, Z. (2020). Development of antioxidant edible films based on mung bean protein enriched with pomegranate peel. *Food Hydrocolloids*, *104*, 105735.

Mohamed, H. M., Mansour, H. A., & Farag, M. D. E. D. H. (2011). The use of natural herbal extracts for improving the lipid stability and sensory characteristics of irradiated ground beef. *Meat Science*, *87*(1), 33–39.

Moraczewski, K., Stepczyńska, M., Malinowski, R., Karasiewicz, T., Jagodziński, B., & Rytlewski, P. (2019). The effect of accelerated aging on polylactide containing plant extracts. *Polymers*, *11*(4), 575.

Muthukumar, M., Naveena, B. M., Vaithiyanathan, S., Sen, A. R., & Sureshkumar, K. (2014). Effect of incorporation of Moringa oleifera leaves extract on quality of ground pork patties. *Journal of Food Science and Technology*, *51*(11), 3172–3180.

Norajit, K., Kim, K. M., & Ryu, G. H. (2010). Comparative studies on the characterization and antioxidant properties of biodegradable alginate films containing ginseng extract. *Journal of Food Engineering*, *98*(3), 377–384.

Nouri, L., & Nafchi, A. M. (2014). Antibacterial, mechanical, and barrier properties of sago starch film incorporated with betel leaves extract. *International Journal of Biological Macromolecules*, *66*, 254–259.

Özünlü, O., Ergezer, H., & Gökçe, R. (2018). Improving physicochemical, antioxidative and sensory quality of raw chicken meat by using acorn extracts. *LWT*, *98*, 477–484.

Özvural, E. B., & Vural, H. (2012). The effects of grape seed extract on quality characteristics of frankfurters. *Journal of Food Processing and Preservation*, *36*(4), 291–297.

Pateiro, M., Vargas, F. C., Chincha, A. A., Sant'Ana, A. S., Strozzi, I., Rocchetti, G., & Lorenzo, J. M. (2018). Guarana seed extracts as a useful strategy to extend the shelf life of pork patties: UHPLC-ESI/QTOF phenolic profile and impact on microbial inactivation, lipid and protein oxidation and antioxidant capacity. *Food Research International*, *114*, 55–63.

Rababah, T. M., Ereifej, K. I., Alhamad, M. N., Al-Qudah, K. M., Rousan, L. M., Al-Mahasneh, M. A., & Yang, W. (2011). Effects of green tea and grape seed and TBHQ on physicochemical properties of Baladi goat meats. *International Journal of Food Properties*, *14*(6), 1208–1216.

Rahman, M. H., Alam, M. S., Monir, M. M., & Ahmed, K. (2021). Comprehensive effects of black cumin (Nigella sativa) and synthetic antioxidant on sensory and physicochemical quality of beef patties during refrigerant storage. *Journal of Agriculture and Food Research*, *4*, 100145.

Rambabu, K., Bharath, G., Banat, F., Show, P. L., & Cocoletzi, H. H. (2019). Mango leaf extract incorporated chitosan antioxidant film for active food packaging. *International Journal of Biological Macromolecules*, *126*, 1234–1243.

Reddy, G. B., Sen, A. R., Nair, P. N., Reddy, K. S., Reddy, K. K., & Kondaiah, N. (2013). Effects of grape seed extract on the oxidative and microbial stability of restructured mutton slices. *Meat Science, 95*(2), 288–294.

Riaz, A., Lagnika, C., Luo, H., Dai, Z., Nie, M., Hashim, M. M., & Li, D. (2020). Chitosan-based biodegradable active food packaging film containing Chinese chive (Allium tuberosum) root extract for food application. *International Journal of Biological Macromolecules, 150*, 595–604.

Rodríguez, G. M., Sibaja, J. C., Espitia, P. J., & Otoni, C. G. (2020). Antioxidant active packaging based on papaya edible films incorporated with Moringa oleifera and ascorbic acid for food preservation. *Food Hydrocolloids, 103*, 105630.

Rojas, M. C., & Brewer, M. S. (2007). Effect of natural antioxidants on oxidative stability of cooked, refrigerated beef and pork. *Journal of Food Science, 72*(4), S282–S288.

Rojas, M. C., & Brewer, M. S. (2008). Effect of natural antioxidants on oxidative stability of frozen, vacuum-packaged beef and pork. *Journal of Food Quality, 31*(2), 173–188.

Sampaio, S. L., Petropoulos, S. A., Alexopoulos, A., Heleno, S. A., Santos-Buelga, C., Barros, L., & Ferreira, I. C. (2020). Potato peels as sources of functional compounds for the food industry: A review. *Trends in Food Science & Technology, 103*, 118–129.

Samsudin, H., Soto-Valdez, H., & Auras, R. (2014). Poly (lactic acid) film incorporated with marigold flower extract (Tagetes erecta) intended for fatty-food application. *Food Control, 46*, 55–66.

Sayari, N., Sila, A., Balti, R., Abid, E., Hajlaoui, K., Nasri, M., & Bougatef, A. (2015). Antioxidant and antibacterial properties of Citrus paradisi barks extracts during turkey sausage formulation and storage. *Biocatalysis and Agricultural Biotechnology, 4*(4), 616–623.

Sebranek, J. G., Sewalt, V. J. H., Robbins, K., & Houser, T. A. (2005). Comparison of a natural rosemary extract and BHA/BHT for relative antioxidant effectiveness in pork sausage. *Meat Science, 69*(2), 289–296.

Shah, M. A., Bosco, S. J. D., & Mir, S. A. (2015). Effect of Moringa oleifera leaf extract on the physicochemical properties of modified atmosphere packaged raw beef. *Food Packaging and Shelf Life, 3*, 31–38.

Shan, B., Cai, Y. Z., Brooks, J. D., & Corke, H. (2009). Antibacterial and antioxidant effects of five spice and herb extracts as natural preservatives of raw pork. *Journal of the Science of Food and Agriculture, 89*(11), 1879–1885.

Shin, D. J., Choe, J., Hwang, K. E., Kim, C. J., & Jo, C. (2019). Antioxidant effects of lotus (Nelumbo nucifera) root and leaf extracts and their application on pork patties as inhibitors of lipid oxidation, alone and in combination. *International Journal of Food Properties, 22*(1), 383–394.

Singh, T. P., Agrawal, R. K., Mendiratta, S. K., & Chauhan, G. (2021). Preparation and characterization of licorice root extract infused bio-composite film and their application on storage stability of chhana balls-a Sandesh like product. *Food Control, 125*, 107993.

Siripatrawan, U., & Harte, B. R. (2010). Physical properties and antioxidant activity of an active film from chitosan incorporated with green tea extract. *Food Hydrocolloids, 24*(8), 770–775.

Sivarooban, T., Hettiarachchy, N. S., & Johnson, M. G. (2008). Physical and antimicrobial properties of grape seed extract, nisin, and EDTA incorporated soy protein edible films. *Food Research International, 41*(8), 781–785.

Song, H. Y., Shin, Y. J., & Song, K. B. (2012). Preparation of a barley bran protein–gelatin composite film containing grapefruit seed extract and its application in salmon packaging. *Journal of Food Engineering, 113*(4), 541–547.

Šulniūtė, V., Jaime, I., Rovira, J., & Venskutonis, P. R. (2016). Rye and wheat bran extracts isolated with pressurized solvents increase oxidative stability and antioxidant potential of beef meat hamburgers. *Journal of Food Science, 81*(2), H519–H527.

Tajik, H., Farhangfar, A., Moradi, M., & Razavi Rohani, S. M. (2014). Effectiveness of clove essential oil and grape seed extract combination on microbial and lipid oxidation characteristics of raw buffalo patty during storage at abuse refrigeration temperature. *Journal of Food Processing and Preservation, 38*(1), 31–38.

Tan, Y. M., Lim, S. H., Tay, B. Y., Lee, M. W., & Thian, E. S. (2015). Functional chitosan-based grapefruit seed extract composite films for applications in food packaging technology. *Materials Research Bulletin*, *69*, 142–146.

Tayel, A. A., & El-tras, W. F. (2012). Plant extracts as potent biopreservatives for Salmonella typhimurium control and quality enhancement in ground beef. *Journal of Food Safety*, *32*(1), 115–121.

Tengilimoglu-Metin, M. M., Hamzalioglu, A., Gokmen, V., & Kizil, M. (2017). Inhibitory effect of hawthorn extract on heterocyclic aromatic amine formation in beef and chicken breast meat. *Food Research International*, *99*, 586–595.

Turan, E., & Şimşek, A. (2021). Effects of lyophilized black mulberry water extract on lipid oxidation, metmyoglobin formation, color stability, microbial quality and sensory properties of beef patties stored under aerobic and vacuum packaging conditions. *Meat Science*, 108522.

Villalobos-Delgado, L. H., González-Mondragón, E. G., Ramírez-Andrade, J., Salazar-Govea, A. Y., & Santiago-Castro, J. T. (2020). Oxidative stability in raw, cooked, and frozen ground beef using Epazote (Chenopodium ambrosioides L.). *Meat Science*, *168*, 108187.

Wambura, P., Yang, W., & Mwakatage, N. R. (2011). Effects of sonication and edible coating containing rosemary and tea extracts on reduction of peanut lipid oxidative rancidity. *Food and Bioprocess Technology*, *4* (1), 107–115.

Wang, K., Lim, P. N., Tong, S. Y., & San Thian, E. (2019). Development of grapefruit seed extract-loaded poly (ε-caprolactone)/chitosan films for antimicrobial food packaging. *Food Packaging and Shelf Life*, *22*, 100396.

Wang, L., Wang, Q., Tong, J., & Zhou, J. (2017). Physicochemical properties of chitosan films incorporated with honeysuckle flower extract for active food packaging. *Journal of Food Process Engineering*, *40*(1), e12305.

Wang, S., Marcone, M., Barbut, S., & Lim, L. T. (2012). The impact of anthocyanin-rich red raspberry extract (ARRE) on the properties of edible soy protein isolate (SPI) films. *Journal of Food Science*, *77*(4), C497–C505.

Wang, Z., Cheng, Y., Zeng, M., Wang, Z., Qin, F., Wang, Y., & He, Z. (2021). Lotus (Nelumbo nucifera Gaertn.) leaf: A narrative review of its Phytoconstituents, health benefits and food industry applications. *Trends in Food Science & Technology*, *112*, 631–650.

Wójciak, K. M., Dolatowski, Z. J., & Okoń, A. (2011). The effect of water plant extracts addition on the oxidative stability of meat products. *Acta Scientiarum Polonorum Technologia Alimentaria*, *10*(2), 175–188.

Yu, J., Ahmedna, M., & Goktepe, I. (2010). Potential of peanut skin phenolic extract as antioxidative and antibacterial agent in cooked and raw ground beef. *International Journal of Food Science & Technology*, *45* (7), 1337–1344.

Zhang, C., Guo, K., Ma, Y., Ma, D., Li, X., & Zhao, X. (2010). Incorporations of blueberry extracts into soybean-protein-isolate film preserve qualities of packaged lard. *International Journal of Food Science & Technology*, *45*(9), 1801–1806.

Zhang, C., Sun, G., Li, J., & Wang, L. (2021). A green strategy for maintaining intelligent response and improving antioxidant properties of κ-carrageenan-based film via cork bark extractive addition. *Food Hydrocolloids*, *113*, 106470.

Zhang, H., Kong, B., Xiong, Y. L., & Sun, X. (2009). Antimicrobial activities of spice extracts against pathogenic and spoilage bacteria in modified atmosphere packaged fresh pork and vacuum packaged ham slices stored at 4 C. *Meat Science*, *81*(4), 686–692.

Zhang, W., Li, X., & Jiang, W. (2020). Development of antioxidant chitosan film with banana peels extract and its application as coating in maintaining the storage quality of apple. *International Journal of Biological Macromolecules*, *154*, 1205–1214.

CHAPTER

Extraction techniques

2

Saqib Farooq[1], Shabir Ahmad Mir[2], Manzoor Ahmad Shah[3] and Annamalai Manickavasagan[4]

[1]Department of Food Technology, Islamic University of Science and Technology, Awantipora, India [2]Department of Food Science & Technology, Govt. College for Women, Srinagar, India [3]Department of Food Science & Technology, Govt. Degree College for Women, Anantnag, India [4]School of Engineering, College of Engineering and Physical Sciences, University of Guelph, Guelph, ON, Canada

2.1 Introduction

Extraction is one of the important operations for obtaining the target compounds from different types of plant materials. This is the first step for separating the compounds from the plant matrix. The extraction from plant materials can be achieved by different methods based on the nature and quantity of compounds to be extracted, and each approach has its own merits and demerits (Table 2.1). The main aim of selecting the method is to maximize the extraction yield of plant compounds without modifying or contaminating its original form. The efficiency and yield of the extraction procedure depend on various factors such as diversification of bioactive compounds, type of extraction procedure, concentration of solvent, time, and temperature. Different protocols and standards have been developed by the researchers and industries as per the type of plant material to enhance its yield and functional properties.

There are three approaches of extraction depending on the phase of the mixture and the extraction agent. In solid–liquid extraction, solvent extracts as solute from the solid phase. In liquid–liquid extraction, the solvent extracts solute from the liquid phase. Third approach is the supercritical extraction, where a fluid under supercritical conditions is used as the solvent. The conventional methods like maceration and soxhlet extraction have been used since ancient times and are still most commonly used for the extraction of different compounds from plant materials mainly based on their simplicity and moderate operating cost. The majority of conventional methods is based on the extraction efficiency of different solvents, homogenization and thermal factors. However, conventional methods have certain disadvantages such as low extraction yield and selectivity, long extraction procedure, and degradation of heat liable plant compounds. New techniques or assisting procedures are continuously being developed to overcome the limitations of traditional methods for the extraction of plant compounds. The assisting methods such as microwaves, ultrasound, pressurized liquid, pulsed electric field, supercritical fluid and enzymes have gained popularity for the extraction of plant extracts due to less use of chemicals, good yield and eco-friendly nature. Furthermore, new techniques minimize the degradation of plant compounds and also eliminate the undesirable components from the extract.

Plant Extracts: ***Applications in the Food Industry*****. DOI: https://doi.org/10.1016/B978-0-12-822475-5.00005-3**

Table 2.1 Advantage and disadvantages of various extraction methods.

Extraction method	Advantages	Disadvantages
Soxhlet	• Simple operation procedure • Low cost technique • Efficient recovery of extract	• Time consuming • Requires large quantity of solvent
Maceration	• Simple extraction method • Inexpensive technique	• Time consuming process • Low extraction yield • Usage of large amount of solvent
Supercritical fluid	• Short extraction time • Suitable for heat sensitive compounds • Low viscosity and higher diffusion coefficient • Eco-friendly	• Not suitable for extraction of polar molecules
Pressurized liquid	• Reduces solvent consumption and extraction time • Eco-friendly	•Decreased selectivity
Pulsed electric field	• Suitable for large volume of samples • Short extraction time • High yield • Facilitation of purified extract	•Process parameter are required to be maintained critically
Ultrasound assisted	• Short processing time • Less chemical usage • Less power and energy consumption • Higher yield	• Extraction yield depends on optimized process parameters
Microwave assisted	• Less extraction time • Higher extraction yield • High selectivity of desired extracts	• Less eco-friendly due to use of organic solvents • Equipments are costly • Poor extraction yield for nonpolar compounds • Unsuitable for heat liable compounds
Enzyme assisted	• Eco-friendly • High extraction rate • Suitable for bound compounds • High selectivity of desired extracts	• High enzyme cost • Not suitable at industrial level • Difficult to maintain suitable conditions during extraction

2.2 Extraction methods

Several methods are applied for the preparation of extracts from various types of plant sources (Fig. 2.1). The selection of suitable procedure is very important for the maximum yield or extraction with high phytochemical quality. Solvent has primary importance for the extraction of compounds irrespective of the method of extraction. The chemical nature of solvent favor the compound solubility, selectivity and extraction yield. The solvent interacts with the compounds depends on its chemical properties and the knowledge of the chemical behavior of the solvent are the basis for selection of solvent for efficient results. Several studies have focused on the selectivity of different solvents for the extraction of targeted compounds. Numerous technologies have been implemented in food industry to enhance the extraction capacity of plant compounds (Farooq et al., 2020).

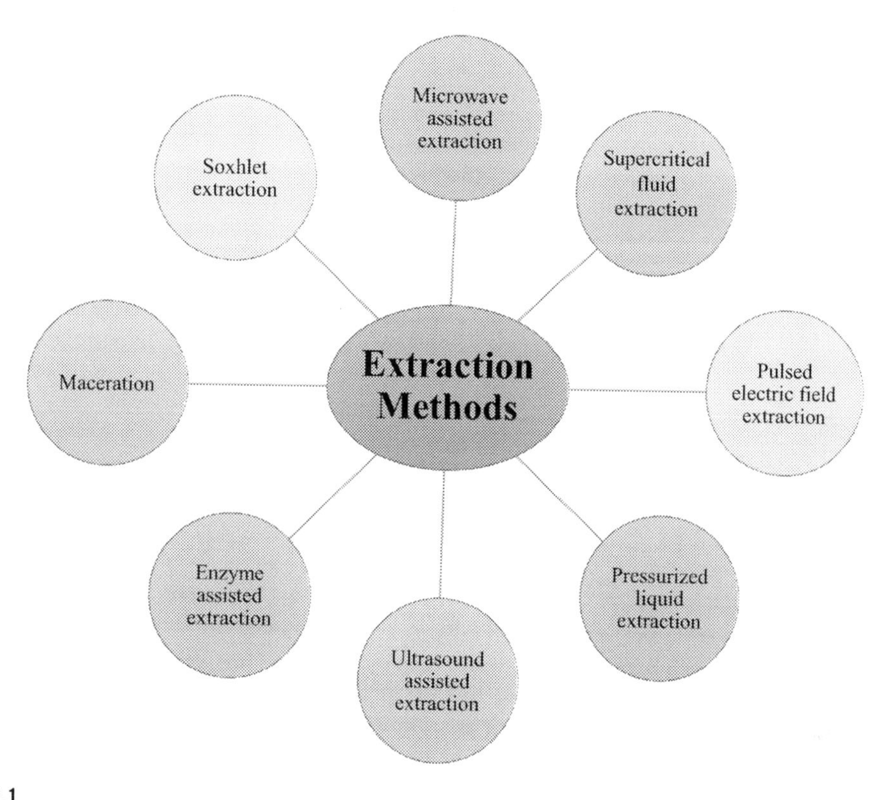

FIGURE 2.1

Extraction methods from plant material.

The development of new methods with considerable advantages over traditional methods for the extraction from the plant material plays an important role in ensuring the availability of high-quality extracts for different applications in food industry.

2.2.1 Maceration

Maceration is one of the simplest extraction techniques in which coarse and powdered plant material is soaked in solvents such as methanol, ethanol, ethyl acetate, acetone, hexane etc. It is one of the popular and inexpensive techniques used for the extraction of different bioactive compounds from plant material. However, maceration procedure has certain limitations such as low extraction yield, lower efficiency and use of large amount of solvents which have some health hazards. Furthermore, the selection of appropriate solvent is important along the methodology for the extraction of particular plant extract. Maceration process consists of grinding of plant material into smaller particles to increase the surface area for easy mixing with solvent and efficient extraction of compounds. Then this mixture of plant material and solvent is kept for longer time, agitated at different intervals and filtered through a filtration medium. The efficiency for the removal of bioactive compounds from the plant material depends on the type of solvent and type of plant material. The polarity of solvent is the important parameter affecting the extraction efficiency. In this method different solvents and time-temperature combinations are used for efficient extraction. Maceration raptures the cell structure and expose the chemical constituents to react with the solvent and helps in removal of different plant components. This method is extensively used for the exaction of different types of bioactive compounds at laboratory scale. Maceration as one of the simplest method was used for obtaining the *Papaver rhoeas* L. flower extracts (Marsoul, Ijjaali, Oumous, Bennani, & Boukir, 2020), *Morus* leaf extracts (Radojković et al., 2016), kinnow peel extract (Safdar et al., 2017) and chokeberry fruit extract (Ćujić et al., 2016). This technique can be operated at both small and large scale and finds application at an industrial scale.

2.2.2 Soxhlet extraction method

Soxhlet extraction is a widely used method for the extraction of different compounds from plant material. The popularity of soxhlet method is due to its simple operation procedure, low processing cost, efficient recovery of extracts, suitable for bulk extraction, and consumes less time as compared to other conventional methods. In this method, powdered sample is placed in a thimble that is positioned in a soxhlet extractor (Fig. 2.2). The solvent of particular interest such as petroleum ether, hexane etc. is placed in distillation flask which is heated by heating mantle and its vapors are condensed by the reflux condenser. The condensed solvent drips into the thimble containing the sample and extracts the bioactive compounds by contact. When the level of the solvent reached to an overflow level, the solvent of the thimble holder (extractor) is aspirated by a siphon and unloads the solvent back into the distillation flask. This method is still used as a reference method to compare the success of new extraction method. Alara, Abdurahman, and Ukaegbu (2018) reported that efficiency of soxhlet extraction depends on various factors including extraction time, solvent

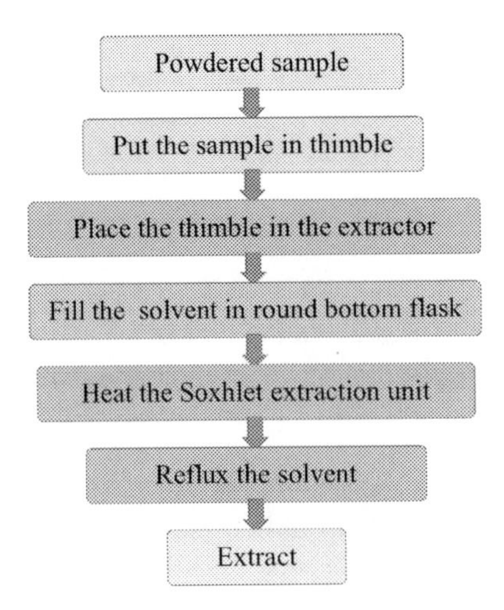

FIGURE 2.2

Flow diagram of Soxhlet extraction method.

concentration and feed to solvent ratio. The soxhlet extraction has been reported as an excellent method for extraction of polyphenols, antioxidants and antimicrobials from *P. rhoeas* flower (Marsoul et al., 2020). Karabegović et al. (2014) reported that soxhlet extraction facilitates the highest extractive yield from cherry laurel leaf and fruit. Soxhlet extraction also showed better performance for the extraction of phenolic compounds from olive leaves (Lama-Muñoz et al., 2020).

2.2.3 Supercritical carbon dioxide extraction

Supercritical carbon dioxide is a state of carbon dioxide at a temperature and pressure above its critical point where distinct liquid and gas phases do not exist (Fig. 2.3). The critical point of carbon dioxide is easily accessible and allowing the fluid to be used at low temperature without leaving the harmful organic residues. The fundamental concept behind the supercritical CO_2 extraction strategy is to replace the conventional organic solvents with a supercritical fluid solvent (Fig. 2.4). Compared to other extraction techniques, supercritical fluid extraction prevents sample oxidation, requires less extraction time, uses nontoxic solvents and is also safe. Supercritical fluid extraction is very effective and efficient technique for extraction of different bioactive compounds from plant material. This technique extracts the compounds near to ambient temperature, thus preventing the degradation of heat liable plant material. There are several solvents such as CO_2, benzene, hexane, pentene, butane, toluene, ethanol, ammonia and nitrous oxide, which can be used as supercritical fluid. However, carbon dioxide is commonly used fluid in supercritical fluid extraction technology. Carbon dioxide is nontoxic, cheap, and chemically stable and absorb a wide variety of organic compounds under supercritical conditions. Being a "green" process, supercritical fluid extraction presents an alternative technology to replace the organic solvents.

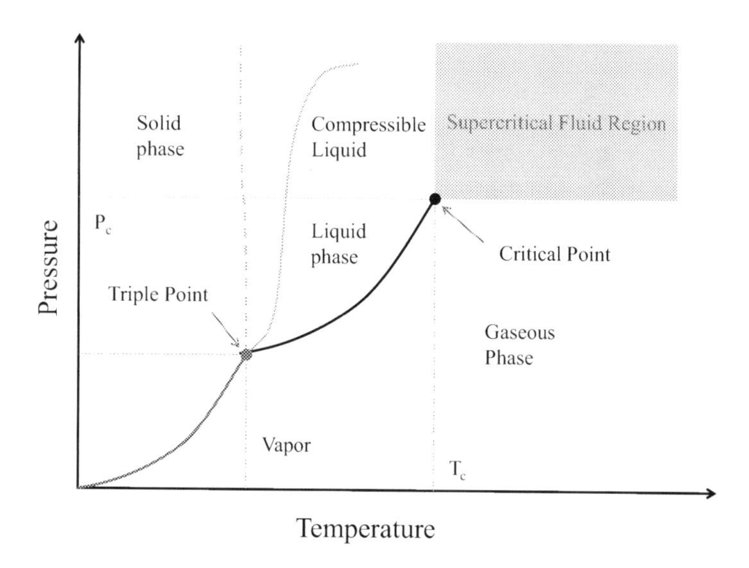

FIGURE 2.3

Phase diagram of a supercritical fluid.

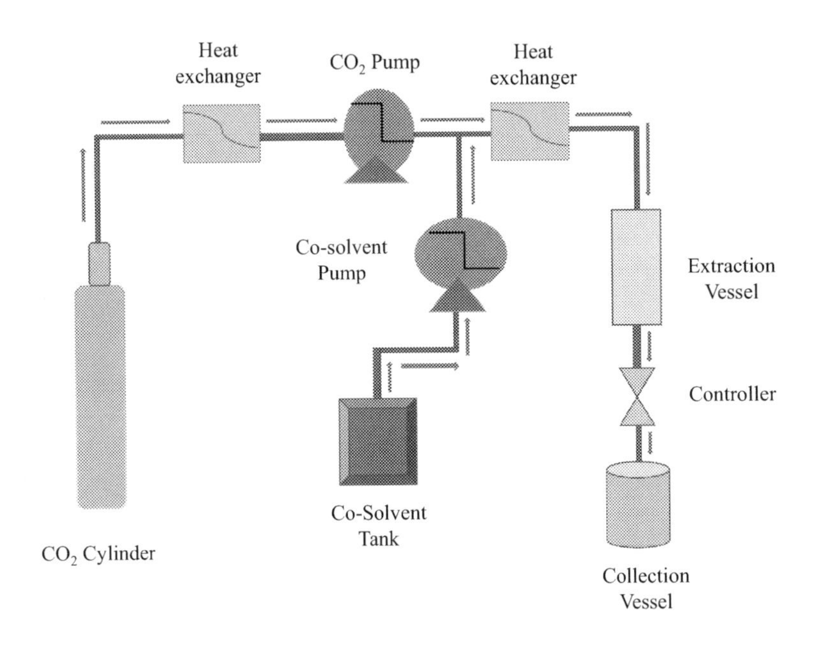

FIGURE 2.4

Schematic representation of supercritical CO_2 equipment.

Supercritical fluid extraction is one of the effective techniques for maximizing the extraction of phenolic compounds and flavonoids with high quality and efficiency. Different process parameters such as temperature, pressure, and solvent or co-solvent flow rate are involved in supercritical fluid extraction. Researchers have optimized the process parameters according to the type of plant material to improve the extraction percentage and selectivity of the recovered compound. Extraction of bioactive compounds from radish leaves were obtained by using supercritical CO_2 (Goyeneche, Fanovich, Rodrigues, Nicolao, & Di Scala, 2018). This technique efficiently extracted the bioactive compounds especially phenolic compounds and flavonoids. The maximum yield was observed at 400 bar and temperature of 40°C. Zhang et al. (2020) reported that supercritical CO_2 extraction was efficient and produced higher yield as compared to the methanol extraction method for extraction of plant compounds form cherry leaf.

Supercritical CO_2 has a high solvent strength because of its high density, which in turn increases its dielectric constant. As a result, this thick gas has the ability to dissolve a wide range of organic solutes, primarily nonpolar ones. The strong diffusivity of supercritical CO_2 together with its conveniently configurable solvent power is it's another advantage. At room temperature and pressure, CO_2 is gaseous in nature which makes the extraction of analytes easy and allowing for solvent-free analytes (Herrero, Mendiola, Cifuentes, & Ibáñez, 2010). Supercritical CO_2 is a special kind of extraction solvent. By changing the pressure and temperature, the solvent properties can be modified according to the protocol. It is sensitive to pressure and temperature changes owing to its compressibility. Density varies with pressure and temperature. It increases with increasing pressure and decreases with increasing temperature. The relationships with the compounds, and therefore their extraction can be altered as a result of density changes. A co-solvent, also known as a modifier, may be applied to increase the potential and selectivity of extraction by attenuating the polarity of supercritical CO_2. The polarity of the extracting solvent can be increased by applying an alcoholic modifier like ethanol to CO_2. Co-solvent can be safely extracted after processing. Modifying the characteristics of supercritical CO_2 using temperature, pressure, and co-solvents is a crucial step in achieving extraction selectivity (Lefebvre, Destandau, & Lesellier, 2020).

2.3 Microwave-assisted extraction

Conventional extraction methods such as organic solvent extraction and soxhlet extraction uses significant energy, time consuming, and pollutes the environment with toxic solvents. Microwave-assisted extraction employs an ecologically sound solvent and takes less energy while lowering the time of extraction. The extraction time and solvent volume are also reduced by using this extraction technique. Nonionizing electromagnetic waves are used in microwave-assisted extraction to extract bioactive chemicals from plant matrixes. Furthermore, microwave-assisted extraction allows for more efficient and selective transfer heating (without heat transfers), reducing temperature gradients, and may take use of different types of solvents to expand the extractions polarity range, boosting the variety and volume of extracted compounds (Esquivel-Hernández et al., 2016).

Microwaves heat molecules by a combination of ionic conduction and dipole rotation. The heating of substances is caused by both ionic conduction and dipole rotation. The minute microscopic residues of moisture that exist in plant cells are the focus for heating in plant material. The microwave action heats up the moisture inside the plant cell, causing evaporation and creating immense pressure on the cell wall. Because of pressure, the cell wall is forced from within and ruptures. As a result, bioactive components are expelled from the burst cells, facilitates the production of phytochemical compounds. Most microwave-assisted extraction techniques use solvents with a high dielectric constant and the ability to absorb a large amount of microwave energy. However, solvent combinations may be used to alter extraction selectivity and the medium's propensity to interact with microwaves. Polar molecules with a high dielectric constant such as water and ethanol may absorb and re-emit this energy, causing the system to heat up, but solvents with a low dielectric constant, such as hexane are insensitive to microwaves (Lefebvre et al., 2020). The addition of water to the solvent results in higher extraction yields. For the extraction of phenolic compounds, microwave transparent solvents such as acetone found to be the most effective (Proestos & Komaitis, 2008). Temperature is the most important parameter in microwave-assisted extraction. The temperature may be adjusted by varying the duration and power of the irradiation (Lefebvre et al., 2020).

Microwave-assisted extraction is a fine choice to get extracts under moderate circumstances. When applied to plant materials, microwaves produce little or no quality degradation, whereas moist heat causes quality degradation (Gupta, Naraniwal, & Kothari, 2012). Plant extracts with potential antibacterial and antioxidant properties have been prepared by microwave-assisted extraction. Microwave assisted extraction showed the most efficient technology for obtaining higher yields of polar and nonpolar bioactive compounds from elder bark and annatto seeds (Bachtler & Bart, 2021).

Microwave assisted extraction technique has been proven efficient and fast extraction method for the extraction of phytochemical compounds from *Ficus racemosa* (Sharma, Kumar, Kumar, & Panesar, 2020). Microwave extraction was an effective technology for obtaining the extracts with high antioxidant potential from avocado peels (Araujo et al., 2021).

Phenolic compounds in peanut skins were extracted using a microwave extraction technique. Under optimal conditions (90% microwave power, 30 s of irradiation duration, and 1.5 g skins), the maximum estimated total phenolic content was 143.6 mg gallic acid equivalent/g skins (Ballard, Mallikarjunan, Zhou, & O'Keefe, 2010). Microwave extraction of phenols such as chlorogenic acids from beans of green coffee was shown to be a viable alternative to standard procedures, and the extracted extracts had excellent radical scavenging activity (Upadhyay, Ramalakshmi, & Rao, 2012).

The extraction of antioxidants from potato peels at the methanol concentration of 67%, extraction period of 15 min, and a microwave power level of 14.67% leads to the production of maximum total phenolics content. The highest levels of caffeic acid, and ferulic acid were achieved at a methanol concentration of 100%, an extraction duration of 15 min, and a microwave power level of 10%. The optimal antioxidant activity was reached at a methanol concentration of 100%, an extraction period of 5 min, and a microwave power level of 10% (Singh et al., 2011). Several studies have been published comparing microwave-assisted extraction to traditional extraction procedures. The extraction of capsaicinoids from peppers, using traditional stirring extraction took 15 min to achieve the extractions of 95% of the capsaicinoids, while as microwave extraction achieved this percentage in 5 min (Barbero, Palma, & Barroso, 2006). Microwave extraction of antioxidants from the exotic *Gordonia axillaris* fruit was studied by Li et al. (2017). The best results were obtained with an ethanol concentration of 36.89%, a solvent/material ratio of 29.56 mL/g, extraction duration of 71.04 min, a temperature of 40°C, and a microwave power of 400 W. Furthermore, the microwave-assisted extraction approach was compared with traditional extraction techniques including soxhlet extraction and maceration. The antioxidant capacity of the microwave-assisted extract was higher than that produced by maceration and soxhlet techniques.

2.4 Pressurized liquid extraction

Pressurized liquid extraction method, also referred as accelerated solvent extraction, is regarded as an environmentally friendly extraction process for the extraction of bioactive compounds from plant sources (Alvarez-Rivera, Bueno, Ballesteros-Vivas, Mendiola, & Ibañez, 2020). This technique reduces the solvent consumption and extraction time, with increased efficiency and analyte recovery precision compared to other methods. Pressurized liquid extraction can be viewed as an extension of supercritical fluid extraction, utilizing organic solvents instead of CO_2. This technique holds solvents near their supercritical region where solvents have elevated extraction properties, while remaining in a liquid state. Pressurized liquid extraction parameters include extraction temperatures (25°C–200°C), pressure (500–3000 psi), solvents (organic to weak acids), number of extraction cycles, duration of static cycles and rinse volume.

This method uses pressure to raise the temperature of the extraction solvent beyond its boiling point. As this method achieves higher temperatures than traditional extraction methods, it leads to higher extraction performance. Higher temperatures result in increased solvent solubility potential, decreased viscosity, improved solvent penetration into plant cells and decreased solute matrix interactions. These effects increases the extraction yield, however results in a decrease in selectivity. Temperature is the primary optimizing parameter during the pressurized liquid extraction as it critically affects the extraction capacity. Depending on the chemical structure of the compounds, temperature may affect their selective extraction in a positive or negative way. Furthermore, even when pressure allows for temperatures higher than the boiling point of the solvent, it is not necessary to use excessive heat for the extraction (Lefebvre et al., 2020). Pressurized liquid extraction promoted the largest recovery of valuable compounds from beetroot leaves and stem (Lasta et al., 2019). The technique provided the higher extraction yield, phenolic content and antioxidant activity than soxhlet and ultrasound assisted extraction.

An optimization analysis was carried out with ethanol as the solvent and independent variables of extraction pressure (5–10 MPa), temperature (313K–393K), and static extraction time (3–15 min) for the extraction of anthocyanins and other phenolic compounds from jabuticaba skins (Santos, Veggi, & Meireles, 2012). Under optimal conditions, traditional low-pressure solvent extraction and pressurized liquid extraction yielded similar extraction yields; however, pressurized liquid extraction extracted 2.15 and 1.66 times more anthocyanins and total phenolic compounds, respectively. Pressurized liquid extraction has been shown to be a successful approach for recovering bioactive compounds from blackberry residues and other food by-products. Machado, Pasquel-Reátegui, Barbero, and Martínez (2015) used pressurized liquid extraction to remove antioxidant compounds from blackberry residues. Temperature has a beneficial effect on yield, overall phenolics, and antioxidant capacity. Authors reported that pressurized liquid extract method is superior for obtaining the extracts from plant as compared to other traditional methods.

2.5 Ultrasound-assisted extraction

The development of effective extraction techniques to minimize extraction time and increase yield is in highly demand. Ultrasound is a term used to describe sonic waves that have frequencies greater than those heard by the human ear. Ultrasound-assisted extraction has the benefit of being easy, taking lesser time and using little amount of solvent in comparison to other processes, and it can be conveniently combined with several other extraction methods. This procedure can resist oxidation and decomposition of target natural substances since it can be done at room temperature. The use of ultrasound-assisted extraction in the extraction of various natural ingredients has become increasingly common. This extraction methodology doesn't require any complex instrumentation and is relatively low cost technique (Fig. 2.5).

The utilization of ultrasound improves the extraction process. At specific frequencies and amplitudes, these waves produce cavitation bubbles, which implode as they reach a nonstable stage, releasing high temperature and pressure. This occurrence has the potential to break down cell walls, allowing metabolites to escape. Ultrasonic waves can be influenced by several factors, the major ones being frequency and amplitude. Power is the amplitude over time, while intensity is the power over surface area. These variables alter ultrasonic waves, causing plant samples to interact accordingly. Ramić et al. (2015) investigated the effect of various factors on the extraction of

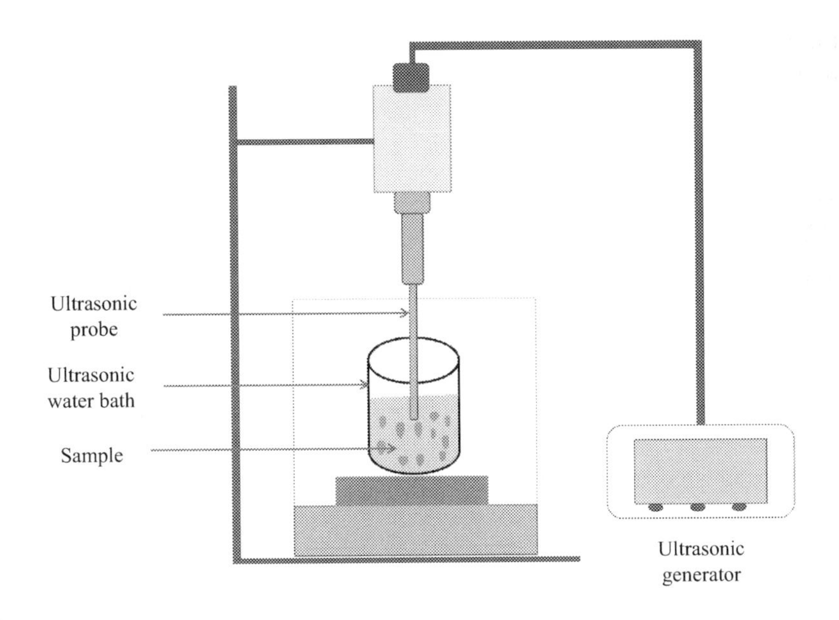

FIGURE 2.5

Schematic representation of basic ultrasound equipment.

polyphenolic compounds from *Aronia melanocarpa* by-products, including temperature (30°C–70°C) and ultrasonic strength (72–216 W). The optimized extraction temperature was 70°C and ultrasonic power of about 200 W. Bimakr, Ganjloo, Zarringhalami, and Ansarian (2017) investigated the impact of ultrasound-assisted extraction variables on the extractive value of bioactive phenolics from *Malva sylvestris* leaves. The concentration of bioactive phenolics showed a significant increase under the optimum ultrasound-assisted extraction conditions. Under the optimal ultrasound extraction conditions (48°C, 110.00 W, and 48.77 min), the experimental extractive value was 279.89 mg/g with 71.12% inhibition of scavenged DPPH, 73.35% inhibition of scavenged ABTS, and a total phenolic content of 152.25 mg GAE/g.

Pan, Qu, Ma, Atungulu, and McHugh (2012) used ultrasound extraction in continuous and pulsed modes for the extraction of antioxidants from the dry peel of pomegranate marc. Pulsed ultrasound-assisted extraction at an intensity level of 59.2 W/cm^2, a pulse duration of 5 s, and a resting interval of 5 s improved antioxidant yield by 22% and minimized time of extraction by 87% as compared to traditional extraction. Similarly, at the same intensity level, continuous ultrasound-assisted extraction enhanced antioxidant yield by 24%, while decreasing extraction time by 90%. As pulsed ultrasound-assisted extraction saved 50% more energy than continuous ultrasound-assisted extraction, the study suggests using pulsed ultrasound-assisted technique for the extraction with 14.5% antioxidant yield and 5.8 g/g DPPH scavenging activity. Vinatoru (2001) compared ultrasonic extraction to traditional methods for extracting bioactive substances from fennel, marigold and mint. The findings showed that the extraction yield (34%) was significantly higher than that of the traditional processes. Chukwumah, Walker, Verghese, and Ogutu (2009) investigated the impacts of ultrasound conditions (frequency and duration of ultrasound) on the extraction efficiency of peanut isoflavones and trans-resveratrol. The findings demonstrate that ultrasound treatment at frequency 80 kHz increased resveratrol, while ultrasound treatment at 25 kHz improved the extraction yields of daidzein and genistein.

2.6 Enzyme-assisted extraction

Enzyme-assisted extraction is one of the promising methods to conventional methods for the preparation of plant extracts. Enzymes have are beneficial to recover the phytochemical compounds from the plant material and their by-products. The main advantage of this method is the decrease of solvent quantity, environmental friendliness, selectivity and ability to catalyze the reaction under mild conditions. Enzyme based extraction method depends on the reaction time, types and concentration of enzyme. Furthermore, this method reduces the extraction time and increases the quality and quality of plant extracts. The enzymes used for the extraction hydrolyzing the plant matrix and the enzymatic reactions disintegrate the cell wall and increases the intracellular release.

The commonly used enzymes evaluated for the extraction of plant extracts are cellulose, pectinase, tannase etc. The enzymes with specific properties rupture the matrix of the plant material in order to gain the access to the bioactive compounds found within the cell matrix and bound to the cell walls. Enzyme assisted extraction resulted in the enhanced phytochemical compounds from bay leaves (Boulila et al., 2015), ginger (Manasa, Srinivas, & Sowbhagya, 2013), red capsicum (Nath, Kaur, Rudra, & Varghese, 2016), sweet cherry pomace (Domínguez-Rodríguez, Marina, & Plaza, 2021). To ensure the complete fragmentation of cellular matrix, it is important to select the appropriate enzyme for the extraction.

2.7 Pulsed electric field extraction

Pulsed electric field process uses the high voltage pulses for different applications in food industry. Pulsed electric field method is one of the promising techniques used for the extraction of various valuable compounds from the plant material. The exposure of plant material to pulsed electric disrupts the cell membrane via electroporation by the high voltage short pulses and promotes the release of intracellular components (Fig. 2.6). The pulsed electric field method obtained extracts from plant material with high yield and purity. This method also helps to shortening the extraction time, decrease the solvent amount and also facilitate the selective extraction of intracellular compounds. The extraction efficiency by pulsed electric field depends on the factors such as properties of plant material, number and duration of electric pulses. The critical factors which effect the processing during this technique are treatment time, electric field intensity, pulse wave form, conductivity and ionic strength of the medium.

The applicability of pulsed electric field for the extraction of bioactive compounds is reported in literature. Pulsed electric field method is a potential alternative for the conventional extraction methods. Pulsed electric field (2 kV/cm, 30 pulses) treatment significantly enhances the extraction of the naringin bioactive from the pomelo peel (Niu, Ren, Li, Zeng, & Li, 2021). The pulsed electric field extraction efficiency increased with increase of intensity from custard apple leaf extraction (Shiekh, Olatunde, Zhang, Huda, & Benjakul, 2021). The increased extraction yield is due to the

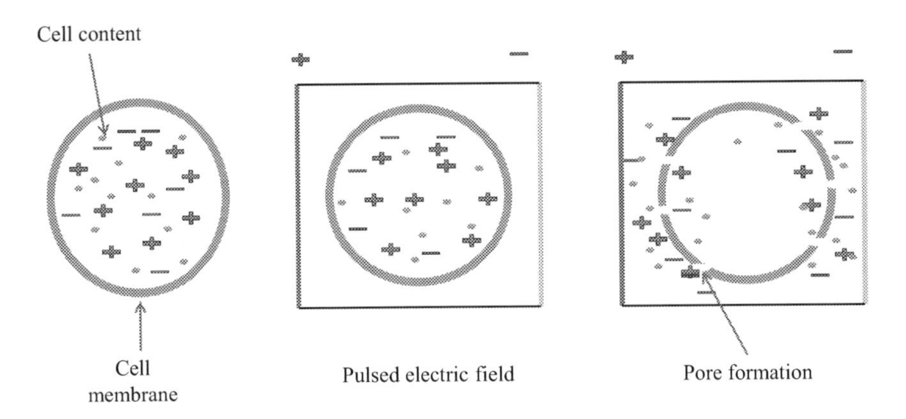

FIGURE 2.6

Schematic representation of electroporation in the cell membrane due to pulsed electric field.

electroporation which facilitates the disintegration of cell. Pashazadeh, Elhamirad, Hajnajari, Sharayei, and Armin (2020) also reported that pulsed electric field increased the extraction of bioactive substance from cinnamon by increasing the plant cell permeability. Pulsed electric field method has shown the enhanced phenolic and antioxidant recover from rosemary and thyme by-products as compared to the ultrasound method (Tzima, Brunton, Lyng, Frontuto, & Rai, 2021). The application of pulsed electric field enhanced the extraction of polyphenols form citrus fruits at 3 kV/cm (El Kantar et al., 2018).

2.8 Conclusion

The exploitation of bioactive compounds in the food industry has urged the researchers and industries to use different methodologies for the extraction of these compounds from plant materials. Many protocols and standards have been used, which affect the yield and composition of the extracted material. Many conventional methods have been used since long time in the industry. However, for some decades, nonconventional extraction methods also gained significant popularity and were commercialized at industrial scale. Emerging methods are environmental friendly extraction techniques due to the decreased use of synthetic chemicals, reduced process time, better yield, and quality of extract. The efficiency of the extraction methods depends on various factors including nature of plant matrix, chemistry of plant extract, and type of solvent used for extraction and methodology.

References

Alara, O. R., Abdurahman, N. H., & Ukaegbu, C. I. (2018). Soxhlet extraction of phenolic compounds from Vernonia cinerea leaves and its antioxidant activity. *Journal of Applied Research on Medicinal and Aromatic Plants*, *11*, 12–17.

Alvarez-Rivera, G., Bueno, M., Ballesteros-Vivas, D., Mendiola, J. A., & Ibañez, E. (2020). *Pressurized liquid extraction. Liquid-Phase Extraction* (pp. 375–398). Elsevier.

Araujo, R. G., Rodríguez-Jasso, R. M., Ruíz, H. A., Govea-Salas, M., Pintado, M., & Aguilar, C. N. (2021). Recovery of bioactive components from avocado peels using microwave-assisted extraction. *Food and Bioproducts Processing, 127*, 152–161.

Bachtler, S., & Bart, H. J. (2021). Increase the yield of bioactive compounds from elder bark and annatto seeds using ultrasound and microwave assisted extraction technologies. *Food and Bioproducts Processing, 125*, 1–13.

Ballard, T. S., Mallikarjunan, P., Zhou, K., & O'Keefe, S. (2010). Microwave-assisted extraction of phenolic antioxidant compounds from peanut skins. *Food Chemistry, 120*(4), 1185–1192.

Barbero, G. F., Palma, M., & Barroso, C. G. (2006). Determination of capsaicinoids in peppers by microwave-assisted extraction–high-performance liquid chromatography with fluorescence detection. *Analytica Chimica Acta, 578*(2), 227–233.

Bimakr, M., Ganjloo, A., Zarringhalami, S., & Ansarian, E. (2017). Ultrasound-assisted extraction of bioactive compounds from Malva sylvestris leaves and its comparison with agitated bed extraction technique. *Food Science and Biotechnology, 26*(6), 1481–1490.

Boulila, A., Hassen, I., Haouari, L., Mejri, F., Amor, I. B., Casabianca, H., & Hosni, K. (2015). Enzyme-assisted extraction of bioactive compounds from bay leaves (Laurus nobilis L.). *Industrial Crops and Products, 74*, 485–493.

Chukwumah, Y. C., Walker, L. T., Verghese, M., & Ogutu, S. (2009). Effect of frequency and duration of ultrasonication on the extraction efficiency of selected isoflavones and trans-resveratrol from peanuts (Arachis hypogaea). *Ultrasonics Sonochemistry, 16*(2), 293–299.

Ćujić, N., Šavikin, K., Janković, T., Pljevljakušić, D., Zdunić, G., & Ibrić, S. (2016). Optimization of polyphenols extraction from dried chokeberry using maceration as traditional technique. *Food Chemistry, 194*, 135–142.

Domínguez-Rodríguez, G., Marina, M. L., & Plaza, M. (2021). Enzyme-assisted extraction of bioactive non-extractable polyphenols from sweet cherry (Prunus avium L.) pomace. *Food Chemistry, 339*, 128086.

El Kantar, S., Boussetta, N., Lebovka, N., Foucart, F., Rajha, H. N., Maroun, R. G., & Vorobiev, E. (2018). Pulsed electric field treatment of citrus fruits: Improvement of juice and polyphenols extraction. *Innovative Food Science & Emerging Technologies, 46*, 153–161.

Esquivel-Hernández, D. A., López, V. H., Rodríguez-Rodríguez, J., Alemán-Nava, G. S., Cuéllar-Bermúdez, S. P., Rostro-Alanis, M., & Parra-Saldívar, R. (2016). Supercritical carbon dioxide and microwave-assisted extraction of functional lipophilic compounds from *Arthrospira platensis*. *International Journal of Molecular Sciences, 17*(5), 658.

Farooq, S., Shah, M. A., Siddiqui, M. W., Dar, B. N., Mir, S. A., & Ali, A. (2020). Recent trends in extraction techniques of anthocyanins from plant materials. *Journal of Food Measurement and Characterization*, 1–12.

Goyeneche, R., Fanovich, A., Rodrigues, C. R., Nicolao, M. C., & Di Scala, K. (2018). Supercritical CO_2 extraction of bioactive compounds from radish leaves: Yield, antioxidant capacity and cytotoxicity. *The Journal of Supercritical Fluids, 135*, 78–83.

Gupta, A., Naraniwal, M., & Kothari, V. (2012). Modern extraction methods for preparation of bioactive plant extracts. *International Journal of Applied And Natural Sciences, 1*(1), 8–26.

Herrero, M., Mendiola, J. A., Cifuentes, A., & Ibáñez, E. (2010). Green processes for the extraction of bioactives from Rosemary: Chemical and functional characterization via ultra-performance liquid chromatography-tandem mass spectrometry and in-vitro assays. *Journal of Chromatography A, 1217*, 2495–2511.

Karabegović, I. T., Stojičević, S. S., Veličković, D. T., Todorović, Z. B., Nikolić, N. Č., & Lazić, M. L. (2014). The effect of different extraction techniques on the composition and antioxidant activity of cherry laurel (Prunus laurocerasus) leaf and fruit extracts. *Industrial Crops and Products, 54*, 142–148.

Lama-Muñoz, A., del Mar Contreras, M., Espínola, F., Moya, M., Romero, I., & Castro, E. (2020). Content of phenolic compounds and mannitol in olive leaves extracts from six Spanish cultivars: Extraction with the Soxhlet method and pressurized liquids. *Food Chemistry*, *320*, 126626.

Lasta, H. F. B., Lentz, L., Rodrigues, L. G. G., Mezzomo, N., Vitali, L., & Ferreira, S. R. S. (2019). Pressurized liquid extraction applied for the recovery of phenolic compounds from beetroot waste. *Biocatalysis and Agricultural Biotechnology*, *21*, 101353.

Lefebvre, T., Destandau, E., & Lesellier, E. (2020). Selective extraction of bioactive compounds from plants using recent extraction techniques: A review. *Journal of Chromatography A*, 461770.

Li, Y., Li, S., Lin, S. J., Zhang, J. J., Zhao, C. N., & Li, H. B. (2017). Microwave-assisted extraction of natural antioxidants from the exotic *Gordonia axillaris* fruit: Optimization and identification of phenolic compounds. *Molecules (Basel, Switzerland)*, *22*(9), 1481.

Machado, A. P. D. F., Pasquel-Reátegui, J. L., Barbero, G. F., & Martínez, J. (2015). Pressurized liquid extraction of bioactive compounds from blackberry (Rubus fruticosus L.) residues: A comparison with conventional methods. *Food Research International*, *77*, 675–683.

Manasa, D., Srinivas, P., & Sowbhagya, H. B. (2013). Enzyme-assisted extraction of bioactive compounds from ginger (Zingiber officinale Roscoe). *Food Chemistry*, *139*(1–4), 509–514.

Marsoul, A., Ijjaali, M., Oumous, I., Bennani, B., & Boukir, A. (2020). Determination of polyphenol contents in Papaver rhoeas L. flowers extracts (soxhlet, maceration), antioxidant and antibacterial evaluation. *Materials Today: Proceedings*, *31*, S183–S189.

Nath, P., Kaur, C., Rudra, S. G., & Varghese, E. (2016). Enzyme-assisted extraction of carotenoid-rich extract from red capsicum (Capsicum annuum). *Agricultural Research*, *5*(2), 193–204.

Niu, D., Ren, E. F., Li, J., Zeng, X. A., & Li, S. L. (2021). Effects of pulsed electric field-assisted treatment on the extraction, antioxidant activity and structure of naringin. *Separation and Purification Technology*, *265*, 118480.

Pan, Z., Qu, W., Ma, H., Atungulu, G. G., & McHugh, T. H. (2012). Continuous and pulsed ultrasound-assisted extractions of antioxidants from pomegranate peel. *Ultrasonics Sonochemistry*, *19*(2), 365–372.

Pashazadeh, B., Elhamirad, A. H., Hajnajari, H., Sharayei, P., & Armin, M. (2020). Optimization of the pulsed electric field-assisted extraction of functional compounds from cinnamon. *Biocatalysis and Agricultural Biotechnology*, *23*, 101461.

Proestos, C., & Komaitis, M. (2008). Application of microwave-assisted extraction to the fast extraction of plant phenolic compounds. *LWT-Food Science and Technology*, *41*(4), 652–659.

Radojković, M., Zeković, Z., Mašković, P., Vidović, S., Mandić, A., Mišan, A., & Đurović, S. (2016). Biological activities and chemical composition of Morus leaves extracts obtained by maceration and supercritical fluid extraction. *The Journal of Supercritical Fluids*, *117*, 50–58.

Ramić, M., Vidović, S., Zeković, Z., Vladić, J., Cvejin, A., & Pavlić, B. (2015). Modeling and optimization of ultrasound-assisted extraction of polyphenolic compounds from Aronia melanocarpa by-products from filter-tea factory. *Ultrasonics Sonochemistry*, *23*, 360–368.

Safdar, M. N., Kausar, T., Jabbar, S., Mumtaz, A., Ahad, K., & Saddozai, A. A. (2017). Extraction and quantification of polyphenols from kinnow (Citrus reticulate L.) peel using ultrasound and maceration techniques. *Journal of Food and Drug Analysis*, *25*(3), 488–500.

Santos, D. T., Veggi, P. C., & Meireles, M. A. A. (2012). Optimization and economic evaluation of pressurized liquid extraction of phenolic compounds from jabuticaba skins. *Journal of Food Engineering*, *108*(3), 444–452.

Sharma, B. R., Kumar, V., Kumar, S., & Panesar, P. S. (2020). Microwave assisted extraction of phytochemicals from Ficus racemosa. *Current Research in Green and Sustainable Chemistry*, *3*, 100020.

Shiekh, K. A., Olatunde, O. O., Zhang, B., Huda, N., & Benjakul, S. (2021). Pulsed electric field assisted process for extraction of bioactive compounds from custard apple (Annona squamosa) leaves. *Food Chemistry*, 129976.

Singh, A., Sabally, K., Kubow, S., Donnelly, D. J., Gariepy, Y., Orsat, V., & Raghavan, G. S. V. (2011). Microwave-assisted extraction of phenolic antioxidants from potato peels. *Molecules (Basel, Switzerland)*, *16*(3), 2218–2232.

Tzima, K., Brunton, N. P., Lyng, J. G., Frontuto, D., & Rai, D. K. (2021). The effect of pulsed electric field as a pre-treatment step in ultrasound assisted extraction of phenolic compounds from fresh rosemary and thyme by-products. *Innovative Food Science & Emerging Technologies*, *69*, 102644.

Upadhyay, R., Ramalakshmi, K., & Rao, L. J. M. (2012). Microwave-assisted extraction of chlorogenic acids from green coffee beans. *Food Chemistry*, *130*(1), 184–188.

Vinatoru, M. (2001). An overview of the ultrasonically assisted extraction of bioactive principles from herbs. *Ultrasonics Sonochemistry*, *8*(3), 303–313.

Zhang, H., Li, Q., Qiao, G., Qiu, Z., Wen, Z., & Wen, X. (2020). Optimizing the supercritical carbon dioxide extraction of sweet cherry (Prunus avium L.) leaves and UPLC-MS/MS analysis. *Analytical Methods*, *12*(23), 3004–3013.

CHAPTER

Chemistry of plant extracts

3

Havalli Bommegowda Rashmi and Pradeep Singh Negi
Department of Fruit and Vegetable Technology, CSIR-Central Food Technological Research Institute, Mysore, India

3.1 Introduction

Plant extracts have been used traditionally to cure and prevent diseases throughout the history. Plant extracts are used as preservatives because of their strong antimicrobial and antioxidant activities. The efficacy of plant extracts lies in chemical substances possessed by them, and these compounds are classified as terpenoids, alkaloids, phenolic compounds, glucosinolates, and various organic acids. Plant extract or pure compounds obtained from various plants offer opportunities for food preservation because of their chemical diversity. For centuries, many plant products have been used to improve the sensory characteristics and extend the shelf life of many foods. Plant extracts extend the storage life of foods by controlling the growth of spoilage and pathogenic bacteria and enhance the quality of foods by inhibiting the oxidative rancidity.

The activity of plant extracts is determined by its chemical properties, the solubility being most important for food application, and its pH being the other important factors (Stratford & Eklund, 2003). Active ingredients in plant extracts differ due to many factors such as parts of plant utilized (roots, stems, leaves, bark, flowers, fruits, and seeds), stage of their maturity, genotype, climatic factors, soil factors, cultivation practices, the time of harvesting, postharvest operations, pretreatments before extraction, and the extraction methods. Plant extracts are widely used in the food industry, and this chapter summarizes the extraction of phytochemicals, their chemical composition, and relationship of various chemical structures in plant extracts with biological activities exerted by them for further widening the application of plant extracts in natural preservation.

3.2 Extraction procedures and chemical composition of plant extracts

The procedure adopted for the extraction of compounds from plants determines the quality and quantity of plant extracts. Various extraction methods followed include solvent extraction (water, ethanol, methanol, and other solvents) (Pinelo, Rubilar, Jerez, Sineiro, & Núñez, 2005; Ye, Liang, Li, & Zhao, 2015), pressurized-liquid extraction (Luthria, 2008), supercritical-fluid extraction (Inczedy, Lengyel, & Ure, 1998; Temelli & Güçlü-Üstündağ, 2005), and various assisted extraction techniques, such as the microwave-assisted extraction (Alupului, Calinescu, & Lavric, 2012), ultrasound assisted extraction (Herrera & De Castro, 2005), and enzyme assisted extraction (Rosenthal,

Plant Extracts: *Applications in the Food Industry*. DOI: https://doi.org/10.1016/B978-0-12-822475-5.00004-1

Pyle, & Niranjan, 1996). Before the extraction of any bioactive compound, it is also important to consider the developmental stage and plant part accumulating the bioactive compound. A distribution of various chemical classes of bioactive compounds present in plants is presented in Table 3.1. The plant extracts can be quantified for the presence of active compounds by various

Table 3.1 Distribution of various chemical classes of bioactive compounds in plants.

Chemical class of bioactive compound	Distribution of bioactive compounds in plants (developmental stage/tissue/plant part)	References
Monoterpenes Sesquiterpenes	Development and growth phase; aerial parts of plants	Aharoni et al. (2003)
Glucosinolates	High content in seeds, siliques, and young leaves; moderate content in roots, stems, and leaves; and low content in senescing leaves	Brown, Tokuhisa, Reichelt, and Gershenzon (2003)
Anthocyanins	Vegetative tissue and embryo	Lepiniec et al. (2006)
Proanthocyanidins	Endothelium of developing seed coats	
Flavonols	Vegetative and reproductive tissues	
Flavan-4-ol 3-deoxyflavonoids apiferol & luteoferol	Pericarp	
Isoflavones	Embryo and seed coat	Halkier and Gershenzon (2006)
Benzylisoquinoline-derived alkaloids	Accumulates in specialized cell present in vascular tissues known as laticifers	Bird, Franceschi, and Facchini (2003), Weid, Ziegler, and Kutchan (2004)
Myrosinase and glucosinolates	Distinct subcellular compartments	Grubb and Abel (2006)
Limonene	*Pinus ponderosa:* bark	Harborne (1986, 1989, 1993, 1997)
Pulegone and carvone	*Satureja douglasii*: leaf	
Lactucin 8-deoxylactucin	*Cichorium intybus*: leaf	
Caryophyllene epoxide	*Melampodium divaricatum:* leaf	
Zingiberene	*Lycopersicon hirsutum:* leaf trichome	
Germacrone	*Ledum groenlandicum*: leaf	
Kaurenoic Trachylobanoic acids	Floret of sunflower (*Helianthus annuus)*	
Camphor	White spruce: leaf	
Alkaloids Cyanogens Monoterpenoids Diterpenoids	Seeds	Harborne and Baxter (1996)
Tannins	Fruits of persimmon (*Diospyros kaki)*	Harborne (1999)
Saponins	Holly (*Ilex opaca*): juvenile leaves Leek (*Allium porrum)*:flowers	
Sesquiterpenes: (*E*)-β-farnesene, β-humulene, and (γ)-muurolene	*Chrysothamnus nauseasus*: leaves	

chromatographic techniques. Plant extracts are available in various forms such as spray-dried powders, pure active ingredients, and encapsulated forms. A general flow diagram for extraction, quantification, and utility of plant extracts is presented in Fig. 3.1.

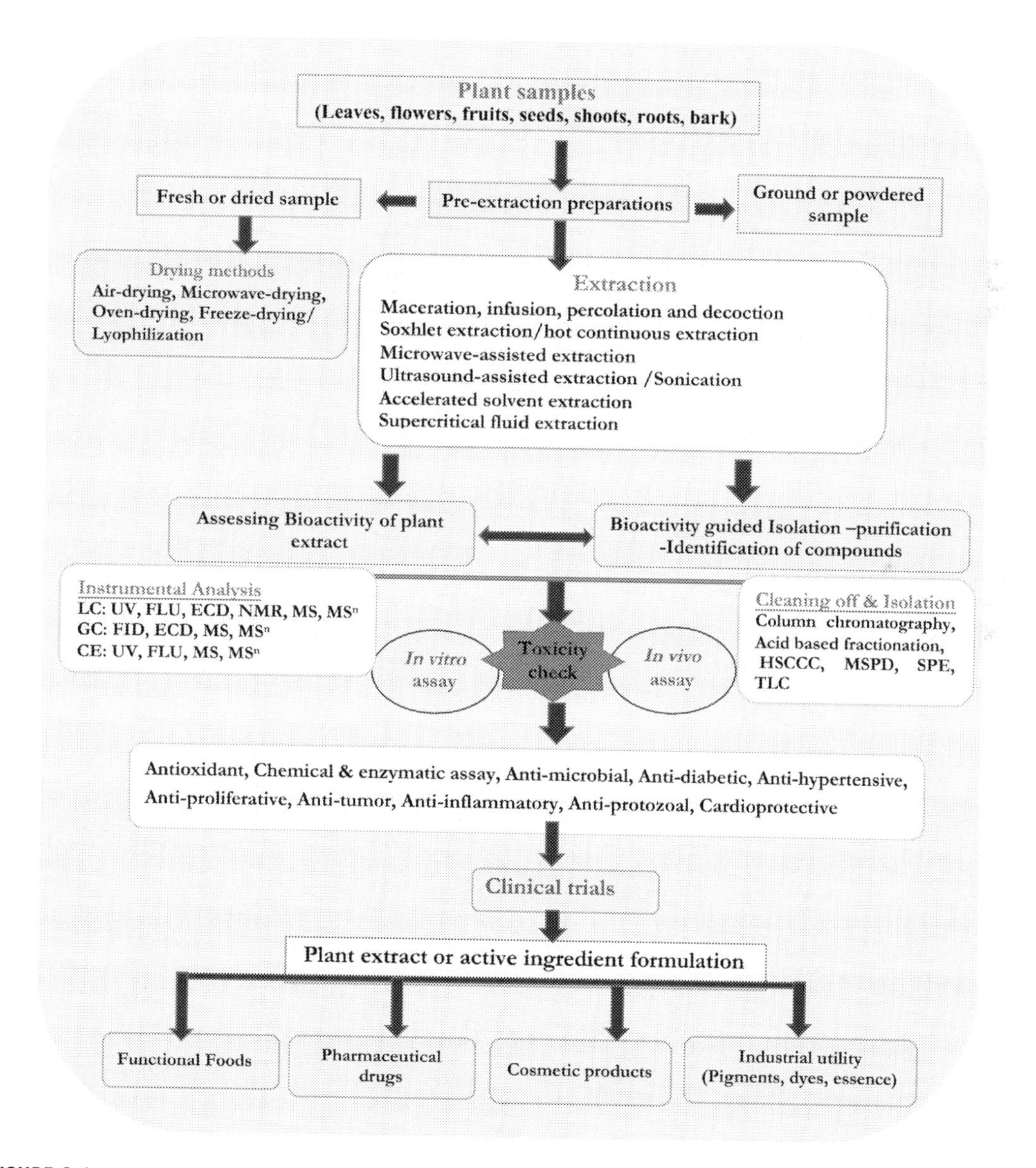

FIGURE 3.1

General scheme of extraction, identification, and utilization of plant extracts.

Various fruits, vegetables, herbs, and spices are used to extract the natural bioactive compounds (Dimitrios, 2006). Several plant extracts or molecules from plants have been commercialized, like moso bamboo (Takeguard) of Takex Labo (Japan) or blend of several natural extracts including green tea (Biovia YM10) by Danisco DuPont, rosemary extract by Naturex's, rosemary extract (Oxikan) by Kancor, and rosemary extract (ExtenFo) by Arjuna Chemicals, India. Members of Punicaceae, Juglandaceae, Rosaceae, Grossulariaceae, Asteraceae, Ericaceae, Empetraceae, and Zingiberaceae families were identified as the main source of plant bioactive compounds (Halvorsen et al., 2002).

3.3 Classification of bioactive compounds in plant extracts

Bioactives in plant extracts are mainly classified as terpenoids, phenolic compounds, glucoinolates, and few miscellaneous compounds. The classification of bioactive compounds based on their chemical structure is discussed in this section, and chemical structures of a few bioactive compounds reported in plant extracts are presented in Fig. 3.2.–3.4.

3.3.1 Terpenoids

Terpenes refer to isoprene polymers and their derivatives with the general formula of (C_5H_8) n, which are most abundantly found in plants and meager in animals. Majority of terpenes form several oxygenated derivatives such as alcohols, aldehydes, ketones, carboxylic acids, esters and glycosides, along with few nitrogen-containing and sulfur-containing derivatives. Terpenoids are hydrocarbons synthesized from isoprene subunits through reactions such as condensation and cyclization, as well they are hydrophobic in nature (Ruchika & Pandey, 2019). Terpenes are categorized depending on their isoprene units, as hemiterpene, monoterpene, sesquiterpene, diterpene, sesterpene, triterpene, tetraterpene and polyterpene (Table 3.2).

Terpenoids compounds have significant role in plants, being a part of important hormones such as gibberellin, abscisic acid and insect juvenile hormone, photosynthetic pigments such as carotenoids and chlorophyll; prominent molecule plastoquinone in photosynthesis and quinone in respiration chain; sterols being the component of the biological membrane. Monoterpene and sesquiterpene can be seen as major component of volatile oil, diterpene in resin; triterpenoid in plant saponins and resins, tetraterpene in some fat-soluble pigments. Few of the bioactivities of terpenoids include roundworm expelling effect by ascaridole and santonin; antimalarial activity by artemisinin, while antibacterial by andrographolidume (Jan & Abbas, 2018; Ruchika & Pandey, 2019).

Carotenoids, the major terpenoids distributed widely in plants, are responsible for different hues and are commonly occurring natural pigments (Namitha & Negi, 2010). Most carotenoids are derived from a 40-carbon structure consisting of eight isoprene units, which includes a system of conjugated double bonds. Carotenoids are classified as carotenes (just carbon and hydrogen atoms) and oxocarotenoids or xanthophylls (minimum one oxygen atom along with carbon and hydrogen). The nature of the specific end groups in carotenoids influences their polarity, and therefore individual carotenoids interact with biological membranes differently (Britton, 1995). Carotenoids

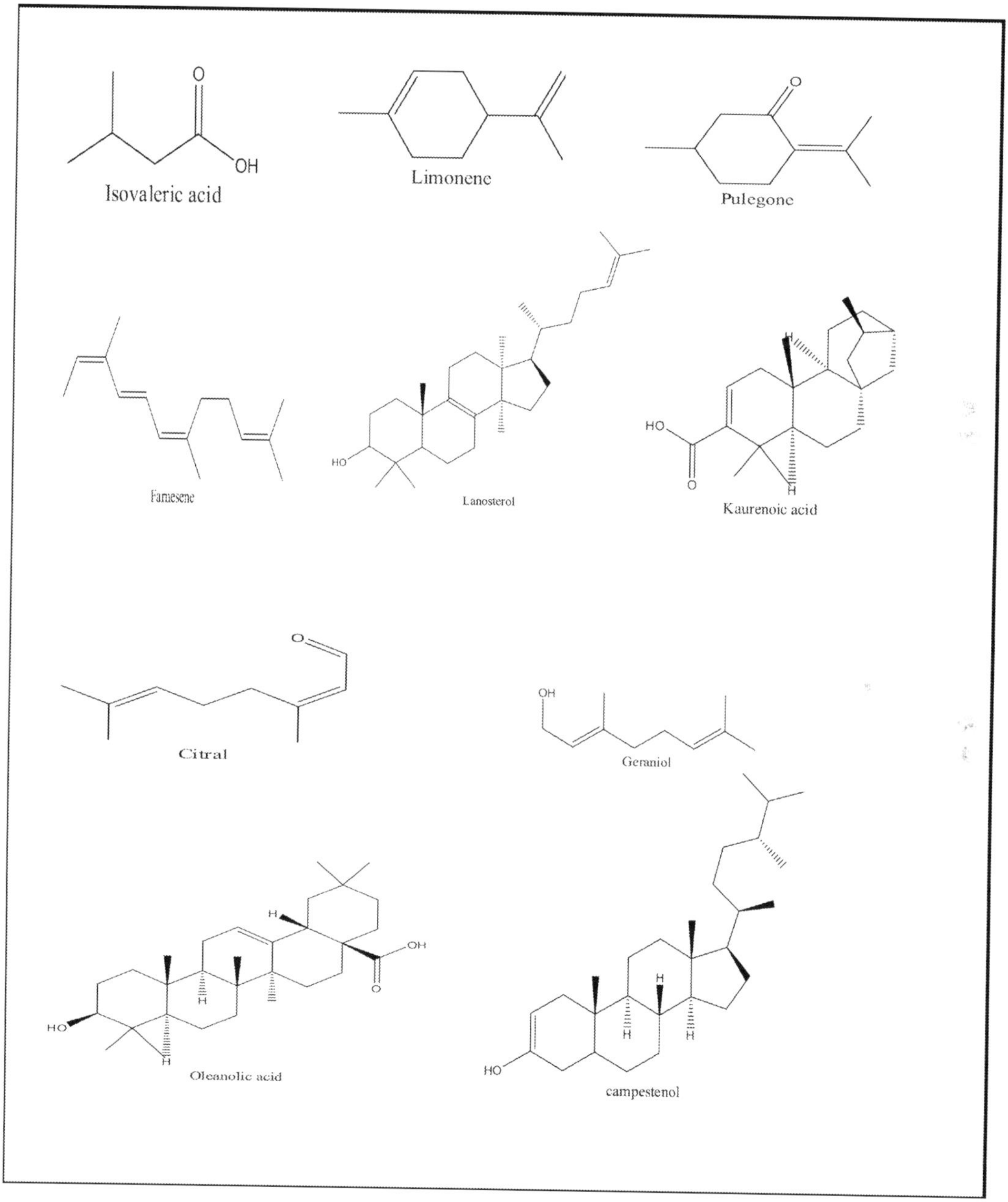

FIGURE 3.2

Chemical structures of terpenoids.

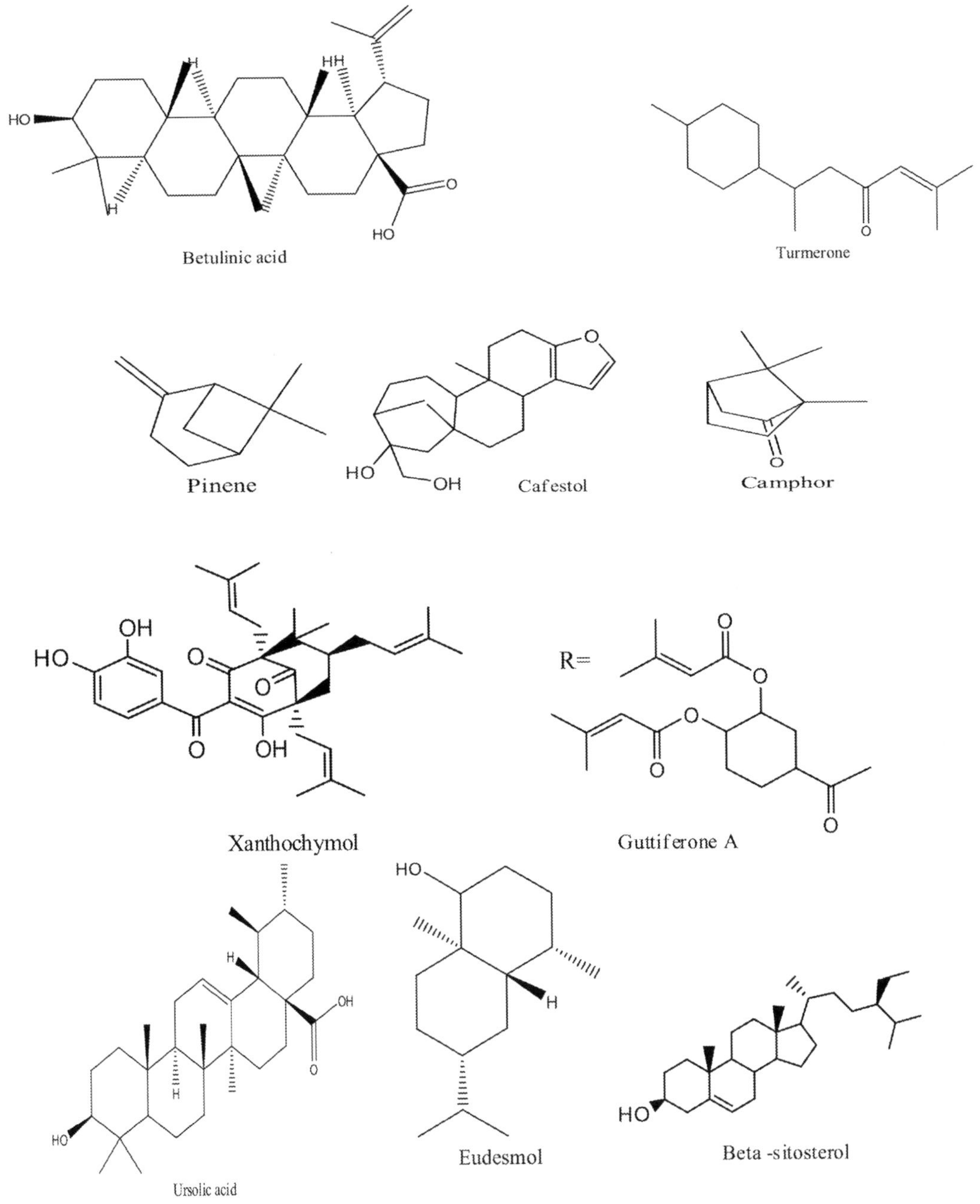

Betulinic acid

Turmerone

Pinene

Cafestol

Camphor

Xanthochymol

R=

Guttiferone A

Ursolic acid

Eudesmol

Beta -sitosterol

FIGURE 3.2

(Continued)

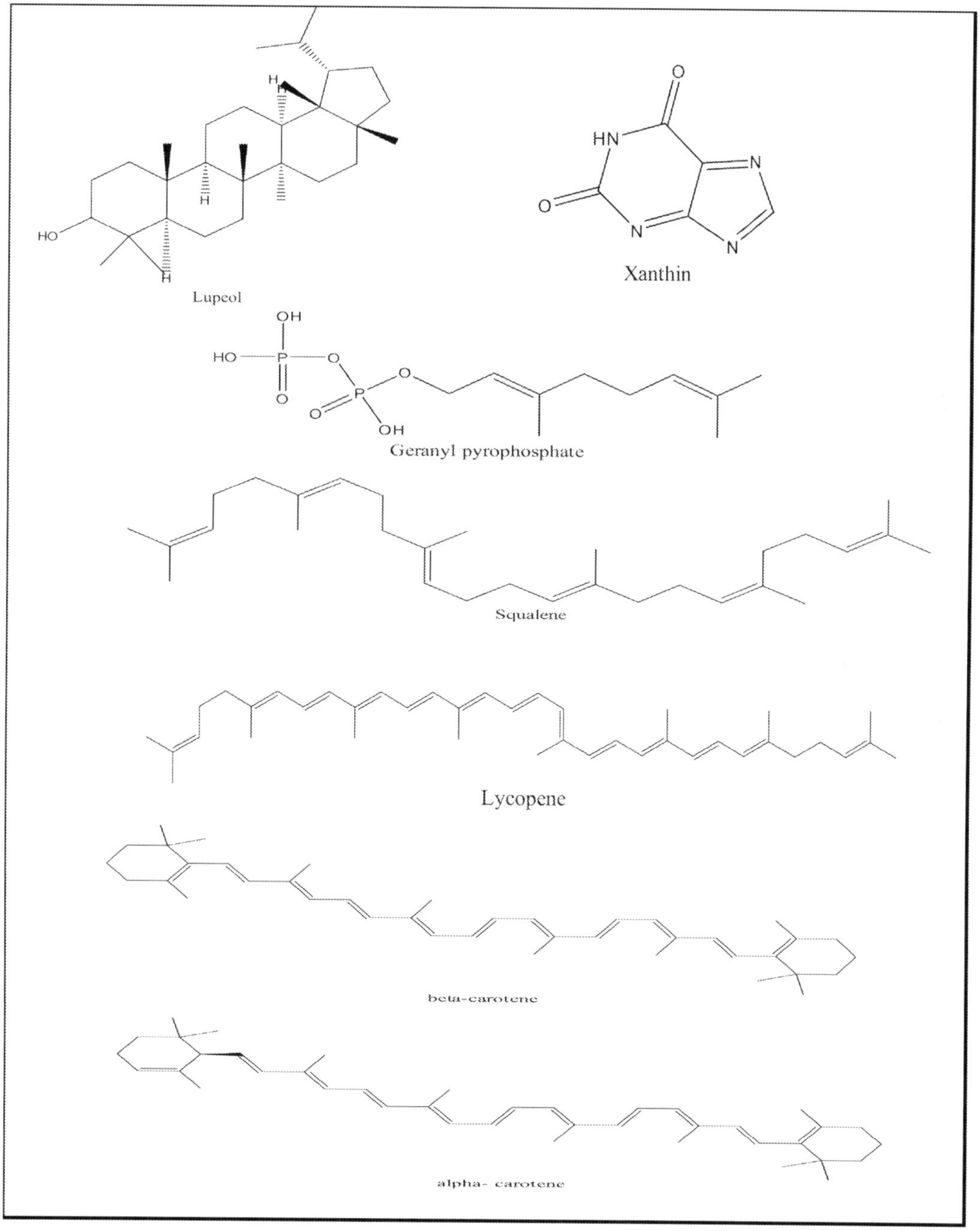

Lupeol

Xanthin

Geranyl pyrophosphate

Squalene

Lycopene

beta-carotene

alpha- carotene

FIGURE 3.2

(Continued)

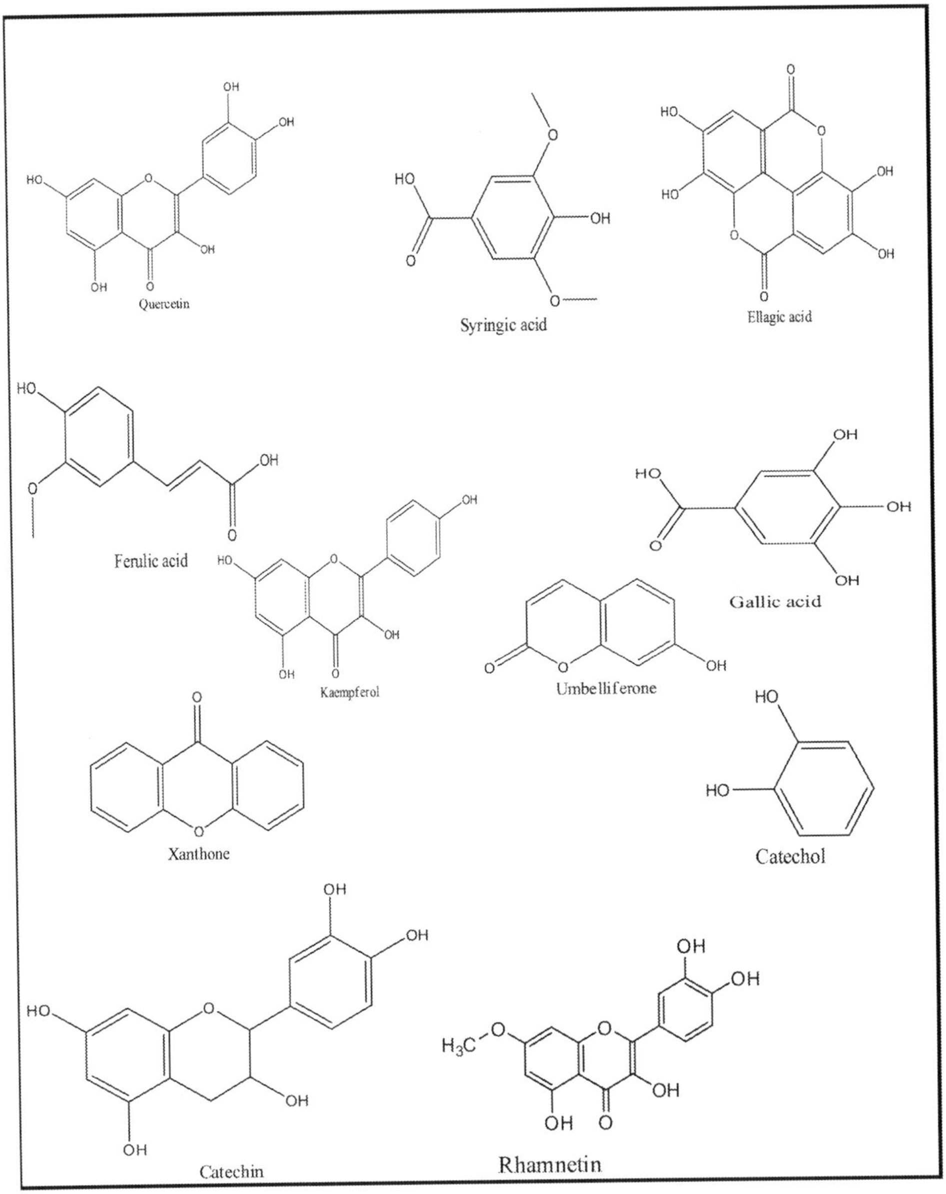

FIGURE 3.3

Chemical structures of phenolic compounds.

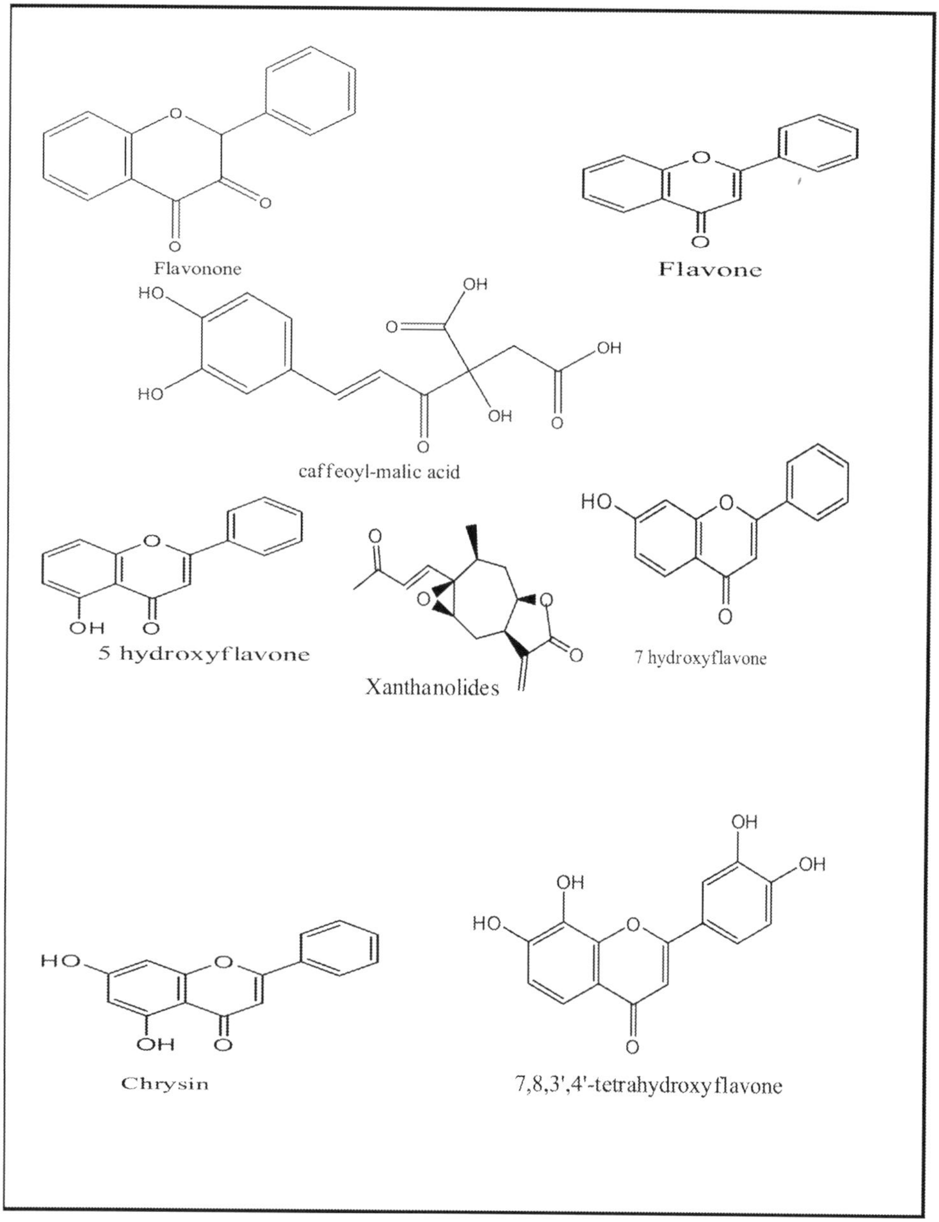

Flavonone

Flavone

caffeoyl-malic acid

5 hydroxyflavone

Xanthanolides

7 hydroxyflavone

Chrysin

7,8,3',4'-tetrahydroxyflavone

FIGURE 3.3

(Continued)

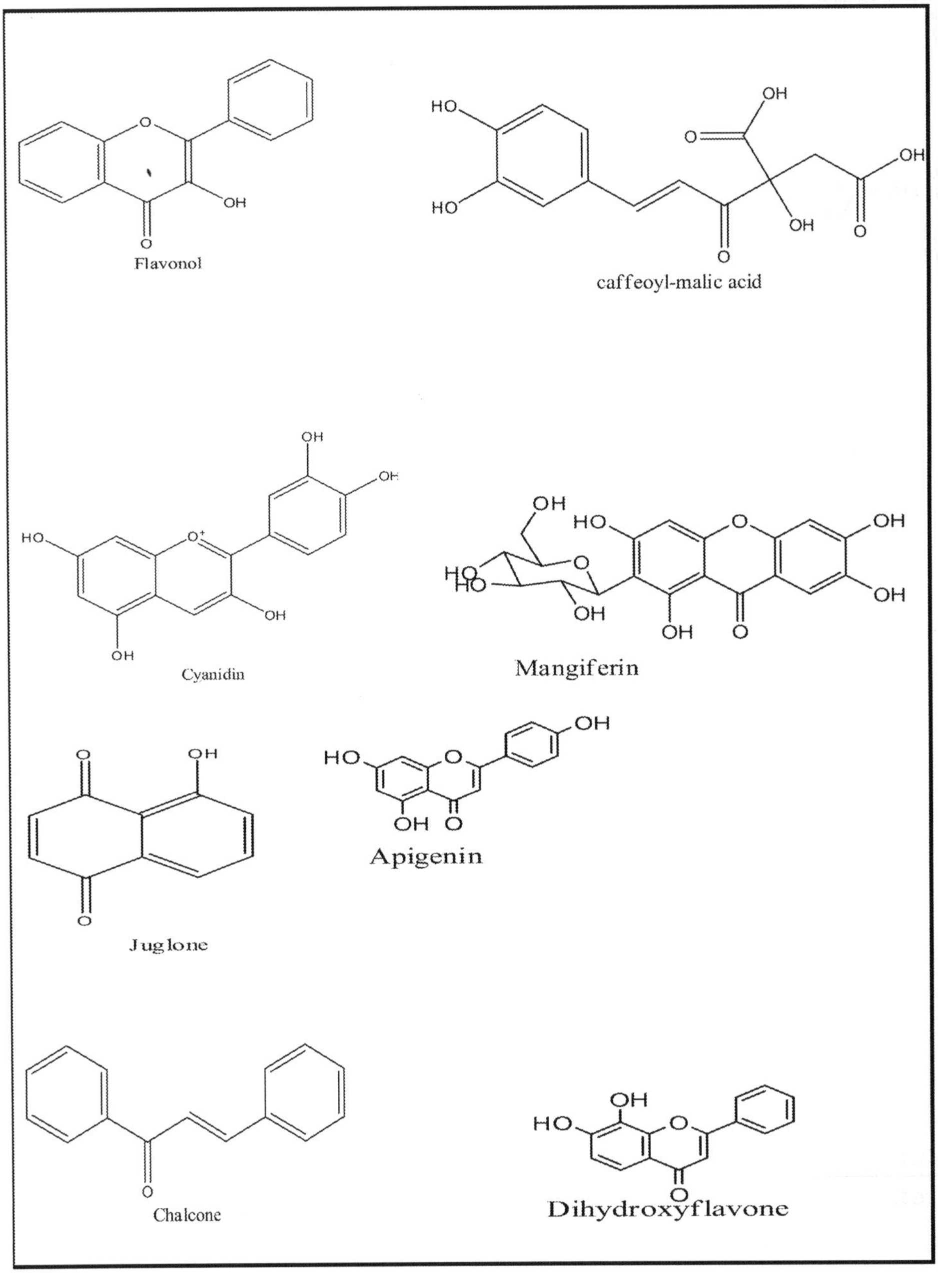

Flavonol

caffeoyl-malic acid

Cyanidin

Mangiferin

Apigenin

Juglone

Chalcone

Dihydroxyflavone

FIGURE 3.3

(Continued)

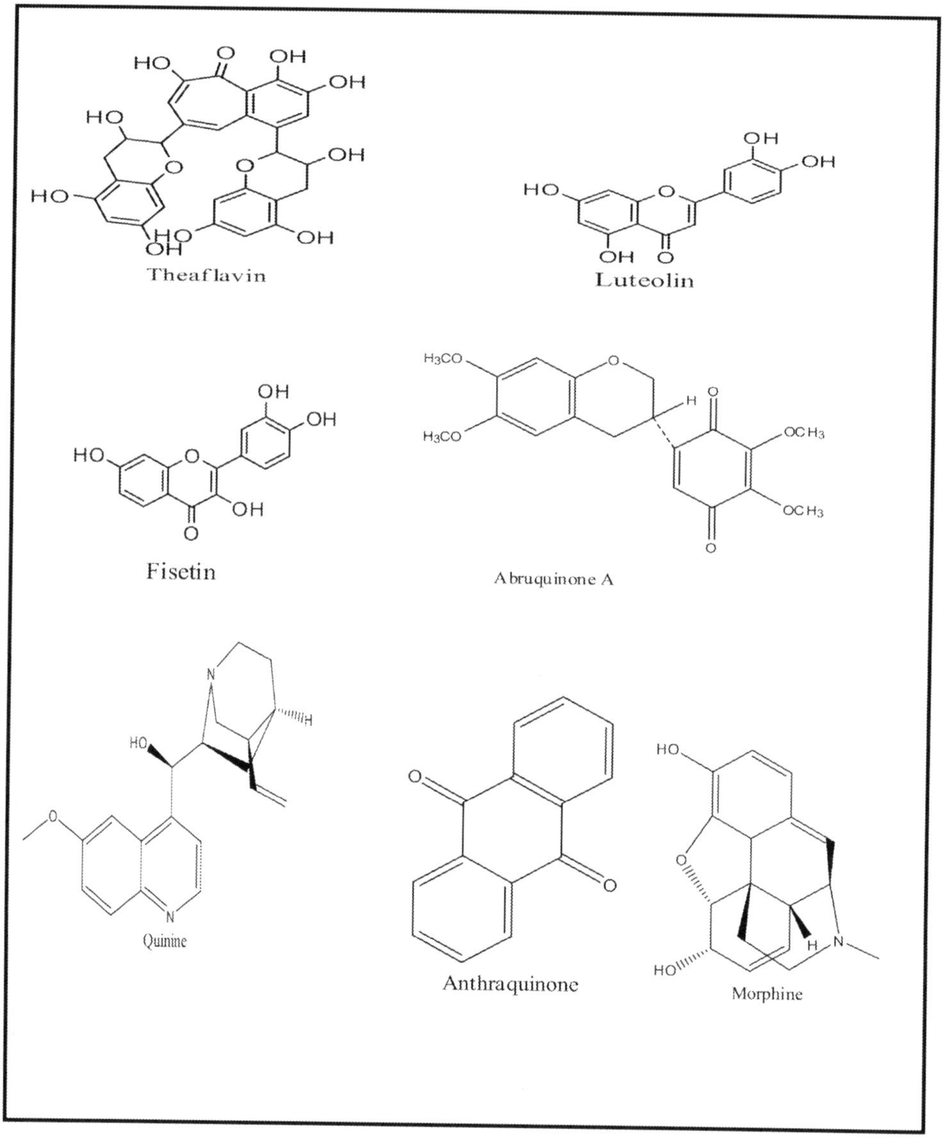

FIGURE 3.3

(Continued)

generally get isomerized and form a mixture of mono- and poly-cis-isomers in addition to the natural all-trans form. Being lipophilic, carotenoids accumulate in lipophilic parts of cells, which influences their absorption, transport and excretion in the organism (Stahl, Schwarz, & Sies, 1993).

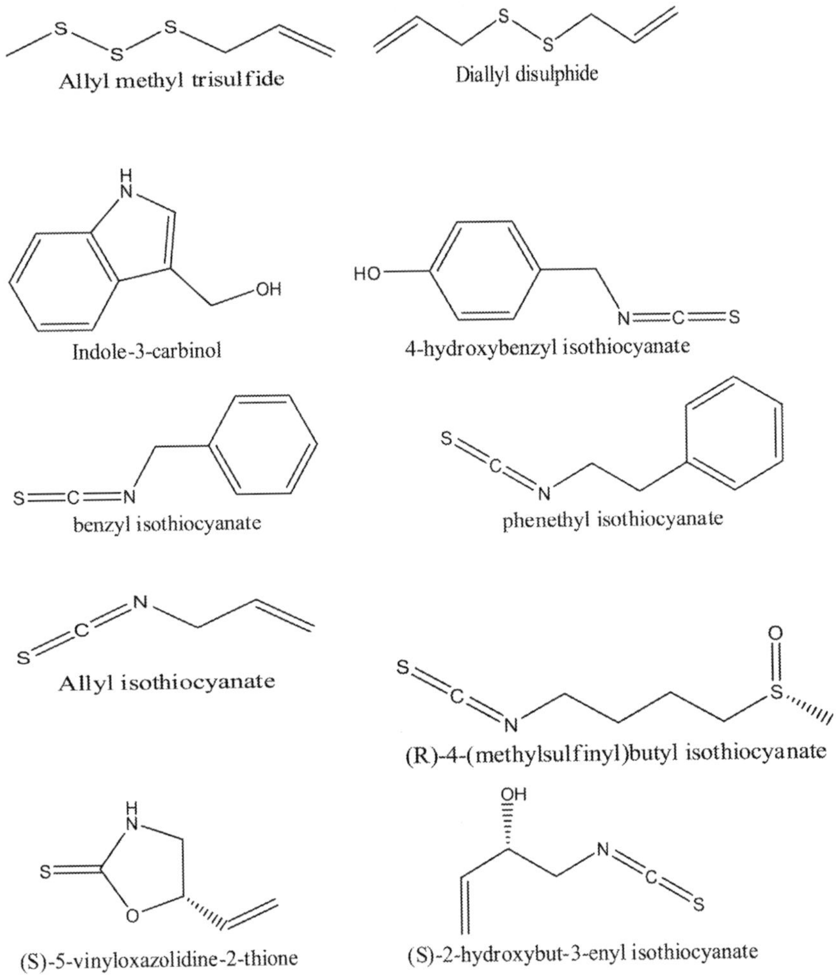

Allyl methyl trisulfide

Diallyl disulphide

Indole-3-carbinol

4-hydroxybenzyl isothiocyanate

benzyl isothiocyanate

phenethyl isothiocyanate

Allyl isothiocyanate

(R)-4-(methylsulfinyl)butyl isothiocyanate

(S)-5-vinyloxazolidine-2-thione

(S)-2-hydroxybut-3-enyl isothiocyanate

FIGURE 3.4

Chemical structures of glucosinolates.

Carotenoids have excellent antioxidant activity (Sies & Stahl, 1995). Carotenoids are the most potent quenchers of singlet oxygen such as hydrogen peroxide, singlet oxygen, nitrogen oxides, super oxide anion, and other reactive oxygen species (ROS) (Boileau, Moore, & Erdman, 1999; Paiva & Russell, 1999). Carotenoids can deactivate the excited sensitizer molecules involved in the generation of radicals and singlet oxygen (Truscott, 1990; Young & Lowe, 2001). Alternative mechanisms of antioxidant activity of carotenoids include chain-breaking antioxidant to terminate lipid oxidation by β-carotene, and decrease in the release of lactate dehydrogenase to protect cells from lipid peroxidation and membrane damage by β-carotene and lutein (Martin, Failla, & Smith, 1996).

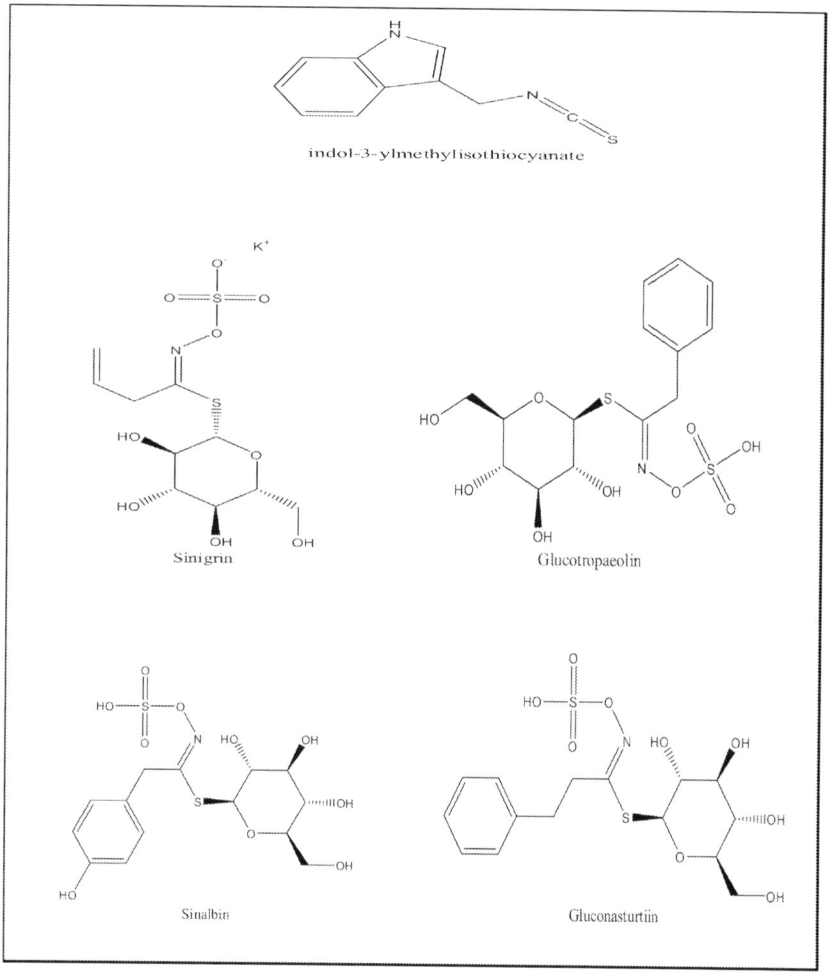

indol-3-ylmethylisothiocyanate

Sinigrin

Glucotropaeolin

Sinalbin

Gluconasturtiin

FIGURE 3.4

(Continued)

3.3.2 Phenolic compounds

Polyphenol is derived primarily from the shikimate-derived phenylpropanoid and/or the polyketide pathway(s), containing more than one phenolic ring. Phenolic compounds are abundantly found in plants and plant derived foods (Rashmi & Negi, 2020a) exhibiting vast variations in their structure, comprising from simple molecules (gallic acid, caffeic acid) to polyphenols (stilbenes, flavonoids, and polymers). Based on the chemical structure, they are grouped into two major groups: flavonoids (flavonols, flavones, flavanols, flavanones, anthocyanidins, isoflavonoids) and nonflavonoids (phenolic acids, stilbenes, coumarins, tannins). The chemical structures of flavonoid show a basic

Table 3.2 Classification of terpenoids.

Class	Structural frame work	Individual compounds
Hemiterpenes	Single isoprene unit	Isovaleric acid Prenol Pulegone
Monoterpenes	C_{10} Two isoprene units	Geraniol Limonene Perillyl alcohol Geranyl pyrophosphate Eucalyptol Citral Camphor Pinene Guttiferone Xanthochymol
Sesquiterpenes	C_{15} Three isoprene units	Artemisinin Bisabolol Fernesol Eudesmol (*E*)-β-farnesene β-humulene (γ)-muurolene Lactone parthenolide
Diterpenes	C_{20} Four isoprene units	Cafestol Cembrene Taxadiene Gibberellins Taxol Kaurenoic acid
Sesterterpenoids	C_{25} Five isoprene units	(2Z,6Z,10E,14E)-Geranylfarnesol C25 analog of transphytol Leucosceptrine Leucosesterterpenone Leucosesterlactone Salvimirzacolide Trinorsesterterpene glycoside Xanthanolides Xanthane
Triterpenoid	C_{30} Six isoprene units	Saponins, Oleanolic acid Ursolic acid Betulinic acid Moronic acid Lanosterol Squalene Azadirachtin b-Amyrin Luperol

Table 3.2 *(Continued)*

Class	Structural frame work	Individual compounds
Sterols	C_{27-30} Six isoprene units	Campesterol Beta Sitosterol Gamma sitosterol Stigmasterol Cholesterol Tocopherols (vitamin E)
Tetraterpenoids/ Carotenoids	C_{40} Eight isoprene units	***Carotenes:*** Orange pigments α-Carotene β- Carotene γ-Carotene δ-Carotene ε-carotene Lycopene ***Xanthophylls:*** yellow pigments Canthaxanthin Cryptoxanthin Zeaxanthin Astaxanthin Lutein Rubixanthin
Polyterpenoids	>C40 More than 8 isoprene units	Rubber

skeleton of diphenyl propane, the two benzene rings (rings A and B) linked with three carbon chains form a pyran ring with benzene ring A. Phenolic compounds are also classified based on number of carbon atoms and carbon skeleton (Table 3.3).

3.3.2.1 Flavonoids and their derivatives

Flavonoids and their derivatives constitute a major class of phytochemicals (Table 3.4). Flavonoids are found in several plants as they impart various color shades to them. They are synthesized from aromatic amino acids by phenylpropanoid pathway. Different classes of flavonoids have same basic structure of benzopyrano moiety and an aromatic ring, existing as glycones or aglycones (Ruchika & Pandey, 2019). Flavonoids are known for their high biological activity such as antimicrobial, antioxidant, anti-carcinogenic and immune enhancing properties.

3.3.2.2 Isoflavonoids

Isoflavonoids class comprises isoflavones, isoflavanones, isoflavans, rotenoids, and pterocarpans, and they are also known as soy flavonoids as they are abundantly present in them. Isoflavonoids exhibit several functional properties such as antioxidant, antimutagenic, anti-carcinogenic, and anti-proliferative (Miadokova, 2009). Isoflavonoids are known as dietary antioxidants, and they protect against free radical damage (Yoon & Park, 2014). The consumption of isoflavonoids is reported to reduce osteoporosis and suppress post-menopausal symptoms (Chen, Ko, & Chen, 2019).

Table 3.3 Classification of phenolic compounds.

No of carbon	Carbon skeleton	Class	Examples
6	C_6	Simple phenol	Catechol
7	$C_6{-}C_1$	Hydroxy benzoate	Phenolic acids: gallic, syringic, hydroxy-benzoic acids
9	$C_6{-}C_3$	1. Hydroxycinnamate 2. Coumarins	Caffeic, p-coumaric, ferulic, isoferulic scopolatin, aesculetin, umbelliferone
10	$C_6{-}C_4$	Napthoquinones	Juglone
13	$C_6{-}C_1{-}C_6$	Xanthones	Mangiferin, mangostein
14	$C_6{-}C_2{-}C_6$	Stilbenes	Resveratrol
15	$C_6{-}C_3{-}C_6$	Flavonoids	Classes of flavonones, anthocyanin
18	$(C_6{-}C_3)_2$	Lignans	Secosiolariciresinol, matatresinol
30	$(C_6{-}C_3{-}C_6)_2$	Biflavonoids	Amentoflavones
N	$(C_6{-}C_3)_n$	Lignins	Guaiacyl lignins
N	$(C_6{-}C_3)_n$-Glu	Hydrolysable tannins	Gallotannins, elagitannins, and chebulagic acid
N	$(C_6{-}C_3{-}C_6)_n$	Condensed tannins	Proanthocyanidins

3.3.2.3 Lignans

Phenylpropane units are linked by the central carbon (C8) of their propyl side chains in lignans. Lignans differs significantly in the chemical structure of their basic carbon frameworks, oxidation levels, and aromatic substitution patterns (Umezawa, Yamamura, Ono, Shiraishi, & Ragamustari, 2019). Lignans has a class of 2 or 3 molecules of benzene in different forms of polymerization present in both angiosperms and gymnosperms. Plant lignans are polyphenols derivative which are digested by intestinal bacteria to produce mammalian lignans (enerodiol and enterolactone). Lignans are referred as phytoestrogens as they exhibit estrogen agonist and antagonist properties (Pathak et al., 2018; Wcislo & Szarlej-Wcislo, 2014).

3.3.2.4 Tannins

Tannins are high molecular weight compounds (upto 30,000 dalton), which are formed by the polymerization of various flavonoids, and they show typical phenolic reactions with the ability to precipitate proteins, alkaloids, and polysaccharides. Earlier tannins were considered nutritionally undesirable, but lately they have been shown to possess high antioxidant, antimicrobial and anticancer properties. Tannins are classified as hydrolysable and non-hydrolysable tannins based on their acid/ alkali degradation. Hydrolysable tannins are Gallotannins (Glucose polyesters of gallic acid: digalloyl glucose, 1,3,6-trigalloyl glucose, tannic acid, coumarin); Ellagitannins- Glucose polyesters of hexa hydroxy diphenic acid (forms ellagic acid on hydrolysis: punicalagins, castalagins, vescalagins, punicalins, roburin As, terflavin Bs); Taragallotannins- Gallic acid and quinic acid; and caffetannins–caffeic acid and quinic acid. The Non-hydrolysable (condensed) tannins include polymer of flavan-3-ols (proanthocyanidins), polymer of flavan-3,4-diols (leucoanthocyanidins) or mixture of both (flavolans). Further, tannin like compounds formed by oxidation reaction catalyzed by PPO are also reported, and theaflavin, theasinensin, thearubigin, and theacitrin A are present in tea,

Table 3.4 Major flavonoid, isoflavonoid, and alkaloids in plant extracts.

Components	Structural frame work	Individual compounds	References
Flavonoids	C6–C3–C6; Two C6 units at Ring A and Ring B		
Flavan-3-ols/ Flavanols	Hydroxylation and variation in chromane ring (Ring C), Ring B attached to C2 position of Ring C; Flavanol—C2 and C3: double bond absent; Ring C: C4 carbonyl absent	Catechin, epicatechin gallate, epigallocatechin, epigallocatechin gallate, proanthocyanidins, theaflavins, thearubigins, fisetin	Prior, Lazarus, Cao, Muccitelli, and Hammerstone (2001), Tsao (2010)
Anthocyanins		Cyanidin, delphinidin, pelargonidin	Anderson and Jordheim (2006)
Flavanones		Prenylated flavanones, furanoflavanones, pyranoflavanones, benzylated flavanones	Tsao (2010)
Flavonols		Quercetin, kaempferol, rhamnetin	Tsao and McCallum (2009), Valant-Vetschera and Wallenweber (2006), Williams (2006)
Neoflavonoids	Ring B linked to ring C at C4 position	Dalbergin	Garazd, Garazd, and Khilya (2003)
Chalcones	Lack heterocyclic ring C	Phloretin and its glucoside phloridzin (phloretin 2′-0-β-glucopyranoside), chalconaringenin	Tsao (2010)
Flavanonol	15 carbon structure, Ring A and B: 2 phenyl rings Ring C: heterocyclic ring	Taxifolin (dihydroquercetin), aromadedrin (dihydrokaempferol), engeletin (dihydrokaempferol-3-rhamnoside)	Grayer and Veitch (2006), Kawaii, Tomono, Katase, Ogawa, and Yano (1999)
Proanthocyanidins	A-type structure: monomers attached by C2–*O*–C7 or C2–*O*–C5 bond, B–type structure: C4–C6 or C4–C8	Procyanidins, prodelphinidins, propelargonidins	Mateos-Martín, Fuguet, Quero, Pérez-Jiménez, and Torres (2012), Souquet, Cheynier, Brossaud, and Moutounet (1996), Tsao (2010)
Polyphenolic Amides	N-containing functional substituents	Capsaicinoids, avenanthramides	Bratt et al. (2003), Davis, Markey, Busch, and Busch (2007)
Isoflavonoids			
Isoflavones	Ring B connected to C3 position of Ring C	Daidzein, formononetin, genistein, biochanin A, glycetein	Mazur, Duke, Wahala, Rasku, and Adlercreutz (1998), Wang and Murphy (1994)

(*Continued*)

Table 3.4 *(Continued)*

Components	Structural frame work	Individual compounds	References
Isoflavanes	Have 3-phenylchroman backbone $C_{15}H_{14}O$	Lonchocarpane, laxiflorane, dalvelutinanes A and B, 3 (S)-3′-hydroxy-8-methoxyvestitol, nitidulan, nitidulin	Kaennakam, Siripong, and Tip-pyang (2017)
Isoflavandiols	$C_{15}H_{14}O_3$	Equol (4′,7-isoflavandiol)	Rufer, Glatt, and Kulling (2006)
Isoflavenes	Related to isoflavanes but with a double bond in ring-B	7-hydroxy-2′-methoxy-4′,5′-methylenedioxyisoflav-3-ene (judaicin), judaicin 7-*O*-glucoside, judaicin 7-*O*-(6″-*O*-malonylglucoside)	Stevenson and Veitch (1996)
Pterocarpans or Coumestans (phytoestrogens)	Polycyclic aromatic compound containing a coumestan moiety	Coumestrol, wedelolactone, demethylwedelolactone, psoralidin, flemicoumestan A 1, glycyrol, erythribyssin N, aureol, tephcalostan, plicadin, sophoracoumestan A, coumestoside C, D, hedysarum, coumestans A, B, D, F	Nehybova, Smarda, and Benes (2014)
Rotenoids	Heterocyclic aromatic compound: supplementary ring carbon atom derived from a methoxy group.	Elliptone, deguelin, malaccol, toxicarol, rotenone, sumatrol, tephrosin, amorphigenin, dolineone, pachyrrhizone, erosone	Uddin and Khanna (1979), Patil and Masand (2018)
Alkaloids			
Pyrrolidine alkaloids	Pyrrolidine or Tetrahydropyrrole ring	hygrine	Jan and Abbas (2018)
Pyridine alkaloids	Piperidine or Hexahydropyridine ring	coniine, piperine, isopelletierine	
Pyrrolidine-pyridine alkaloids	Pyrrolidine-pyridine	myosmine, nicotine	
Pyridine-piperidine alkaloids	Pyridine ring joined to piperidine ring	Anabasine	
Quinoline alkaloids	Quinolone	Quinine, anthraquinone, abruquinone	
Isoquinoline alkaloids	Isoquinoline	Opium alkaloids like narcotine, papaverine, morphine, codeine, and heroine	

whereas complex polymerized compounds (flavanols and anthocyanins) are present in wine (Fennema, 1996).

3.3.2.5 Phenolic acids

The principal phenolic acids in plants are derivatives of hydroxybenzoic acids and hydroxycinnamic acids. Hydroxybenzoic acid derivatives exist as glucoside forms, and hydroxycinnamic acids occur as esters with quinic acid or sugars (Rashmi & Negi, 2020b).

3.3.2.5.1 Derivatives of hydroxybenzoic acids

4-hydroxybenzoic acid is a monohydroxybenzoic acid in which a benzoic acid carries a hydroxy substituent at C-4 of the benzene ring, which is a conjugate acid of 4-hydroxybenzoate. Gallic acid is a trihydroxyl derivative, which may also be present in esterified form as hydrolysable or condensed tannins and their monomers. Ellagic acid exists in ellagitannins as esters of diphenic acid analog along with glucose. It is produced due to the hydrolysis of tannins such as ellagitannin and geraniin. Gentisic acid (2,5-dihydroxybenzoic acid) is a biosynthetic derivative and metabolite of salicylic acid (2-hydroxybenzoic acid) (Belles et al., 1999). It is produced by carboxylation of hydroquinone, whereas along with oxygen and enzyme gentisate 1,2-dioxygenase, gentisic acid yields maleylpyruvate (Hudnall, 2000).

Protocatechuic acids, also known as 3,4-dihydroxybenzoic acid, belonging to catechols and a dihydroxybenzoic acid, as well as conjugate acid of 3,4-dihydroxybenzoate. Syringic acid is a chemical compound naturally found as *O*-methylated trihydroxybenzoic acid. It is a derivative from gallic acid and conjugate acid of a syringate. It is synthesized by hydrolyzing eudesmic acid along with sulfuric acid. Vanillic acid (4-hydroxy-3-methoxy benzoic acid) is a dihydroxybenzoic acid derivative. Vanillic acid is the oxidized form of vanillin, and it is an intermediate in the formation of vanillin from ferulic acid (Civolani, Barghini, Roncetti, Ruzzi, & Schiesser, 2000; Lesage-Meessen et al., 1996).

3.3.2.5.2 Derivatives of hydroxycinnamic acids

Hydroxycinnamic acids include several compounds comprising a quinic acid moiety with a cyclohexane ring having four hydroxyl groups at different positions and a carboxylic acid (Alam et al., 2016; http://www.hmdb.ca/metabolites/HMDB0041641). The 3,4-Dicaffeoylquinic acid is a natural product, which is an ester of two polyphenolic caffeic acids and one cyclitol (−)-quinic acid. 3,5-Dicaffeoylquinic acid (3,5-DCQA) is a carboxylic ester that is obtained by the condensation of the hydroxy groups at positions 3 and 5 of (−)-quinic acid with the carboxy group of *trans*-caffeic acid. The 3,4-diferuloylquinic acid and 3,5-diferuloylquinic acid are insoluble in water and a weakly acidic compound (Bezerra et al., 2017; http://foodb.ca/compounds/FDB000277), whereas 3-p-coumaroylquinic acid is slightly soluble in water and a weak acidic compound. Caffeoylquinic acids (CQA) are compounds composed of a quinic acid core, acylated with one or more caffeoyl groups (Miyamae, Kurisu, Han, Isoda, & Shigemori, 2011). This class includes compounds such as Chlorogenic acid (3-*O*-caffeoylquinic acid or 3-CQA), 4-*O*-caffeoylquinic acid (crypto-chlorogenic acid or 4-CQA) and 5-*O*-caffeoylquinic acid (neo-chlorogenic acid or 5-CQA) (Wianowska & Gil, 2019). The 4-*O*-Caffeoylquinic acid (Cryptochlorogenic acid), is an isomer of chlorogenic acid, which possesses antioxidant properties. The 5-caffeoylquinic acids, an isomer of chlorogenic acid are the most important groups of phenolic secondary metabolites, which are the esters formed between cinnamic acid derivatives and quinic acid (Clifford, 2000; Clifford, Johnston, Knight, & Kuhnert, 2004). The 4-feruloylquinic acid (*O*-feruloylquinate), 3-p-coumaroylquinic acid and 1-*O*-feruloyl glucose are slightly soluble in water and weak acidic compound (http://foodb.ca/compounds/FDB000248; http://foodb.ca/compounds/FDB000866; Li & Bet, 2013). The 1-*O*-feruloylglucose belong to the class of hydroxycinnamic acid glycosides, which are glycosylated hydoxycinnamic acids derivatives (http://foodb.ca/compounds/FDB015907). The 5-p-coumaroylquinic acid is a cinnamate ester obtained by formal condensation of the carboxy group of 4-coumaric acid with the 5-hydroxy group of (−)-quinic acid (http://foodb.ca/compounds/FDB000236). Caffeic acid (3,4-dihydroxycinnamic) is the hydroxycinnamate and phenylpropanoid metabolites more widely distributed in plant tissues (Clifford, 2000; Mattila, Hellström, & Törrönen, 2006). Cinnamic acid is the precursor for the biosynthesis of lignins, phenyl-propanoids, coumarins, tannins, flavonoids, pigments, isoflavonoids, flavonoids, stilbenes, aurones, anthocyanins, spermidines, flavor components of spices and various alkaloids such as morphine and colchicines (Vogt, 2010). Cinnamic acids are ester conjugates with quinic acid, or with other acids, sugars or lipids, or form amides with aromatic and aliphatic amines (De, Baltas & Bedos-Belval, 2011). Ferulic acid is chemically defined as ([E]-3-[4-hydroxy-3-methoxy-phenyl] prop-2-enoic acid), which are available in the plant tissues (Mattila & Kumpulainen, 2002). Sinapic acid (3,5-dimethoxy-4-hydroxycinnamic acid) is available both in the free and ester form. Hydroxycinnamic esters occur as sugar esters (glycosides), or as esters of a variety of organic compounds. There are two types of sinapoyl esters, the sinapoyl malate present in leaves, and sinapine (sinapoylcholine) stored in roots. Sinapine is an alkaloidal amine, which is a choline ester of sinapinic acid (Chapple, Vogt, Ellis, & Somerville, 1992; Shirley & Chapple, 2003).

3.3.3 Glucosinolates

Glucosinolates are sulfur- and nitrogen-containing glycosides commonly found in Brassicaceae family. The type and concentration of glucosinolates accumulated in plants depends on genotype, growing conditions, developmental stage, type of plant tissue and postharvest handling. Approximately 140 different glucosinolates have been identified in plants, which also include glucosinolate degradation products such as isothiocyanates, indoles, dithiothiols and other organosulfur compounds. Glucosinolate degradation products are well known for their antimicrobial activity (Barbieri et al., 2017) and anti-carcinogenic effects (Hayes, Kellesher, & Eggleston, 2008).

Glucosinolates consist of a β-D-glycopyranose residue linked to a hydroximinosulfate ester by sulfur bridge, and an R-group. Glucosinolates are broadly classified in three classes based on the structure of different amino acids precursors linked to the R-group, as aliphatic glucosinolates (derived from alanine, leucine, isoleucine, methionine, or valine), aromatic glucosinolates (derived from phenylalanine or tyrosine), or indole glucosinolates (derived from tryptophan). The R chains may also contain double bonds, oxo, hydroxyl, methoxy, carbonyl or di-sulfide linkages. The major glucosinolates include progoitrin, sinigrin, glucobrassican, neoglucobrassican, and glucoraphanin. The other glucosinolates metabolites include thiocyanates, nitriles, sulfates, and goitrins (Bischoff, 2006). Major glucosinolates and their derivatives are given in Table 3.5, and their chemical structures are presented in Fig. 3.4.

Table 3.5 Major glucosinolates and their derivatives in plant extracts.

Glucosinolates	Derivatives	Plant source	References
Isothiocyanate	Sinigrin precursor to allyl isothiocyanate	Mustard, broccoli, cabbage, cauliflower, kale, mustard, radish, brussels sprout, watercress	McNaughton and Marks (2003)
Gluconasturtiin (phenethylglucosinolate)	Phenethyl isothiocyanate	Cruciferous vegetables-cabbage, mustard or rapeseed	Ishida, Hara, Fukino, Kakizaki, and Morimitsu (2014), Li and Kushad (2004)
Glucobrassicin	1-Methoxyglucobrassicin (neoglucobrassicin) 4-Hydroxyglucobrassicin 4-Methoxyglucobrassicin 1,4-Dimethoxyglucobrassicin 1-Sulfoglucobrassicin 6′-Isoferuloylglucobrassicin	Horseradish, cabbage, mustard, broccoli	Agerbirk, De Vos, Kim, and Jander (2009), Galletti, Barillari, Iori, and Venturi (2006), Ishida et al. (2014)
Benzylglucosinolate (Glucotropaeolin)	Benzyl isothiocyanate (BITC)	Cruciferous vegetables, particularly garden cress	Higdon (2005), Ishida et al. (2014)
(*R*)-4-(methylsulfinyl) butylglucosinolate (Glucoraphanin)	Sulforaphane	Cruciferous vegetables mustard, broccoli and red cabbage	Ishida et al. (2014), Leicach and Chludil (2014)
(*R*)-2-hydroxybut-3-enylglucosinolate (progoitrin)	(*S*)-2-hydroxybut-3-enyl isothiocyanate, which is expected to be unstable and immediately cyclize to form (*S*)-5-vinyloxazolidine-2-thione (goitrin)	Mustard, broccoli, cabbage, cauliflower, kale, radish, brussels sprout, watercress, peanut, kohlrabi and spinach	Ishida et al. (2014), Rossiter and James (1990)
Sinalbin	4-Hydroxybenzyl isothiocyanate	Seeds of white mustard	Borek and Morra (2005), Ishida et al. (2014)

3.3.4 Other phytochemicals

Other phytochemicals in various plant extracts include Betalains, Betacyanins, Betanin, Isobetanin, Probetanin, Neobetanin, and Betaxanthins such as Indicaxanthin and Vulgaxanthin. Alkaloids contain one nitrogen atom in the form of primary, secondary, or tertiary amine. Sometimes the number of nitrogen bases may go up to five. Alkaloids have been reported to possess antimicrobial activity since ancient time. Several organic acids such as Phytic acid (Inositol hexaphosphate), Quinic acid, Oxalic acid, Tartaric acid, Anacardic acid, and Malic acid are also present in various extracts. Various plant extracts also contain Chlorophylls and Chlorophyllin, Amines such as Betaine, Choline, Carnitine, Coenzyme Q_{10}, Ubiquinone, and Ubidecarenone.

3.4 Structure-activity relationship of plant extract

The diverse nature of chemical structures in plants is related to their multifaceted properties, which is linked to their specific biological activity (Cheynier, 2012). The nature of chemicals present in plant extracts determines their biological activity. Several plant extracts obtained from grape seeds, green tea, rosemary, pomegranates and cinnamon, are found to be similar or better antioxidant compared to synthetic antioxidants (Shah, Bosco, & Mir, 2014). Some plant extracts, in addition to being antioxidant, also exhibit antimicrobial activity, and influence the textures and flavor of food (Soto-Vaca, Gutierrez, Losso, Xu, & Finley, 2012). The chemical compounds present in plant extracts varies in their structure (Figs. 3.2–3.4), and these structures influence bioactivities of plant extracts.

3.4.1 Structure-activity relationship of carotenoid

The antioxidant properties of carotenoids are related to their chemical structure. The nature of the specific end groups in carotenoids influences their polarity, which affects their interaction with biological membranes and thus antioxidant activity. Arrangement of conjugated double bonds in the carotenoids also impacts antioxidant activity. The opening of the β-ionone ring in the carotenoid structure and addition of oxygenated functional groups increase their ROO· scavenging capacity. The increase in polyene chain length also affects the quenching of singlet oxygen. Isolated double bond and lack of oxygen substituents also influences their antioxidant activity (Kobayashi & Sakamoto, 1999). The size and shape of carotenoids are important for functionality. The tendency of *cis*-isomers to crystallize is much faster that *all trans*-isomers, and are readily solubilized and transported. Therefore, the *trans* isomers are more efficient than their corresponding *cis* ones (Britton, 1995). Table 3.6 summarizes the structure activity relationship of carotenoids present in the plant extracts.

3.4.2 Structure-activity relationship of phenolics

The biological activity of phenolics is related to the presence of a 3-hydroxyl substituent, a 3′4′-dihydroxy (catechol or B ring) moiety, and the C_4 oxo group and C_2C_3 double bond. The hydroxyl groups confer antioxidant and metal chelating activity, whereas methoxy groups increase lipophilicity and membrane transport. A double bond and carbonyl functional group in the heterocycle or polymerization increases biological activity (Heim, Tagliaferro, & Bobilya, 2002). Table 3.7 summarizes the structure activity relationship of various phenolic compounds present in the plant extracts.

3.4.2.1 Antioxidant activity

Increased antioxidant potential of flavonoids is correlated to their enhanced number of hydroxyl groups. The most important structure for scavenging actions is the presence of two hydroxyl groups in the B ring at ortho position. Existence of hydroxyl groups at the positions 5, 6, and 7 in the ring A are crucial for scavenging activity. The presence of metal complexing domains between the

Table 3.6 Structural changes in carotenoids present in plant extracts and their influence on biological activities.

Structural changes	Changes in bioactivity	References
Pattern of conjugated double bonds in the polyene backbone	Antioxidant activity increases with increase in double bonds	Kobayashi and Sakamoto (1999)
Opening of the β-ionone ring	Increase of ROO· scavenging capacity	Rodrigues, Mariutti, Chiste, and Mercadante (2012)
Increase of chromophore extension	Increase of ROO· scavenging capacity	Rodrigues et al. (2012)
Addition of oxygenated functional groups	Increase of ROO· scavenging capacity	Rodrigues et al. (2012)
Increase in polyene chain length	Increase in quenching of the singlet oxygen	Krinsky (1998)
Cis-trans isomerization	*All-trans* form are more efficient ROO· scavenger than their corresponding *cis* ones	Britton (1995)
Linearity of structure	Reduced bioavailability of trans-isomers as compared to corresponding *cis* ones	Levin and Mokady (1995), Ferruzzi, Lumpkin, Schwartz, and Failla (2006)

5-hydroxyl and 4-carbonyl group, the 3-hydroxyl and 4-carbonyl group, and between the 3′,4′-hydroxyl groups are responsible for metal chelating ability of flavonoids (Sordon et al., 2016).

In general, it is considered that a higher number of hydroxyl substituents in flavonoids results in a higher antioxidant activity (Burda & Oleszek, 2001; Pietta, 2000; Rice-Evans, Miller, & Paganga, 1996). The presence of two hydroxyl groups in the ortho position of ring B is confirmed as the most important factor, although adjacent hydroxyl groups at positions 5 and 6 (and 7) in ring A may replace ring B hydroxyl groups scavenging function (Heim et al., 2002). The 3′,4′-dihydroxy-phenolic structure of ring B molecules may result in a stable radical after interaction with ROS. The removal or derivatization of one of the hydroxyl groups in the ortho position of ring B markedly decreases the ROS-scavenging activity. An additional hydroxyl group in ring B (pyrogallol structure) does not greatly influence the activity of flavonols or anthocyanins, but may increase the activity of flavanols (Rice-Evans et al., 1996).

Free radical scavenging by flavonoids is enhanced by the presence of both (2–3 double bond and 4-oxo). Conjugation between the A- and B-rings permits a resonance effect of the aromatic nucleus that stabilizes the flavonoid radical. The research reports indicate that the flavonoids devoid of 2–3 double bond and/or 4-oxy group have lower antioxidant activity than those with both the structural features (Wolfe & Liu, 2008).

Aglycones are more potent antioxidants than their corresponding glycosides. The total number, the position, and structure of the sugar play an important role. Sugar in A-ring shows higher reduction of antioxidant activity than 3-glycosylation in the heterocycle. *O*-glycosylation at carbon 7 weakens the antioxidant effect of flavonoids, however *O*-glycosylation at carbon 3 has no effect. The type of the sugar moiety also plays an important role in determining the antioxidant effect

Table 3.7 Structural changes in phenolic compounds present in plant extracts and their influence on biological activities.

Bioactive compound (s)	Structural changes	Changes in bioactivity	References
Phenolic compounds	Number of hydroxyl groups in relation to carboxyl functional group	Antioxidant activity increases with increase in hydroxyl groups	Afanasev, Dcrozhko, Brodskii, Kostyuk, and Potapovitch (1989), Amarowicz, Pegg, Rahimi-Moghaddam, Barl, and Weil (2004)
Monohydroxybenzoic acids	−OH moiety at ortho- or para-position to −COOH	No antioxidant activity	Rice-Evans et al. (1996)
Trihydroxylated gallic acid	Increase in degree of hydroxylation	Antioxidant activity increases	
Syringic acid	Replacement of hydroxyl groups at 3- and 5-position with methoxyl groups	Antioxidant potential decreases	
Hydroxycinnamic acids	CH = CH−COOH group confirmshigher H-donating capacity and radical stabilization than −COOH group	Antioxidant activity increases	Andreasen, Landbo, Christensen, Hansen, and Meyer (2001), Rice-Evans et al. (1996)
Hydroxybenzoic acids	CH = CH−COOH group endorsesreduced H-donating capability and radical stabilization than −COOH group	Antioxidant activity decreases	Andreasen et al. (2001), Rice-Evans et al. (1996)
Flavonoids	Structural features and nature of substitutions on rings B and C; Double bond between C-2 and C-3, conjugated with 4-oxo group in ring C	Antioxidant activity increases	Pietta (2000)
Catechol group	Degree of hydroxylation and sites of −OH groups in B ring (ortho-dihydroxyl structure of ring B)	Higher antioxidant activity	Pietta (2000), Van Acker et al. (1996a, 1996b)
Pyrogallol group	Presence of hydroxyl groups at 3-O-, 4-O-, and 5-O-positions of ring B	Antioxidant activity increases	Van Acker et al. (1996a, 1996b)
Anthocyanidins	Conversion of 3-O, 4-O-dihydroxyphenyl to 3-O, 4-O, 5-O−trihydroxylphenyl	Antioxidant activity increases	Seeram and Nair (2002)
Catechins	Conversion of 3-O, 4-O-dihydroxyphenyl to 3-O, 4-O, 5-O−trihydroxylphenyl	Antioxidant activity decreases	Seeram and Nair (2002)

Table 3.7 *(Continued)*

Bioactive compound (s)	Structural changes	Changes in bioactivity	References
	B-ring catechol	Antioxidant activity increases due to peroxynitrite scavenging	Kerry and Rice-Evans (1999)
Flavones	Catechol or *o*-trihydroxyl (pyrogallol)—Absent	Weak antioxidant property	Burda and Oleszek (2001), Gao, Huang, Yang, and Xu (1999), Pannala, Chan, O'Brien, and Rice-Evans (2001)
Kaempferol	Double bond between C-2 and C-3, combined with a 3-OH, in ring C	Antioxidant potential increases	Van Acker et al. (1996a, 1996b)
	Substitution of 3-OH	Antioxidant activity reduces	Seeram and Nair (2002)
Quercetin	1. o-diphenolic group (in ring B), 2. 2−3 double bond conjugated with 4-Oxo function, 3. hydroxyl groups at 3 and 5 position	High Antioxidant activity	Bravo (1998)
Luteolin	3,4-catechol structure in B-ring	Enhanced peroxyl radical scavenging ability	Van Acker et al. (1996a, 1996b)
Naringenin (4, 5, 7-trihydroxyflavanone),	Three OH substitution on structure but no 3′,4′-di-OH-structure	No influence on antioxidant activity (Heridictyol vs Naringenin)	Di Majo et al. (2005)
Heridictyol (5,7, 3,4 tetrahydroxyflavanone),	Four hydroxyl groups substitutions with 3′,4′-di-OH-structure		
Neoeriocitrin (Heridictyol-7-neohesperidoside),	Glycosylated with a neohesperidose of 7th OH group, 3′,4′-catechol structure	Increase in antioxidant activity (Neoeriocitrin vs Naringenin)	
Hesperitin (3′, 5, 7-trihydroxy-4′ methoxyflavanone)	*O*-glycosylation at hydroxyl position	Higher antioxidant activity than Neohesperidin due to steric effect disturbs the planarity and ability to delocalize electrons	
Neohesperidin (Hesperitin-7-neohesperidoside) Neoeriocitrin (Heridictyol-7-neohesperidoside)	Replace with a neohesperidoside molecule in 7th position, a methoxylation in 4th position	Decreases antioxidant power-(Neohesperidin vs Neoeriocitrin)	

(*Continued*)

Table 3.7 ***(Continued)***

Bioactive compound (s)	Structural changes	Changes in bioactivity	References
Chalcones	−SCH3 and −OCH3 in para position of A-ring and −OH in B-ring	Antioxidant activity increases	Sivakumar, Prabhakar, and Doble (2011)
Neoechinulin A	C-8/C-9 double bond forms conjugate with indole and diketopiperazine moieties of Neoechinulin A	Antioxidant activity increases	Kuramochi (2013)
	Stereo chemistry of C12	No influence on antioxidant activity	
	Presence of intact diketopiperazine moieties is requirement	No influence on antioxidant activity	
Caffeic acid derivatives	Caffeic acid anilides with electron donating groups at *p*-position	Higher inhibitory activities against *B. subtilis*	Fu, Cheng, Zhang, Fang, and Zhu (2010)
Chalcones	Lipophilicity of ring A of hydroxyl chalcones	Higher inhibitory activities against *S. aureus* and *E. coli*	Batovska et al. (2009)
4-Methoxy phenylpropanone	Introduction of keto group in place of double bond	Increased inhibitory effect against various bacterial strains	Raj, Narayana, Ashalatha, Kumari, and Sarojini (2007)
4-Methoxy phenylpropanol	Addition of alcohol group in place of double bond	Enhanced antifungal activity	Raj et al. (2007)
Polymethoxylated flavones (PMFs),	Four or more methoxyl groups on basic benzo-γ-pyrone (15 carbon, C-6 − C-3 − C-6) skeleton with a carbonyl group at C-4	Enhanced antifungal activity against *Aspergillus niger*	Liu, Xu, Cheng, Yao, and Pan (2012)
Kaempferol,	Hydroxyl group substitutions at C-3, C-5, C-7, and C-4′	Higher Anti-*E. coli* activity with substitution	Wu, Zang, He, Pan, and Xu (2013)
Nobiletin (PMF)	Methoxyl group substitutions at C-5, C-6, C-7, C-8, C-3′, and C-4′	Lowest Anti-*E. coli* activity with substitution	Constantinou et al. (1995)
Quercetin, Myricetin, and Kaempferol	C-4 keto group and hydroxyl group substitutions at C-3, C-7, and C-4′	Higher Anti-*E. coli* activity with substitution	
Flavonoids	4′-OH in B ring	Higher inhibition of influenza virus	Liu et al. (2012)
5-Hydroxyflavanones 5-Hydroxyisoflavanones	One, two, or three additional hydroxyl groups at 7, 2′, and 4′ positions	Higher inhibition of *Streptococcus mutans* and *Streptococcus sobrinus*	Osawa et al. (1992)

Table 3.7 *(Continued)*

Bioactive compound (s)	Structural changes	Changes in bioactivity	References
Cyclic C5-curcuminoids	Changes in benzylidene group and nitrogen heteroatom	Better antiproliferative activity	Huber et al. (2020)
Seco-pseudoguaianolides paulitin and isopaulitin	Two α β unsaturated (C–O–CH = CH2) systems	Better antiproliferative activity	Chen, Liu, and Wang (2011)
Psilostachyin	Single C–O–CH = CH2 moiety in the molecule	No effect on antiproliferative activity	
Sintenin	No α β unsaturated (C–O–CH = CH2)	No effect on antiproliferative activity	
Gallic acid	Three hydroxyls and one carboxylic acid group	Higher antioxidant ability and neuroprotective effect	Lu, Nie, Belton, Tang, and Zhao (2006), Phonsatta et al. (2017), Rajan and Muraleedharan (2017)
Luteolin	OH groups at C-5, C-7, C-3′, and C-4′	Higher anti-leishmanial potential	Tasdemir et al. (2006)
7,8-Dihydroxyflavone	Basic structure	Higher anti-leishmanial potential	
7,8-Dihydroxyflavone	Addition of a catechol structure into the B ring	Diminished anti-leishmanial activity upto fivefold	
Apigenin and Luteolin	Sugars (one or more) at C-5 or the C-7 position	Reduction in anti-leishmanial potency	
3-Hydroxyflavone	Pattern of hydroxylation on ring B	Higher anti-leishmanial potential	
Fisetin	OH groups at C-3, C-7, C-3′, and C-4′	Higher anti-leishmanial potential	
Quercetin	Catechol moiety in ring B	Higher anti-leishmanial potential	
Kaempferol	*p*-Hydroxyphenyl ring	Lesser anti-leishmanial activity	
Morin	OH in *meta*-position at C-2′ and C-4′	Lesser anti-leishmanial activity	
7,8-dihydroxyflavone and 6,7-dihydroxyflavone	Hydroxylation on ring B absent, however, catechol function at side chain	Higher anti-*Trypanosoma brucei rhodesiense* activities	
Catechol, Pyrogallol, Gallic acid, and 3,4-dihydroxybenzoic acid	Two or three OH groups positioned *ortho* to each other	Significant trypanocidal activities	
Phenolic esters; Methyl, propyl and octyl esters of caffeic and gallic acids	Size, degree of ring hydroxyl substitution and length of alkyl chain, lipophilicity	Antiproliferative and/or cytotoxic activity higher than phenolic acids	Fiuza et al. (2004)

(Fennema, 1996). *O*-methylation can change hydrophobicity and impart steric effects (at the cost of OH). The suppression of antioxidant activity by *O*-methylation is attributed to the steric effects (Fennema, 1996). The protection of lipids against oxidative damage is achieved by scavenging of hydroxyl, and peroxyl, or termination of chain reactions or chelation of divalent cations, which help in reducing or recycling the flavonoid radical (Fennema, 1996).

3.4.2.2 Antimicrobial activity

The antibacterial activity of flavonoids is due to the inhibition of nucleic acid synthesis, cytoplasmic membrane function, energy metabolism, biofilm formation, and alteration of the membrane permeability. Hydroxyl groups at the special sites on the aromatic rings of flavonoids improve the antibacterial activity. The methylation of the active hydroxyl groups generally decrease the activity. The hydrophobic substituents such as prenyl groups, alkylamino chains, alkyl chains, and nitrogen or oxygen containing heterocyclic moieties usually enhance the antimicrobial activity for all the flavonoids (Xie, Yang, Tang, Chen, & Ren, 2015).

3.5 Conclusions

The plant extracts are rich in chemicals that impart them various biological activities. Plant extracts show lot of variability as the chemical compound in them are affected by cultivar of plant, growing conditions, plant part, maturity of plants, method of extraction and storage. There is a need to standardize the extract composition. The extract or dried preparation from plants may contain several chemical compounds in varied proportion, which pose a challenge for standardization of plant extracts. The plant extracts contain a diverse range of active chemicals, which include terpenoids, phenolics, flavonoids, tannins, glucosinolates, alkaloids, etc., among others. These chemical compounds are known to exert several bioactivities, such as antimicrobial, antioxidant, anticancer, antidiabetic and cardioprotective effects, and have potential for food and medicinal applications. Therefore, optimization of isolation, purification, and identification procedures that yield better functional properties of active chemical compounds from various plants are required.

References

Afanasev, I. B., Dcrozhko, A. I., Brodskii, A. V., Kostyuk, V. A., & Potapovitch, A. I. (1989). Chelating and free radical scavenging mechanisms of inhibitory action of rutin and quercetin in lipid peroxidation. *Biochemical Pharmacology*, *38*(11), 1763–1769.

Agerbirk, N., De Vos, M., Kim, J. H., & Jander, G. (2009). Indole glucosinolate breakdown and its biological effects. *Phytochemistry Reviews*, *8*(1), 101.

Aharoni, A., Giri, A. P., Deuerlein, S., Griepink, F., de Kogel, W. J., Verstappen, F. W., & Bouwmeester, H. J. (2003). Terpenoid metabolism in wild-type and transgenic Arabidopsis plants. *The Plant Cell*, *15*(12), 2866–2884.

Alam, M. A., Subhan, N., Hossain, H., Hossain, M., Reza, H. M., Rahman, M. M., & Ullah, M. O. (2016). Hydroxycinnamic acid derivatives: A potential class of natural compounds for the management of lipid metabolism and obesity. *Nutrition and Metabolism*, *13*(1), 27.

Alupului, A., Calinescu, I., & Lavric, V. (2012). Microwave extraction of active principles from medicinal plants. *UPB Science Bulletin, Series B*, *74*(2), 129–142.

Amarowicz, R., Pegg, R. B., Rahimi-Moghaddam, P., Barl, B., & Weil, J. A. (2004). Free-radical scavenging capacity and antioxidant activity of selected plant species from the *Canadian prairies*. *Food Chemistry*, *84* (4), 551–562.

Anderson, O. M., & Jordheim, M. (2006). The anthocyanins. In O. M. Anderson, & K. R. Markham (Eds.), *Flavonoids: Chemistry, biochemistry and applications* (pp. 472–551). Boca Raton, FL: CRC Press/Taylor & Francis Group.

Andreasen, M. F., Landbo, A. K., Christensen, L. P., Hansen, Å., & Meyer, A. S. (2001). Antioxidant effects of phenolic rye (*Secale cereale* L.) extracts, monomeric hydroxycinnamates, and ferulic acid dehydrodimers on human low-density lipoproteins. *Journal of Agricultural and Food Chemistry*, *49*(8), 4090–4096.

Barbieri, R., Coppo, E., Marchese, A., Daglia, M., Sobarzo-Sanchez, E., Nabavi, S. F., & Nabavi, S. M. (2017). Phytochemicals for human disease: An update on plant-derived compounds antibacterial activity. *Microbiological Research*, *196*, 44–68.

Batovska, D., Parushev, S., Stamboliyska, B., Tsvetkova, I., Ninova, M., & Najdenski, H. (2009). Examination of growth inhibitory properties of synthetic chalcones for which antibacterial activity was predicted. *European Journal of Medicinal Chemistry*, *44*, 2211–2218.

Belles, J. M., Garro, R., Fayos, J., Navarro, P., Primo, J., & Conejero, V. (1999). Gentisic acid as a pathogen-inducible signal, additional to salicylic acid for activation of plant defenses in tomato. *Molecular Plant-Microbe Interactions*, *12*(3), 227–235.

Bezerra, G. S. N., Pereira, M. A. V., Ostrosky, E. A., Barbosa, E. G., de Moura, M. D. F. V., Ferrari, M., & Gomes, A. P. B. (2017). Compatibility study between ferulic acid and excipients used in cosmetic formulations by TG/DTG, DSC and FTIR. *Journal of Thermal Analysis and Calorimetry*, *127*(2), 1683–1691.

Bird, D. A., Franceschi, V. R., & Facchini, P. J. (2003). A tale of three cell types: Alkaloid biosynthesis is localized to sieve elements in opium poppy. *The Plant Cell*, *15*(11), 2626–2635.

Bischoff, K. L. (2006). Glucosinolates. In R. C. Gupta (Ed.), *Nutraceuticals: Efficacy, safety and toxicity* (pp. 551–554). Cambridge, MA: Academic Press.

Boileau, T. W., Moore, A. C., & Erdman, J. W. (1999). Carotenoids and vitamin. In A. M. A. Papas (Ed.), *Antioxidant status, diet, nutrition and health* (pp. 133–158). Boca Raton, FL: CRC Press.

Borek, V., & Morra, M. J. (2005). Ionic thiocyanate (SCN-) production from 4-hydroxybenzyl glucosinolate contained in *Sinapis alba* seed meal. *Journal of Agricultural and Food Chemistry*, *53*(22), 8650–8654.

Bratt, K., Sunnerheim, K., Bryngelsson, S., Fagerlund, A., Engman, L., Andersson, R. E., & Dimberg, L. H. (2003). Avenanthramides in oats (*Avena sativa* L.) and structure – antioxidant activity relationships. *Journal of Agricultural and Food Chemistry*, *51*(3), 594–600.

Bravo, L. (1998). Polyphenols: Chemistry, dietary sources, metabolism, and nutritional significance. *Nutrition Reviews*, *56*(11), 317–333.

Britton, G. (1995). Structure and properties of carotenoids in relation to function. *The FASEB Journal*, *9*(15), 1551–1558.

Brown, P. D., Tokuhisa, J. G., Reichelt, M., & Gershenzon, J. (2003). Variation of glucosinolate accumulation among different organs and developmental stages of *Arabidopsis thaliana*. *Phytochemistry*, *62*(3), 471–481.

Burda, S., & Oleszek, W. (2001). Antioxidant and antiradical activities of flavonoids. *Journal of Agricultural and Food Chemistry*, *49*, 2774–2779.

Chapple, C. C. S., Vogt, T., Ellis, B. E., & Somerville, C. R. (1992). An Arabidopsis mutant defective in the general phenylpropanoid pathway. *The Plant Cell*, *4*(11), 1413–1424.

Chen, L. R., Ko, N. Y., & Chen, K. H. (2019). Isoflavone supplements for menopausal women: A systematic review. *Nutrients*, *11*(11), 2649.

Chen, Q. F., Liu, Z. P., & Wang, F. P. (2011). Natural sesquiterpenoids as cytotoxic anticancer agents. *Mini Reviews in Medicinal Chemistry*, *11*(13), 1153–1164.

Cheynier, V. (2012). Phenolic compounds: From plants to foods. *Phytochemistry Reviews*, *11*, 153–177.

Civolani, C., Barghini, P., Roncetti, A. R., Ruzzi, M., & Schiesser, A. (2000). Bioconversion of ferulic acid into vanillic acid by means of a vanillate-negative mutant of *Pseudomonas fluorescens* strain BF13. *Applied Environment Microbiology*, *66*(6), 2311–2317. Available from https://doi.org/10.1128/AEM.66.6.2311-2317.2000, PMC 110519. PMID 10831404.

Clifford, M. N. (2000). Chlorogenic acids and other cinnamates–nature, occurrence, dietary burden, absorption and metabolism. *Journal of the Science of Food and Agriculture*, *80*(7), 1033–1043.

Clifford, M. N., Johnston, K. L., Knight, S., & Kuhnert, N. (2004). Hierarchical scheme for LC-MS in identification of chlorogenic acids. *Journal of Agricultural and Food Chemistry*, *51*(10), 2900–2911. Available from https://doi.org/10.1021/jf026187q, PMID 12720369.

Constantinou, A., Mehta, R., Runyan, C., Rao, K., Vaughan, A., & Moon, R. (1995). Flavonoids as DNA topoisomerase antagonists and poisons: Structure-activity relationships. *Journal of Natural Products*, *58*(2), 217–225.

Davis, C. B., Markey, C. E., Busch, M. A., & Busch, K. W. (2007). Determination of capsaicinoids in habanero peppers by chemometric analysis of UV spectral data. *Journal of Agricultural and Food Chemistry*, *55*(15), 5925–5933.

De, P., Baltas, M., & Bedos-Belval, F. (2011). Cinnamic acid derivatives as anticancer agents-a review. *Current Medicinal, Chemistry,18,*, 1672–1703.

Di Majo, D., Giammanco, M., La Guardia, M., Tripoli, E., Giammanco, S., & Finotti, E. (2005). Flavanones in Citrus fruit: Structure–antioxidant activity relationships. *Food Research International*, *38*(10), 1161–1166.

Dimitrios, B. (2006). Sources of natural phenolic antioxidants. *Trends in Food Science & Technology*, *17*(9), 505–512.

Fennema, O.R. (1996). Food Chemistry, Third Edition (Food Science & Technology), CRC Press.

Ferruzzi, M. G., Lumpkin, J. L., Schwartz, S. J., & Failla, M. L. (2006). Digestive stability, micellarization, and uptake of beta-carotene isomers by Caco-2 human intestinal cells. *Journal of Agricultural and Food Chemistry*, *54*(7), 2780–2785.

Fiuza, S. M., Gomes, C., Teixeira, L. J., Da Cruz, M. G., Cordeiro, M. N. D. S., Milhazes, N., & Marques, M. P. M. (2004). Phenolic acid derivatives with potential anticancer properties—A structure–activity relationship study. Part 1: Methyl, propyl and octyl esters of caffeic and gallic acids. *Bioorganic andMedicinal Chemistry*, *12*(13), 3581–3589.

Fu, J., Cheng, K., Zhang, Z. M., Fang, R. Q., & Zhu, H. L. (2010). Synthesis, structure and structure-activity relationship analysis of caffeic acid amides as potential antimicrobials. *European Journal of Medicinal Chemistry*, *45*, 2638–2643.

Galletti, S., Barillari, J., Iori, R., & Venturi, G. (2006). Glucobrassicin enhancement in woad (*Isatis tinctoria*) leaves by chemical and physical treatments. *Journal of the Science of Food and Agriculture*, *86*(12), 1833–1838.

Gao, Z., Huang, K., Yang, X., & Xu, H. (1999). Free radical scavenging and antioxidant activities of flavonoids extracted from the radix of *Scutellaria baicalensis* Georgi. *Biochimica et Biophysica Acta (BBA)-General Subjects*, *1472*(3), 643–650.

Garazd, M. M., Garazd, Y. L., & Khilya, V. P. (2003). Neoflavones. 1. Natural distribution and spectral and biological properties. *Chemistry of Natural Compounds*, *39*(1), 54–121.

Grayer, R. J., & Veitch, N. C. (2006). Flavanones and dihydroflavonols. In O. M. Anderson, & K. R. Markham (Eds.), *Flavonoids: Chemistry, biochemistry and applications* (pp. 918–1002). Boca Raton, FL: CRC Press/Taylor & Francis Group.

Grubb, C. D., & Abel, S. (2006). Glucosinolate metabolism and its control. *Trends in Plant Science, 11*(2), 89–100.

Halkier, B. A., & Gershenzon, J. (2006). Biology and biochemistry of glucosinolates. *Annual Review of Plant Biology, 57*, 303–333.

Halvorsen, B. L., Holte, K., Myhrstad, M. C., Barikmo, I., Hvattum, E., Remberg, S. F., & Moskaug, O. (2002). A systematic screening of total antioxidants in dietary plants. *The Journal of Nutrition, 132*(3), 461–471.

Harborne, J. B. (1986). Recent advances in ecological chemistry. *Natural Products Reports, 3*, 323–344.

Harborne, J. B. (1989). Recent advances in chemical ecology. *Natural Product Reports, 6*(1), 85–109.

Harborne, J. B. (1993). Advances in chemical ecology. *Natural Product Reports, 10*(4), 327–348.

Harborne, J. B. (1997). Recent advances in chemical ecology. *Natural Product Reports, 14*(2), 83–98.

Harborne, J. B. (1999). Plant chemical ecology. *Comprehensive Natural Products Chemistry*, 137–196. Available from https://doi.org/10.1016/b978-0-08-091283-7.00051-5.

Harborne, J. B., & Baxter, H. (1996). *Dictionary of plant toxins*. John Wiley and Sons.

Hayes, J. D., Kellesher, M. O., & Eggleston, I. M. (2008). The cancer chemopreventive action of phytochemicals derived from glucosinolates. *European Journal of Nutrition, 47*, 73–88.

Heim, K. E., Tagliaferro, A. R., & Bobilya, D. J. (2002). Flavonoid antioxidants: Chemistry, metabolism and structure-activity relationships. *The Journal of Nutritional Biochemistry, 13*(10), 572–584.

Herrera, M. C., & De Castro, M. L. (2005). Ultrasound-assisted extraction of phenolic compounds from strawberries prior to liquid chromatographic separation and photodiode array ultraviolet detection. *Journal of Chromatography A, 1100*(1), 1–7.

Higdon, J. (2005). *Isothiocyanates*. Corvallis, OR: Linus Pauling Institute, Oregon State University.

Huber, I., Rozmer, Z., Gyöngyi, Z., Budán, F., Horváth, P., Kiss, E., & Perjési, P. (2020). Structure activity relationship analysis of antiproliferative cyclic C5-curcuminoids without DNA binding: Design, synthesis, lipophilicity and biological activity. *Journal of Molecular Structure, 1206*, 127661.

Hudnall, P. M. (2000). *Hydroquinone. Ullmann's encyclopedia of industrial chemistry*. Weinheim: Wiley-VCH. Available from http://doi.org/10.1002/14356007.a13_499.

Inczedy, J., Lengyel, T., & Ure, A. M. (1998). *Supercritical fluid chromatography and extraction. Compendium of analytical nomenclature (Definitive Rules 1997)* (3rd ed.). Oxford: Blackwell Science.

Ishida, M., Hara, M., Fukino, N., Kakizaki, T., & Morimitsu, Y. (2014). Glucosinolate metabolism, functionality and breeding for the improvement of Brassicaceae vegetables. *Breeding Science, 64*(1), 48–59.

Jan, S., & Abbas, N. (2018). Chemistry of Himalayan phytochemicals. *Himalayan Phytochemicals*, 121–166. Available from https://doi.org/10.1016/b978-0-08-102227-6.00004-8.

Kaennakam, S., Siripong, P., & Tip-pyang, S. (2017). Cytotoxicities of two new isoflavanes from the roots of *Dalbergia velutina*. *Journal of Natural Medicines, 71*(1), 310–314.

Kawaii, S., Tomono, Y., Katase, E., Ogawa, K., & Yano, M. (1999). Quantitation of flavonoid constituents in citrus fruits. *Journal of Agricultural and Food Chemistry, 47*(9), 3565–3571.

Kerry, N., & Rice-Evans, C. (1999). Inhibition of peroxynitrite-mediated oxidation of dopamine by flavonoid and phenolic antioxidants and their structural relationships. *Journal of Neurochemistry, 73*(1), 247–253.

Kobayashi, M., & Sakamoto, Y. J. B. L. (1999). Singlet oxygen quenching ability of astaxanthin esters from the green alga *Haematococcus pluvialis*. *Biotechnology Letters, 21*(4), 265–269.

Krinsky, N. I. (1998). The antioxidant and biological properties of the carotenoids. *Annals of the New York Academy of Sciences, 854*(1), 443–447.

Kuramochi, K. (2013). Synthetic and structure-activity relationship studies on bioactive natural products. *Bioscience, Biotechnology, and Biochemistry, 77*(3), 446–454.

Leicach, S. R., & Chludil, H. D. (2014). *Plant secondary metabolites: Structure–activity relationships in human health prevention and treatment of common diseases, . Studies in natural products chemistry* (Vol. 42, pp. 267–304). Elsevier.

Lepiniec, L., Debeaujon, I., Routaboul, J. M., Baudry, A., Pourcel, L., Nesi, N., & Caboche, M. (2006). Genetics and biochemistry of seed flavonoids. *Annual Review of Plant Biology*, *57*, 405–430.

Lesage-Meessen, L., Delattre, M., Haon, M., Thibault, J. F., Ceccaldi, B. C., Brunerie, P., & Asther, M. (1996). A two-step bioconversion process for vanillin production from ferulic acid combining *Aspergillus niger* and *Pycnoporus cinnabarinus*. *Journal of Biotechnology*, *50*(2–3), 107–113.

Levin, G., & Mokady, S. (1995). Incorporation of all-trans- or 9-cis-beta-carotene into mixed micelles in vitro. *Lipids*, *30*(2), 177–179.

Li, W., & Bet, T. (2013). Food sources of phenolics compounds. *Natural Products*, 2527–2558. Available from https://doi.org/10.1007/978-3-642-22144-6_68.

Li, X., & Kushad, M. M. (2004). Correlation of glucosinolate content to myrosinase activity in horseradish (*Armoracia rusticana*). *Journal of Agricultural and Food Chemistry*, *52*(23), 6950–6955.

Liu, L., Xu, X., Cheng, D., Yao, X., & Pan, S. (2012). Structure–activity relationship of citrus polymethoxylated flavones and their inhibitory effects on *Aspergillus niger*. *Journal of Agricultural and Food Chemistry*, *60*(17), 4336–4341.

Lu, Z., Nie, G., Belton, P. S., Tang, H., & Zhao, B. (2006). Structure–activityrelationship analysis of antioxidant ability and neuroprotective effect of gallic acid derivatives. *Neurochemistry International*, *48*(4), 263–274.

Luthria, D. L. (2008). Influence of experimental conditions on the extraction of phenolic compounds from Parsley (*Petroselinum crispum*) flakes using a pressurized liquid extractor. *Food Chemistry*, *107*, 745–752. Available from http://dx.doi.org/10.1016/j.foodchem.2007.08.074.

Martin, K. R., Failla, M. L., & Smith, J. C., Jr (1996). β-Carotene and lutein protect HepG2 human liver cells against oxidant-induced damage. *The Journal of Nutrition*, *126*(9), 2098–2106.

Mateos-Martín, M. L., Fuguet, E., Quero, C., Pérez-Jiménez, J., & Torres, J. L. (2012). New identification of proanthocyanidins in cinnamon (*Cinnamomum zeylanicum* L.) using MALDI-TOF/TOF mass spectrometry. *Analytical and Bioanalytical Chemistry*, *402*(3), 1327–1336.

Mattila, P., & Kumpulainen, J. (2002). Determination of free and total phenolic acids in plant-derived foods by HPLC with diode-array detection. *Journal of Agricultural and Food Chemistry*, *50*, 3660–3667.

Mattila, P., Hellström, J., & Törrönen, R. (2006). Phenolic acids in berries, fruits, and beverages. *Journal of Agricultural and Food Chemistry*, *54*(19), 7193–7199. Available from https://doi.org/10.1021/jf0615247.

Mazur, W. M., Duke, J. A., Wahala, K., Rasku, S., & Adlercreutz, H. (1998). Isoflavonoids and lignans in legumes: Nutritional and health aspects in humans. *The Journal of Nutritional Biochemistry*, *9*(4), 193–200.

McNaughton, S. A., & Marks, G. C. (2003). Development of a food composition database for the estimation of dietary intakes of glucosinolates, the biologically active constituents of cruciferous vegetables. *British Journal of Nutrition*, *90*(3), 687–697.

Miadokova, E. (2009). Isoflavonoids—An overview of their biological activities and potential health benefits. *Interdisciplinary Toxicology*, *2*(4), 211–218.

Miyamae, Y., Kurisu, M., Han, J., Isoda, H., & Shigemori, H. (2011). Structure–activity relationship of caffeoylquinic acids on the accelerating activity on ATP production. *Chemical and Pharmaceutical Bulletin*, *59*(4), 502–507.

Namitha, K. K., & Negi, P. S. (2010). Chemistry and biotechnology of carotenoids. *Critical Reviews in Food Science and Nutrition*, *50*(8), 728–760.

Nehybova, T., Smarda, J., & Benes, P. (2014). Plant coumestans: Recent advances and future perspectives in cancer therapy. *Anti-cancer Agents in Medicinal Chemistry*, *14*(10), 1351–1362.

Osawa, K., Yasuda, H., Maruyama, T., Morita, H., Takeya, K., & Itokawa, H. (1992). Isoflavanones from the heartwood of *Swartzia polyphylla* and their antibacterial activity against cariogenic bacteria. *Chemical and Pharmaceutical Bulletin*, *40*(11), 2970–2974.

Paiva, S. A., & Russell, R. M. (1999). β-Carotene and other carotenoids as antioxidants. *Journal of the American College of Nutrition*, *18*(5), 426–433.

Pannala, A. S., Chan, T. S., O'Brien, P. J., & Rice-Evans, C. A. (2001). Flavonoid B-ring chemistry and antioxidant activity: Fast reaction kinetics. *Biochemical and Biophysical Research Communications*, *282*(5), 1161–1168.

Pathak, S., Kesavan, P., Banerjee, A., Banerjee, A., Celep, G. S., Bissi, L., & Marotta, F. (2018). *Metabolism of dietary polyphenols by human gut microbiota and their health benefits. Polyphenols: Mechanisms of action in human health and disease* (pp. 347–359). Academic Press.

Patil, V. M., & Masand, N. (2018). Anticancer potential of flavonoids: Chemistry, biological activities, and future perspectives. *Studies in Natural Products Chemistry*, *59*, 401–430.

Phonsatta, N., Deetae, P., Luangpituksa, P., Grajeda-Iglesias, C., Figueroa-Espinoza, M. C., Le Comte, J., & Panya, A. (2017). Comparison of antioxidant evaluation assays for investigating antioxidative activity of gallic acid and its alkyl esters in different food matrices. *Journal of Agricultural and Food Chemistry*, *65* (34), 7509–7518.

Pietta, P. G. (2000). Flavonoids as antioxidants. *Journal of Natural Products*, *63*, 1035–1042.

Pinelo, M., Rubilar, M., Jerez, M., Sineiro, J., & Núñez, M. J. (2005). Effect of solvent, temperature, and solvent-to-solid ratio on the total phenolic content and antiradical activity of extracts from different components of grape pomace. *Journal of Agricultural and Food Chemistry*, *53*(6), 2111–2117.

Prior, R. L., Lazarus, S. A., Cao, G., Muccitelli, H., & Hammerstone, J. F. (2001). Identification of procyanidins and anthocyanins in blueberries and cranberries (*Vaccinium* spp.) using high-performance liquid chromatography/mass spectrometry. *Journal of Agricultural and Food Chemistry*, *49*(3), 1270–1276.

Raj, K. K. V., Narayana, B., Ashalatha, B. V., Kumari, N. S., & Sarojini, B. K. (2007). Synthesis of some bioactive 2-bromo-5-methoxy-N'-[4-(aryl)-1, 3-thiazol-2-yl] benzohydrazide derivatives. *European Journal of Medicinal Chemistry*, *42*, 425–429.

Rajan, V. K., & Muraleedharan, K. (2017). A computational investigation on the structure, global parameters and antioxidant capacity of a polyphenol, gallic acid. *Food Chemistry*, *220*, 93–99.

Rashmi, H. B., & Negi, P. S. (2020a). *Health benefits of bioactive compounds from vegetables. Plant-derived bioactives* (pp. 115–166). Singapore: Springer.

Rashmi, H. B., & Negi, P. S. (2020b). Phenolic acids from vegetables: A review on processing stability and health benefits. *Food Research International*, *136*, 109298. Available from https://doi.org/10.1016/j.foodres.2020.109298.

Rice-Evans, C. A., Miller, N. J., & Paganga, G. (1996). Structure–antioxidant activity relationships of flavonoids and phenolic acids. *Free Radical Biology & Medicine*, *20*, 933–956.

Rodrigues, E., Mariutti, L. R. B., Chiste, R. C., & Mercadante, A. Z. (2012). Development of a novel micro-assay for evaluation of peroxyl radical scavenger capacity: Application to carotenoids and structure–activity relationship. *Food Chemistry*, *135*, 2103–2111.

Rosenthal, A., Pyle, D. L., & Niranjan, K. (1996). Aqueous and enzymatic processes for edible oil extraction. *Enzyme and Microbial Technology*, *19*(6), 402–420.

Rossiter, J. T., & James, D. C. (1990). Biosynthesis of (R)-2-hydroxybut-3-enylglucosinolate (progoitrin) from [3, 4-3 H] but-3-enylglucosinolate in *Brassica napus*. *Journal of the Chemical Society, Perkin Transactions*, *1*(7), 1909–1913.

Ruchika, N. J., & Pandey, A. (2019). Synthetic metabolism and its significance in agriculture. *Current Developments in Biotechnology and Bioengineering*, 365–391. Available from https://doi.org/10.1016/b978-0-444-64085-7.00015-0.

Rufer, C. E., Glatt, H., & Kulling, S. E. (2006). Structural elucidation of hydroxylated metabolites of the isoflavan equol by gas chromatography-mass spectrometry and high-performance liquid chromatography-mass spectrometry. *Drug Metabolism and Disposition*, *34*(1), 51–60.

Seeram, N. P., & Nair, M. G. (2002). Inhibition of lipid peroxidation and structure–activity related studies of the dietary constituents anthocyanins, anthocyanidins and catechins. *Journal of Agricultural and Food Chemistry*, *50*, 5308–5312.

Shah, M. A., Bosco, S. J. D., & Mir, S. A. (2014). Plant extracts as natural antioxidants in meat and meat products. *Meat Science*, *98*(1), 21–33.

Shirley, A. M., & Chapple, C. (2003). Biochemical characterization of sinapoylglucose: Choline sinapoyltransferase, a serine carboxypeptidase-like protein that functions as an acyltransferase in plant secondary metabolism. *Journal of Biological Chemistry*, *278*(22), 19870–19877.

Sies, H., & Stahl, W. (1995). Vitamins E and C, beta-carotene, and other carotenoids as antioxidants. *The American Journal of Clinical Nutrition*, *62*(6), 1315S–1321S.

Sivakumar, P. M., Prabhakar, P. K., & Doble, M. (2011). Synthesis, antioxidant evaluation, and quantitative structure–activity relationship studies of chalcones. *Medicinal Chemistry Research*, *20*(4), 482–492.

Sordon, S., Madej, A., Popłonski, J., Bartmanska, A., Tronina, T., Brzezowska, E., ... Huszcza, E. (2016). Regioselective ortho-hydroxylations of flavonoids by yeast. *Journal of Agricultural and Food Chemistry*, *64*(27), 5525–5530.

Soto-Vaca, A., Gutierrez, A., Losso, J. N., Xu, Z., & Finley, J. W. (2012). Evolution of phenolic compounds from color and flavor problems to health benefits. *Journal of Agricultural and Food Chemistry*, *60*(27), 6658–6677.

Souquet, J. M., Cheynier, V., Brossaud, F., & Moutounet, M. (1996). Polymeric proanthocyanidins from grape skins. *Phytochemistry*, *43*(2), 509–512.

Stahl, W., Schwarz, W., & Sies, H. (1993). Human serum concentrations of all-trans β-and α-carotene but not 9-cis β-carotene increase upon ingestion of a natural isomer mixture obtained from *Dunaliella salina* (Betatene). *The Journal of Nutrition*, *123*(5), 847–851.

Stevenson, P. C., & Veitch, N. C. (1996). Isoflavenes from the roots of Cicer judaicum. *Phytochemistry*, *43*(3), 695–700.

Stratford, M., & Eklund, T. (2003). Organic acids and esters. In N. J. Russell, & G. W. Gould (Eds.), *Food Preservatives* (pp. 48–84). London: Kluwer Academic/Plenum Publishers.

Tasdemir, D., Kaiser, M., Brun, R., Yardley, V., Schmidt, T. J., Tosun, F., & Ruedi, P. (2006). Antitrypanosomal and anti-leishmanial activities of flavonoids and their analogues: In vitro, in vivo, structure-activity relationship, and quantitative structure-activity relationship studies. *Antimicrobial Agents and Chemotherapy*, *50*(4), 1352–1364. Available from https://doi.org/10.1128/aac.50.4.1352-1364.2006.

Temelli, F., & Güçlü-Üstündağ, O. (2005). *Supercritical technologies for further processing of edible oils. Bailey's industrial oil and fat products*. London: Wiley Edible oils and fat products: processing technologies. Available from https://doi.org/10.1002/047167849X.

Truscott, T. G. (1990). The photophysics and photochemistry of the carotenoids. *Journal of Photochemistry and Photobiology B*, *6*, 359–371.

Tsao, R. (2010). Chemistry and biochemistry of dietary polyphenols. *Nutrients*, 2(12), 1231–1246.

Tsao, R., & McCallum, J. (2009). Chemistry of flavonoids. In L. A. de la Rosa, E. Alvarez-Parrilla, & G. Gonzalez–Aguilar (Eds.), *Fruit and vegetable phytochemicals: Chemistry, nutritional value and stability* (pp. 131–153). Ames, IA: Blackwell Publishing, Chapter 5.

Uddin, A., & Khanna, P. (1979). Rotenoids in tissue cultures of *Crotalaria burhia. Planta Medica*, *36*(06), 181–183.

Umezawa, T., Yamamura, M., Ono, E., Shiraishi, A., & Ragamustari, S. K. (2019). Recent advances in lignan OMT studies. *Mokuzai Gakkaishi*, *65*(1), 1–12.

Valant-Vetschera, K. M., & Wallenweber, E. (2006). Flavones and flavonols. In O. M. Anderson, & K. R. Markham (Eds.), *Flavonoids: Chemistry, biochemistry and applications* (pp. 618–748). Boca Raton, FL: CRC Press/Taylor & Francis Group.

Van Acker, S. A., de Groot, M. J., van den Berg, D. J., Tromp, M. N., Donné-Op den Kelder, G., van der Vijgh, W. J., & Bast, A. (1996a). A quantum chemical explanation of the antioxidant activity of flavonoids. *Chemical Research in Toxicology*, *9*(8), 1305–1312.

Van Acker, S. A., Tromp, M. N., Griffioen, D. H., Van Bennekom, W. P., Van Der Vijgh, W. J., & Bast, A. (1996b). Structural aspects of antioxidant activity of flavonoids. *Free Radical Biology and Medicine*, *20* (3), 331–342.

Vogt, T. (2010). Phenylpropanoid biosynthesis. *Molecular Plant*, *3*, 2–20.

Wang, H. J., & Murphy, P. A. (1994). Isoflavone content in commercial soybean foods. *Journal of Agricultural and Food Chemistry*, *42*(8), 1666–1673.

Wcislo, G., & Szarlej-Wcislo, K. (2014). *Colorectal cancer prevention by wheat consumption: A three-valued logic–true, False, or otherwise? Wheat and rice in disease prevention and health* (pp. 91–111). Academic Press.

Weid, M., Ziegler, J., & Kutchan, T. M. (2004). The roles of latex and the vascular bundle in morphine biosynthesis in the opium poppy, *Papaver somniferum. Proceedings of the National Academy of Sciences*, *101* (38), 13957–13962.

Wianowska, D., & Gil, M. (2019). Recent advances in extraction and analysis procedures of natural chlorogenic acids. *Phytochemistry Reviews*, *18*(1), 273–302.

Williams, C. A. (2006). Flavone and flavonol O-glycosides. In O. M. Anderson, & K. R. Markham (Eds.), *Flavonoids: Chemistry, biochemistry and applications* (pp. 749–856). Boca Raton, FL: CRC Press/Taylor & Francis Group.

Wolfe, K. L., & Liu, R. H. (2008). Structure – activity relationships of flavonoids in the cellular antioxidant activity assay. *Journal of Agricultural and Food Chemistry*, *56*(18), 8404–8411. Available from https://doi.org/10.1021/jf8013074.

Wu, T., Zang, X., He, M., Pan, S., & Xu, X. (2013). Structure–activity relationship of flavonoids on their anti-Escherichia coli activity and inhibition of DNA gyrase. *Journal of Agricultural and Food Chemistry*, *61*(34), 8185–8190.

Xie, Y., Yang, W., Tang, F., Chen, X., & Ren, L. (2015). Antibacterial activities of flavonoids: Structure-activity relationship and mechanism. *Current Medicinal Chemistry*, *22*(1), 132–149.

Ye, F., Liang, Q., Li, H., & Zhao, G. (2015).). Solvent effects on phenolic content, composition, and antioxidant activity of extracts from florets of sunflower (*Helianthus annuus* L.). *Industrial Crops and Products*, *76*, 574–581.

Yoon, G., & Park, S. (2014). Antioxidant action of soy isoflavones on oxidative stress and antioxidant enzyme activities in exercised rats. *Nutrition Research and Practice*, *8*(6), 618–624.

Young, A. J., & Lowe, G. M. (2001). Antioxidant and prooxidant properties of carotenoids. *Archives of Biochemistry and Biophysics*, *385*(1), 20–27.

CHAPTER

4 Encapsulation techniques for plant extracts

Chagam Koteswara Reddy[1,2], Ravindra Kumar Agarwal[2], Manzoor Ahmad Shah[3] and M. Suriya[4]

[1]*Department of Biochemistry and Bioinformatics, Institute of Science, GITAM (Deemed to be University), Visakhapatnam, India* [2]*Department of Food Technology, Centre for Health and Applied Sciences, Ganpat University, Mehsana, India* [3]*Department of Food Science & Technology, Govt. Degree College for Women, Anantnag, India* [4]*Centre for Food Technology, Anna University, Chennai, India*

4.1 Introduction

The intake of functional foods has been increasing in the daily human diet. This has drawn the interest of researchers to formulate and use bioactive compounds, including plant extract compounds in different food formulations (Comunian, Silva, & Souza, 2021). Efforts to make functional foods have been increased because of the health benefits of these foods. Functional foods are edible products fortified with superior compounds/agents, including essential oils, antioxidants, minerals, vitamins, flavors, and bioactive compounds (Fang & Bhandari, 2010; Granato et al., 2020; Mir, Shah, Ganai, Ahmad, & Gani, 2019).

Plants are widely known as a natural source of numerous bioactive compounds having several biological activities. Plant extracts are complex mixtures of chemical compounds with numerous biological properties and are mostly derived from the different parts of medicinal plants and herbs (Fang & Bhandari, 2010; Muñoz-Shugulí, Vidal, Cantero-López, & Lopez-Polo, 2021). Bioactive compounds, including carotenoids, alkaloids, flavonoids, terpenoids, anthocyanins, polyphenols, saponins, and essential oils, are substances extracted from different plant parts and have numerous health benefits (Jia, Dumont, & Orsat, 2016; Muñoz-Shugulí et al., 2021). Some of these compounds have antioxidant, antiviral, antibacterial, antifungal, anticancer, anti-inflammatory, hypoglycemic, antihypertensive, and immunomodulatory activities (Shishir, Xie, Sun, Zheng, & Chen, 2018).

However, these plant extract compounds are easily susceptible to degradation under adverse factors during processing and storage, and these are sensitive to air, temperature, pH, and humidity (Fang & Bhandari, 2010). Also, bioactive compounds extracted from plants with polar and nonpolar solvents are commonly insoluble or sparingly soluble in water (Akolade, Oloyede, & Onyenekwe, 2017). Furthermore, plant extract compounds are inadequately applied in the novel food formulations because of their chemical instability, low thermal stability, bitter taste, and sensitivity to oxidation (Shishir et al., 2018). Therefore, to maintain the physicochemical and biological properties of plant extract compounds or to improve their sustained release properties and applicability to food formulations, encapsulation is considered a viable alternative.

Plant Extracts: *Applications in the Food Industry*. DOI: https://doi.org/10.1016/B978-0-12-822475-5.00008-9

Encapsulation is a physicochemical process where an active molecule or compound is trapped into an immiscible substance, which can be either liquid or solid (Rehman et al., 2019). Furthermore, in this technology, a physical barrier is employed to protect and maintain the physicochemical properties, bioavailability, and biological properties of the bioactive compounds against any adverse environmental conditions (Fang & Bhandari, 2010; Muñoz-Shugulí et al., 2021). The applications of encapsulation have recently expanded in the food industry since encapsulated bioactive compounds can be protected from processing and storage conditions (Saifullah, Shishir, Ferdowsi, Rahman, & Van Vuong, 2019). Furthermore, encapsulation enhances the stability and biological properties, sustained release properties, and prolonged shelf life of plant extract compounds (Fang & Bhandari, 2010). A wide range of techniques have been established to encapsulate plant extract compounds, including spray drying, spray chilling, freeze drying, coacervation, emulsion, liposome entrapment, and inclusion complexation.

This chapter presents techniques that have recently been employed for encapsulating plant extract compounds, as well as demonstrate recent developments in encapsulation techniques and their application in food industries.

4.2 Encapsulation

Encapsulation is a rapidly growing technique in which molecules are uniquely packed in the form of micro- and nanoparticles and is described as a physicochemical process for entrapping one substance (guest molecule) within another (host molecule) (Mahdavi, Jafari, Ghorbani, & Assadpoor, 2014). In this process, the guest molecules are also labeled as the core molecule, while the host molecules are known as the carrier material, matrix, or wall material (Devi, Sarmah, Khatun, & Maji, 2017). Encapsulation is widely used in food industries to protect and maintain plant extract compounds against processing, and other adverse factors. Encapsulation improves the physicochemical properties, water solubility, bioavailability, biological properties and sustained the release properties of bioactive compounds. Use of encapsulated bioactive compounds instead of free compounds can overcome the drawbacks of bioactive compounds that is, their instability, unpleasant taste, or flavor, as well as expand their physicochemical properties, biological activities, and shelf life (Fang & Bhandari, 2010; Muñoz-Shugulí et al., 2021). Therefore, encapsulation contributes by preserving as well as improving the functionality of plant-extracted compounds.

Wall materials (host or carrier materials) are vital for encapsulation and must be selected based on the purpose of encapsulation and the compatibility of the material with the food system. Furthermore, some key features of perfect wall materials are high solubility, low hygroscopy, as well as the ability of developing a highly stable emulsion and providing highly safe and secure environment (Shishir et al., 2018). The most commonly used wall materials for encapsulation in food industries are starch and its starch derivatives, chitosan, gums, phospholipids, soy proteins, milk and whey proteins (Fang & Bhandari, 2010; Wandrey, Bartkowiak, & Harding, 2010).

4.3 Encapsulation techniques

In the food industry, several techniques are used for encapsulating plant extract compounds (Fig. 4.1). Even though there is no standard system of encapsulation, different basic parameters, including core (guest molecule) and carrier (host material) molecular weight, structure, shape, polarity, and encapsulation efficiency, need to be considered before selecting a proper technique; and they are continuously optimized to secure the quality characteristics of diverse core molecules. According to their mechanism, encapsulation techniques are categorized into physical procedures,

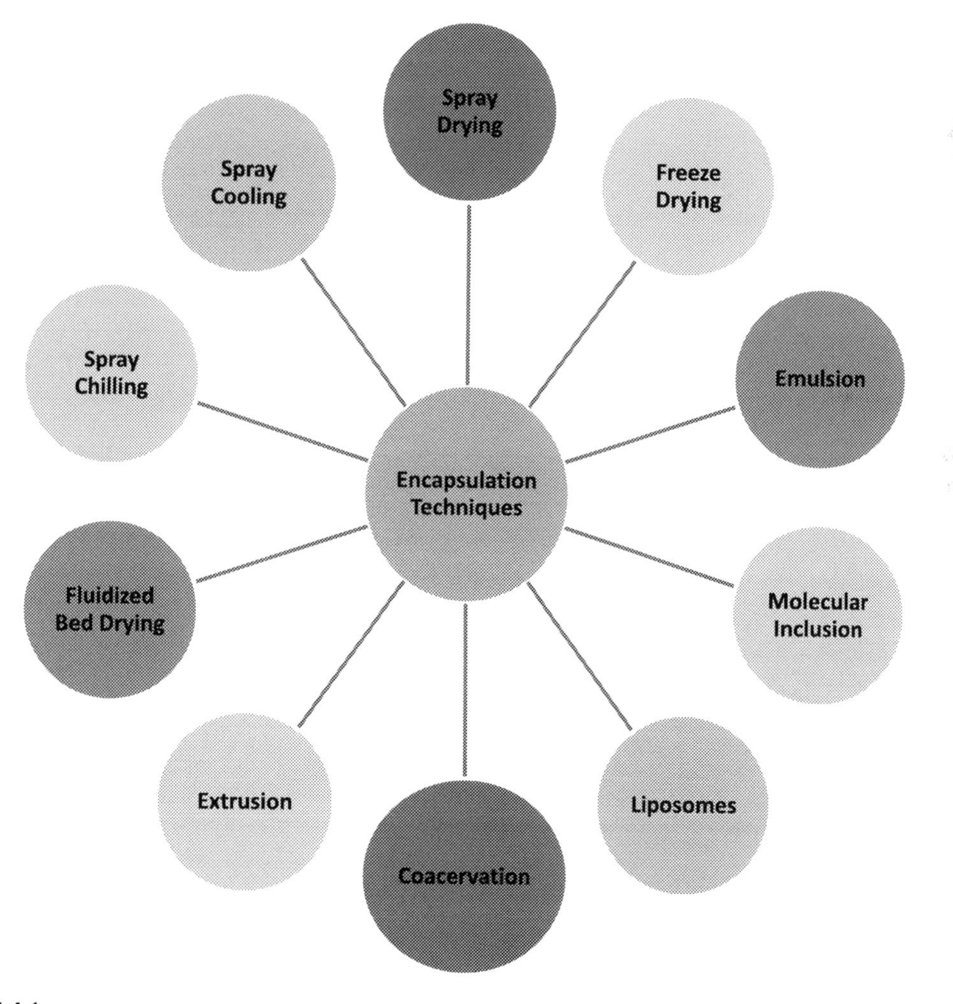

FIGURE 4.1

Techniques used for encapsulation of plant extracts.

based on the distribution of emulsion, or dispersion using different mechanical practices (de Moura, Berling, Germer, Alvim, & Hubinger, 2018), and chemical procedures, where a carrier material is formed around a core compound in liquid (Nedovic, Kalusevic, Manojlovic, Levic, & Bugarski, 2011).

4.3.1 Spray drying

Spray drying has been widely used in the food industries for encapsulation and stabilization of plant extract compounds (Fig. 4.2). In spray drying, during atomization of an emulsion, atomized droplets encounter hot gas, leading to micro-particle formation (Gadkari & Balaraman, 2015). Spray drying involves some steps, namely selection of suitable coating material, formulation of emulsion between guest and host materials, and the drying process that affects the morphological characteristics of developed capsules; the encapsulation efficiency, retention rate, and bioactivity of guest molecules are also affected during drying (Gadkari & Balaraman, 2015). Further, the coating material should be hydrophilic because water-based solutions or emulsions are employed in spray drying. In spray drying, carbohydrate molecules including as starch and its derivatives, edible gums, or other substances can be used as coating materials (Fang & Bhandari, 2010)), while hydrophobic

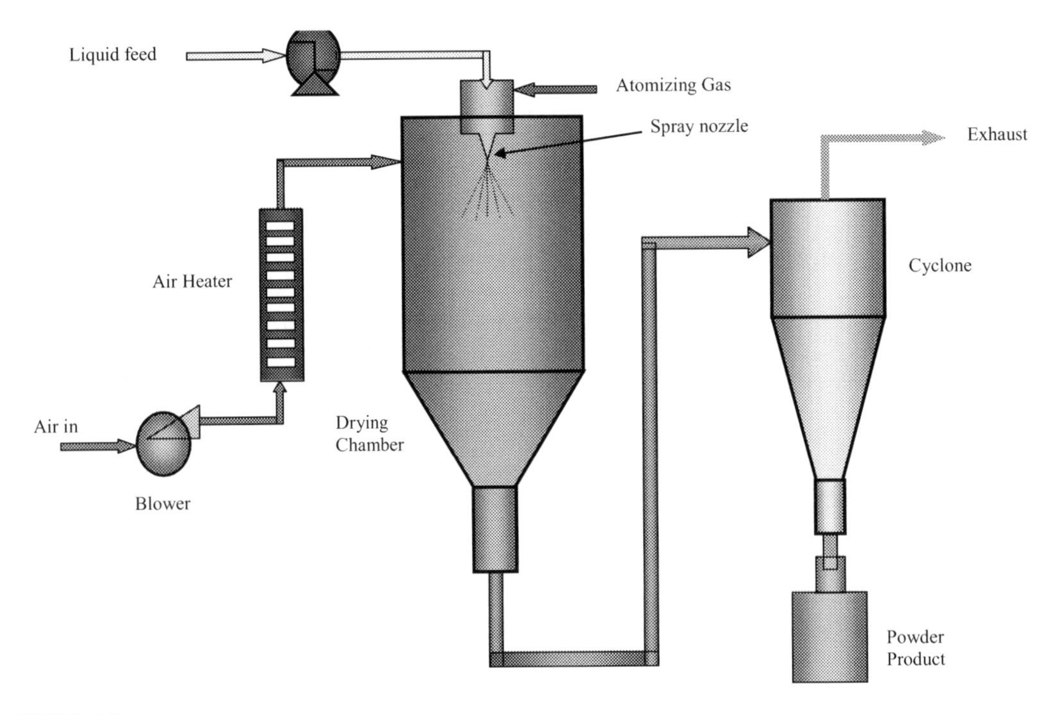

FIGURE 4.2

Schematic diagram showing the spray drying process.

and/or hydrophilic plant extract compounds can be applied as core materials (Desai & Jin Park, 2005). Moreover, nitrogen and air gas are applied to dry the dispersions as they are inert toward the material to be dried.

Spray drying is a low-cost technique for encapsulating plant extract compounds. Initially, a liquid solution comprising a host material and the guest molecule in a solvent is atomized into droplets using compressed gas. Subsequently, heated carrier gas is carried in contact with the atomized dispersion feed through a gas disperser, leading to solvent evaporation. As the solvent swiftly evaporates from the droplet, a micro-particle forms and falls to the bottom of the chamber. The resultant powder is recovered from the exhaust gases using a cyclone separator. It produces good quality particles of size $<40\,\mu m$ (Gadkari & Balaraman, 2015; Zuidam & Heinrich, 2010). The size of the formulated particles is desirable for the sensorial characteristics of final food products.

Besides the advantages of encapsulation, chemical compounds, including polyphenols, flavonoids, carotenoids, saponins, anthocyanins, essential oils, and terpenoids are successfully extracted from the byproducts of food industries, including pomegranate peel, grape pomace, and citrus processing byproducts. Rosemary essential oil was encapsulated using spray drying, with hydrocolloid and maltodextrin protecting the major compounds (Fernandes et al., 2017). The encapsulated essential oil acted as a potent bio-preservative inhibiting the growth of microbes and extending the shelf life of cheese. In another study, Santhalakshmy, Bosco, Francis, and Sabeena (2015) encapsulated jamun fruit juice using spray drying. They used maltodextrin as the wall material to expand the physicochemical properties, and shelf-life of anthocyanins. Sablania, Bosco, Rohilla, and Shah (2018) prepared encapsulated powder from curry leaf extracts by spray drying using different wall materials. These encapsulated powders may be used as functional additives in various food systems because of their antioxidant potential.

4.3.2 Freeze drying

Freeze drying is commonly used for the drying up of all heat-sensitive plant extract compounds. In this process, the material is frozen, the adjacent pressure is decreased, and sufficient heat is maintained to allow the frozen water molecule in the material to transfer clearly from the solid phase to the gas phase (Nedovic et al., 2011). Encapsulation through freeze drying is accomplished when core compounds homogenize in the matrix or carrier solution and then co-lyophilize, typically ensuing in diverse forms (Desai & Jin Park, 2005). This encapsulation process produces high-quality products, does not alter their sensorial characteristics, retains their biological properties, and offers a longer shelf life for core molecules (Tarone, Cazarin, & Junior, 2020). The main drawbacks of freeze drying are long processing time, high energy use, and formation of an open porous structure, which is usually not a very good barrier between core molecule and its storage conditions (Ezhilarasi, Karthik, Chhanwal, & Anandharamakrishnan, 2013).

Among all encapsulation techniques, freeze drying is a superior system for plant extract compounds because of higher encapsulation efficiency, better retention of stability, sustained release properties, and biological activities (Fang & Bhandari, 2010). Through freeze drying, Laine, Kylli, Heinonen, and Jouppila (2008) encapsulated cloudberry extract by using maltodextrin as the wall material. The encapsulated cloudberry extract shown improved protection of polyphenols during processing and storage. Furthermore, Elsebaie and Essa (2018) employed freeze drying to encapsulate bioactive compounds from red onion peels by using soy protein isolates and maltodextrin as wall materials to improve polyphenol stability. Using freeze drying, Reddy, Jung, Son, and Lee (2020) recently encapsulated catechin-rich green tea extract. They used β-cyclodextrin as the wall material to expand the stability and antioxidant activity of catechins.

4.3.3 Spray chilling and spray cooling

Spray chilling and spray cooling are very similar to spray drying, but the air temperature used for drying is different. Spray chilling is a promising encapsulation procedure that is widely used for plant extract compounds, particularly flavors and flavonoids. In spray chilling, the dispersion of core molecule and wall materials are atomized into cooled air, which initiates the solidification of the wall material around the core molecule (Desai & Jin Park, 2005). In spray cooling, the wall material is usually vegetable oil or its derivatives. In spray chilling, the formulated microparticles are maintained at a low temperature in a set-up, and lipid droplets adhere to already hard lipid particles before solidification. The melting point of the lipid used should be in the range of 32°C–42°C for spray chilling and 45°C–122°C for spray cooling (Risch, 1995).

Further, microcapsules formulated by spray cooling or chilling are water insoluble due to lipid layer coating. These two practices, which vary only in the melting point of the wall material used, are most often applied to encapsulate solid materials (Saifullah et al., 2019). Because one can choose the melting point of wall materials, these encapsulation systems can be utilized for the controlled release of guest molecules. Thus spray cooling and spray chilling cannot be employed for encapsulating water-soluble core materials (Desai & Jin Park, 2005). Oriani et al. (2016) produced oleoresin flavor-loaded microparticles through spray chilling by using saturated fatty acids. The study revealed that highly volatile and pungent compounds can be encapsulated through spray chilling. In another study, crystalline particles were formulated for the encapsulation of different flavor compounds through spray chilling by using erythritol as a carrier material (Sillick & Gregson, 2012).

4.3.4 Fluidized bed coating

Fluidized bed coating is one of the most proficient encapsulation systems and is finding ever-growing applications in food industries. In fluidized bed coating process, a coating is applied onto powder particles in a batch or continuous set-up. The powder particles are suspended at a precise temperature by using air stream and sprayed with an atomized coating material (Nedović, Kalušević, Manojlović, Petrović, & Bugarski, 2013). Aqueous solutions of gums, starch, cellulose, and proteins are employed as coating materials for fluidized bed coating. The fluidized bed coating process comprises three basic steps: (1) fluidization of powder particles to be coated in the coating chamber with the assistance of an air stream, (2) spraying of a coating material through a nozzle onto the particles, and (3) evaporation of the solvent of the coating material by hot air, and consequently, the coating material stick to the particles . The drawbacks of fluidized bed coating are direct exposure to high temperature, which can cause guest molecule degradation, and probable cluster formation of developed particles. The authors, Anwar and Kunz (2011), investigated the effect of fluidized bed coating on the stabilization of fish oil microcapsules with maltodextrin, cyclodextrin, and starch. Fish oil stability and bioactivity improved significantly through encapsulation. Using fluidized bed coating, Oehme, Valotis, Krammer, Zimmermann, and Schreier (2011) developed hydroxypropyl methyl cellulose-coated anthocyanin amidated pectin beads as dietary colonic delivery systems.

4.3.5 Extrusion

Encapsulation of plant-derived bioactive compounds by extrusion is relatively new compared with other encapsulation techniques. In extrusion, droplets are produced by extruding a liquid mixture of core and matrix materials. For the extrusion process, the solution-containing core and matrix materials is loaded into a syringe and passed through a needle into gelling conditions to form a gel (Saifullah et al., 2019). During extrusion, the temperature and pressure are kept at below 118°C and 100 psi, respectively. Furthermore, microcapsules prepared by extrusion have a hard, dense, and glassy structure, which protects them from adverse conditions (Nedović et al., 2013). Based on their mechanism, extrusion methods are categorized as jet cutting, spinning disk atomization, and electrostatic simple dripping extrusion (Đorđević et al., 2015). Among all extrusion methods, jet cutting is the best method for industrial applications. The coextrusion process is employed to formulate spherical microbeads with a hydrophobic guest compound and a hydrophilic or hydrophobic matrix material (Nedovic et al., 2011).

Recently, polyphenol-rich chokeberry extract was encapsulated through electrostatic extrusion by using alginate and inulin. Polyphenol-loaded microcapsules displayed higher functionality, and bioavailability of polyphenols (Ćujić et al., 2016). Encapsulation led to higher temperature stability of the plant extract compounds compared with the free compounds. Furthermore, C-phycocyanin isolated from *Arthrospira platensis* was encapsulated through extrusion by using alginate and calcium chloride. The C-phycocyanin-loaded microcapsules presented higher functionality and thermal stability of C-phycocyanin (Pan-utai & Iamtham, 2019).

Electrospinning is an effective approach for formulating sub-micron or nano-scale polymer fibers. In electrospinning, an appropriate polymer solution is spun by a high potential electric field to obtain fibers or particles. In this process, the polymer solution must be highly soluble and have suitable concentration, dielectric properties, and electrical conductivity (Isik, Altay, & Capanoglu, 2018). Tampau, González-Martinez, and Chiralt (2017) reported that carvacrol can be well encapsulated in maize starch-sodium caseinate matrices by electrospinning. Coextrusion is a process that produces beads with the guest compound being enclosed by a matrix (Waterhouse, Wang, & Sun-Waterhouse, 2014). Chew and Nyam (2016) reported that kenaf seed oil can be encapsulated using coextrusion, and encapsulation is an efficient, stable, and reproducible process. In another study, antioxidant-fortified canola oil was encapsulated through coextrusion by using alginate as the encapsulant. Coextrusion significantly changed oil bead characteristics, improved canola oil stability, and maintained the levels of polyphenols (Wang, Waterhouse, & Sun-Waterhouse, 2013).

4.3.6 Emulsion

Emulsion is a standard encapsulation system and is widely used as a delivery system for diverse plant extract compounds because of its better encapsulation efficiency and controlled release properties (Lu, Kelly, & Miao, 2016). Basically, an emulsion system comprises of two immiscible liquids, with one liquid being dispersed as small spherical droplets in the other (McClements, 2015). Moreover, based on the relative spatial distribution of water and oil phases, emulsions can be categorized as follows: water/oil (W/O) emulsions, oil/water (O/W) emulsions, and an alternative to these two emulsions, that is, a water/oil/water (W/O/W) double emulsion (Đorđević et al., 2015). Several emulsion systems with a desired structure and features have recently been established to ensure encapsulation and delivery of numerous types of plant extracts with significant health benefits.

Furthermore, the procedures for emulsion development are diverse, and few of them are injection and gelation of a particular biopolymer dispersion; phase separation and gelation of particular biopolymer dispersion; aggregative separation and gelation of a mixed biopolymer solution (Đorđević et al., 2015). After the formation of an emulsion, a gelling agent can be added to obtain microbeads. Moreover, in emulsions, stabilizers may develop films and offer a barrier to release of core molecule at the internal interface and performance as a steric stabilizer on the external aqueous phase (Zuidam & Heinrich, 2010). Further, an emulsion can be dried by spray or freeze drying to obtain a powder. Finally, the resultant powder form of emulsions have been used as encapsulates in the preparation of diverse functional foods.

Recently, emulsions have been used as a delivery vehicle for plant extract compounds. Usually, plant extract compounds have several health benefits. However, the applications of plant extract compounds in industries are limited because of their major drawbacks, including low water solubility and high sensitivity to processing conditions. Ye et al. (2020) developed a tea polyphenol-loaded emulsion using zein as a stabilizer to expand polyphenol stability as well as its bioactivity during storage. Moreover, emulsions developed using both natural and synthetic emulsifiers expand the stability of plant extract compound, especially resveratrol (Sessa, Tsao, Liu, Ferrari, & Donsì, 2011).

4.3.7 Coacervation

Coacervation, a relatively simple technique, is the partition of two liquid phases in a colloidal solution. At its molecular level, one phase contains polymer (coacervate phase) and the other does not have a polymer (equilibrium solution). The coacervation system is categorized as simple and complex depending on the number of polymer types (Dias, Ferreira, & Barreiro, 2015). In single coacervation, inorganic salts or water miscible non-solvents are typically added to the reaction mixture to initiate phase partition (Xiao, Li, & Zhu, 2015). In complex coacervation, oppositely charged polymers are mixed in a solution to develop a coating around the guest molecule, primarily under the effect of pH and temperature adjustments or in the presence of electrolytic elements (Nori et al., 2011). In complex coacervation, different polymers are utilized for encapsulation of guest molecules, and the typical polymeric pairs consist of gelatin and gums or proteins and polysaccharides (Timilsena, Adhikari, Barrow, & Adhikari, 2016). In food industries, complex coacervation is extensively employed for aroma encapsulation; however, it can only encapsulate water-soluble vitamins and phenolic compounds (Comunian et al., 2013).

Complex coacervation is primarily used to generate favorable encapsulation products with a high loading capacity. Moreover, capsules developed through coacervation are heat resistant and have better controlled release qualities under mechanical stress and temperature and pH shifts (Nori et al., 2011). The major limitation of complex coacervation is that it cannot be used for encapsulating hydrophilic core compounds. Also, the encapsulated particles are not fully spherical, and this process is more expensive. However, complex coacervation is useful for polyphenol encapsulation.

The authors, Nori et al. (2011) encapsulated polyphenol-rich propolis extract through complex coacervation by using pectin and soy protein isolates as encapsulants to enhance the antioxidant activity and stability of polyphenols during storage.

4.3.8 Liposomes

Liposomes are spherical and nano and/or micro-sized colloidal vesicles comprising one or more lipid bilayers. They are nontoxic, biodegradable, biocompatible, and non-immunogenic. Liposomes are widely used for encapsulating lipophilic, hydrophilic, and amphiphilic compounds, which makes them attractive carriers for food industries. Moreover, they are a constructive encapsulation system for water-soluble phenolic compounds as they allow them to expand their stability and maintain their biological activities (Akgün et al., 2020). In addition, liposomes offer pH and ionic strength stability to phenolic and flavonoid compounds. Usually, liposomes form instinctively when phospholipid molecules, such as lecithin and cephalin, are dispersed in water. The basic principle for liposome development is the hydrophilic–hydrophobic interaction between water and phospholipids (Fang & Bhandari, 2010). Moreover, liposomes can entrap hydrophilic compounds in aqueous compartments, hydrophobic compounds within lipid bilayers, and amphiphilic compounds at the aqueous–lipid interface (Hupfeld, Holsaeter, Skar, Frantzen, & Brandl, 2006). Liposomes provide a delivery platform for lipophilic and/or hydrophilic plant extract compounds. However, water soluble liposomal formulations have low kinetic and thermal stability.

Liposomes are yet widely used in food applications; they can be used to deliver functional ingredients or compounds that offer several health benefits to foods. Different practices have been used to formulate liposomes. The major practices are the solvent injection procedure, bubble method, thin-film hydration procedure, and heating-based procedure (Tarone et al., 2020). In recent times, some novel practices have been developed for liposome formation, such as membrane contactor-based process, pro-liposome process, and freeze drying of double emulsion procedure (Emami, Azadmard-Damirchi, Peighambardoust, Valizadeh, & Hesari, 2016).

Liposome encapsulation has recently triggered significant interest in food applications. Moreover, numerous studies have been performed to formulate liposomes as a delivery system for different plant extract compounds. Liposomes stabilize the encapsulated guest compounds against environmental variations during processing and thereafter. Plant extract compounds including polyphenols and flavonoids encapsulated in liposomes can escape digestion in the stomach and shows considerable levels of absorption in the gastrointestinal tract, leading to an increase in bioavailability (Fang & Bhandari, 2010). Gibis, Vogt, and Weiss (2012) formulated soy lecithin liposomes encapsulating polyphenol-rich grape seed extract through high-pressure homogenization. Encapsulation of grape seed polyphenols significantly enhanced their shelf life, and thermal and oxidative stability. Further, Rashidinejad, Birch, Sun-Waterhouse, and Everett (2014) described that liposome formation between green tea catechin and soy lecithin can expand the antioxidant activity of catechins.

However, liposomes are inadequately applied in food industries because of their instability in biological fluids, low storage stability, and poor reproducibility. For liposome formation, formulations need to be maintained in fairly dilute water dispersions, and this may be a main limitation for the large-scale production, and distribution of encapsulated plant extract compounds (Đorđević et al., 2015). To enhance liposome functions, Lu, Li, and Jiang (2011) used the thin-film ultrasonic dispersion method to encapsulate tea polyphenols by using nanoliposomes and observed an increase in tea polyphenol stability and bioavailability. Furthermore, Silva-Weiss et al. (2018) encapsulated quercetin in dipalmitoyl lecithin-based liposomes to maintain and regulate their release in the edible film, thereby improving its bioactivity and expanding food shelf life.

4.3.9 Molecular inclusion

Inclusion complexes are biopolymer-based encapsulation systems that can offer a safe and secure environment for different plant extract compounds. At its molecular level, inclusion complexation involves encapsulation of a core compound into a cavity- or helix-bearing substrate (host material), which was achieved through hydrogen bonding, van der Waals forces, and electrostatic interactions between core and carrier materials (Reddy et al., 2020). Inclusion complexes can be made with starch and its derivatives, including amylose, cyclodextrins (α-, β-, and γ-), chitosan, and β-glucan molecules, which may offer secure and maintain the biological properties of plant extract compounds from hydration, oxidation, and thermal processes.

Cyclodextrins are cyclic oligosaccharides comprising α-D-glucopyranose units joined by α-1,4-glycosidic bonds, which are generated from starch by enzymatic hydrolysis (Periasamy, Nayaki, Sivakumar, & Ramasamy, 2020). At its molecular level, a cyclodextrin molecule consist of a hydrophilic outer surface with several OH-groups, and a hydrophobic central cavity. This unique feature allows cyclodextrin to turn as a matrix or carrier that may formulate inclusion complexes with diverse hydrophobic compounds including flavors, fatty acids, and polyphenols through hydrogen bonding and van der Waals interactions (Numata & Shinkai, 2011). For inclusion complexation, three main forms of cyclodextrins are used, namely α-, β-, and γ-cyclodextrins, which can be made up of six, seven, and eight glucopyranose units, respectively. Among these, β-cyclodextrin is the most commonly used in food and pharmaceutical applications because of its suitable cavity size, accessibility, non-toxicity, and biocompatibility (Reddy et al., 2020). Moreover, β-cyclodextrin consists of a truncated cone-shaped structure with a flexible hydrophobic tridimensional cavity, which permits the development of non-covalent inclusion complexes with diverse guest molecules (Fernandes et al., 2018).

Inclusion complexes of host and guest molecules are formed through co-precipitation, and these inclusion complexes can be dried using spray- or freeze-drying techniques. By forming inclusion complexes with diverse plant extract compounds, cyclodextrins may expand the physical and bio-functional characteristics of plant extract compounds. Also, inclusion complexation increases the stability, aqueous solubility, and biological properties of core compounds against hydration, oxidation, and thermal processes (Ho, Thoo, Young, & Siow, 2017). Several studies have recently been conducted to enhance the water solubility, thermal and chemical stability of plant extract compounds through inclusion complexation with cyclodextrins. Kong, Su, Zhang, Qin, and Zhang (2019) prepared inclusion complexes of grape seed extract with β-cyclodextrin to improve aqueous solubility and reduce sensitivity to extreme environment conditions. Recently, Reddy et al. (2020) described that molecular inclusion of catechin-rich green tea extract by using β-cyclodextrin can expand the thermal stability and antioxidant activity of catechins. Moreover, inclusion complexation can overcome the adverse qualities of polyphenols and protect their antioxidative properties.

4.4 Conclusions

At present, the potential of different plant extracts as the resource of several chemical compounds with crucial biological activities is very well known. Plant extract compounds might be used in distinct areas, including food industries. However, commercial formulations of plant extract compounds are limited owing to their instability, easy degradability, unpleasant taste, and high oxidation sensitivity. This chapter demonstrates an overview of the recent accomplishments in encapsulation of plant extract compounds.

Through encapsulation, plant extract compounds can be incorporated without dropping their functional and quality characteristics. Encapsulation of these compounds also increases their chemical stability, aqueous solubility, biological activities, and shelf life of functional foods. Furthermore, encapsulation delivers a broad range of solutions in this respect—enhancement of sensory attributes, nutrient value, textural properties, and/or health aspects of functional foods.

References

Akgün, D., Gültekin-Özgüven, M., Yücetepe, A., Altin, G., Gibis, M., Weiss, J., & Özçelik, B. (2020). Stirred-type yoghurt incorporated with sour cherry extract in chitosan-coated liposomes. *Food Hydrocolloids*, *101*, 105532.

Akolade, J. O., Oloyede, H. O. B., & Onyenekwe, P. C. (2017). Encapsulation in chitosan-based polyelectrolyte complexes enhances antidiabetic activity of curcumin. *Journal of Functional Foods*, *35*, 584–594.

Anwar, S. H., & Kunz, B. (2011). The influence of drying methods on the stabilization of fish oil microcapsules: Comparison of spray granulation, spray drying, and freeze drying. *Journal of Food Engineering*, *105*(2), 367–378.

Chew, S.-C., & Nyam, K.-L. (2016). Microencapsulation of kenaf seed oil by co-extrusion technology. *Journal of Food Engineering*, *175*, 43–50.

Comunian, T. A., Silva, M. P., & Souza, C. J. F. (2021). The use of food by-products as a novel for functional foods: Their use as ingredients and for the encapsulation process. *Trends in Food Science & Technology*, *108*, 269–280.

Comunian, T. A., Thomazini, M., Alves, A. J. G., de Matos, F. E., Junior, de Carvalho Balieiro, J. C., & Favaro-Trindade, C. S. (2013). Microencapsulation of ascorbic acid by complex coacervation: Protection and controlled release. *Food Research International*, *52*(1), 373–379.

Ćujić, N., Trifković, K., Bugarski, B., Ibrić, S., Pljevljakušić, D., & Šavikin, K. (2016). Chokeberry (*Aronia melanocarpa* L.) extract loaded in alginate and alginate/inulin system. *Industrial Crops and Products*, *86*, 120–131.

de Moura, S. C., Berling, C. L., Germer, S. P., Alvim, I. D., & Hubinger, M. D. (2018). Encapsulating anthocyanins from *Hibiscus sabdariffa* L. calyces by ionic gelation: Pigment stability during storage of microparticles. *Food Chemistry*, *241*, 317–327.

Desai, K. G. H., & Jin Park, H. (2005). Recent developments in microencapsulation of food ingredients. *Drying Technology*, *23*(7), 1361–1394.

Devi, N., Sarmah, M., Khatun, B., & Maji, T. K. (2017). Encapsulation of active ingredients in polysaccharide–protein complex coacervates. *Advances in Colloid and Interface Science*, *239*, 136–145.

Dias, M. I., Ferreira, I. C. F. R., & Barreiro, M. F. (2015). Microencapsulation of bioactives for food applications. *Food & Function*, *6*(4), 1035–1052.

Đorđević, V., Balanč, B., Belščak-Cvitanović, A., Lević, S., Trifković, K., Kalušević, A., ... Nedović, V. (2015). Trends in encapsulation technologies for delivery of food bioactive compounds. *Food Engineering Reviews*, *7*(4), 452–490.

Elsebaie, E. M., & Essa, R. Y. (2018). Microencapsulation of red onion peel polyphenols fractions by freeze drying technicality and its application in cake. *Journal of Food Processing and Preservation*, *42*(7), e13654.

Emami, S., Azadmard-Damirchi, S., Peighambardoust, S. H., Valizadeh, H., & Hesari, J. (2016). Liposomes as carrier vehicles for functional compounds in food sector. *Journal of Experimental Nanoscience*, *11*(9), 737–759.

Ezhilarasi, P., Karthik, P., Chhanwal, N., & Anandharamakrishnan, C. (2013). Nanoencapsulation techniques for food bioactive components: A review. *Food and Bioprocess Technology*, *6*(3), 628–647.

Fang, Z., & Bhandari, B. (2010). Encapsulation of polyphenols—A review. *Trends in Food Science & Technology*, *21*(10), 510–523.

Fernandes, A., Rocha, M. A. A., Santos, L., Brás, J., Oliveira, J., Mateus, N., & de Freitas, V. (2018). Blackberry anthocyanins: β-Cyclodextrin fortification for thermal and gastrointestinal stabilization. *Food Chemistry*, *245*, 426–431.

Fernandes, R. V. D. B., Guimarães, I. C., Ferreira, C. L. R., Botrel, D. A., Borges, S. V., & de Souza, A. U. (2017). Microencapsulated rosemary (*Rosmarinus officinalis*) essential oil as a biopreservative in minas frescal cheese. *Journal of Food Processing and Preservation*, *41*(1), e12759.

Gadkari, P. V., & Balaraman, M. (2015). Catechins: Sources, extraction and encapsulation: A review. *Food and Bioproducts Processing*, *93*, 122–138.

Gibis, M., Vogt, E., & Weiss, J. (2012). Encapsulation of polyphenolic grape seed extract in polymer-coated liposomes. *Food & Function*, *3*(3), 246–254.

Granato, D., Barba, F. J., Bursać Kovačević, D., Lorenzo, J. M., Cruz, A. G., & Putnik, P. (2020). Functional foods: Product development, technological trends, efficacy testing, and safety. *Annual Review of Food Science and Technology*, *11*, 93–118.

Ho, S., Thoo, Y. Y., Young, D. J., & Siow, L. F. (2017). Inclusion complexation of catechin by β-cyclodextrins: Characterization and storage stability. *LWT*, *86*, 555–565.

Hupfeld, S., Holsaeter, A. M., Skar, M., Frantzen, C. B., & Brandl, M. (2006). Liposome size analysis by dynamic/static light scattering upon size exclusion-/field flow-fractionation. *Journal of Nanoscience and Nanotechnology*, *6*(9–10), 3025–3031.

Isik, B. S., Altay, F., & Capanoglu, E. (2018). The uniaxial and coaxial encapsulations of sour cherry (*Prunus cerasus* L.) concentrate by electrospinning and their in vitro bioaccessibility. *Food Chemistry*, *265*, 260–273.

Jia, Z., Dumont, M.-J., & Orsat, V. (2016). Encapsulation of phenolic compounds present in plants using protein matrices. *Food Bioscience*, *15*, 87–104.

Kong, F., Su, Z., Zhang, L., Qin, Y., & Zhang, K. (2019). Inclusion complex of grape seeds extracts with sulfobutyl ether β-cyclodextrin: Preparation, characterization, stability and evaluation of α-glucosidase and α-amylase inhibitory effects in vitro. *LWT*, *101*, 819–826.

Laine, P., Kylli, P., Heinonen, M., & Jouppila, K. (2008). Storage stability of microencapsulated cloudberry (*Rubus chamaemorus*) phenolics. *Journal of Agricultural and Food Chemistry*, *56*(23), 11251–11261.

Lu, Q., Li, D. C., & Jiang, J. G. (2011). Preparation of a tea polyphenol nanoliposome system and its physicochemical properties. *Journal of Agricultural and Food Chemistry*, *59*(24), 13004–13011.

Lu, W., Kelly, A. L., & Miao, S. (2016). Emulsion-based encapsulation and delivery systems for polyphenols. *Trends in Food Science & Technology*, *47*, 1–9.

Mahdavi, S. A., Jafari, S. M., Ghorbani, M., & Assadpoor, E. (2014). Spray-drying microencapsulation of anthocyanins by natural biopolymers: A review. *Drying technology*, *32*(5), 509–518.

McClements, D. J. (2015). *Food emulsions: Principles, practices, and techniques*. CRC press.

Mir, S. A., Shah, M. A., Ganai, S. A., Ahmad, T., & Gani, M. (2019). Understanding the role of active components from plant sources in obesity management. *Journal of the Saudi Society of Agricultural Sciences*, *18* (2), 168–176.

Muñoz-Shugulí, C., Vidal, C. P., Cantero-López, P., & Lopez-Polo, J. (2021). Encapsulation of plant extract compounds using cyclodextrin inclusion complexes, liposomes, electrospinning and their combinations for food purposes. *Trends in Food Science & Technology*, *108*, 177–186.

Nedovic, V., Kalusevic, A., Manojlovic, V., Levic, S., & Bugarski, B. (2011). An overview of encapsulation technologies for food applications. *Procedia Food Science*, *1*, 1806–1815.

Nedović, V., Kalušević, A., Manojlović, V., Petrović, T., & Bugarski, B. (2013). *Encapsulation systems in the food industry. Advances in food process engineering research and applications* (pp. 229–253). Springer.

Nori, M. P., Favaro-Trindade, C. S., Matias de Alencar, S., Thomazini, M., de Camargo Balieiro, J. C., & Contreras Castillo, C. J. (2011). Microencapsulation of propolis extract by complex coacervation. *LWT - Food Science and Technology*, *44*(2), 429–435.

Numata, M., & Shinkai, S. (2011). 'Supramolecular wrapping chemistry' by helix-forming polysaccharides: A powerful strategy for generating diverse polymeric nano-architectures. *Chemical Communications*, *47*(7), 1961–1975.

Oehme, A., Valotis, A., Krammer, G., Zimmermann, I., & Schreier, P. (2011). Preparation and characterization of shellac-coated anthocyanin pectin beads as dietary colonic delivery system. *Molecular Nutrition & Food Research*, *55*(S1), S75–S85.

Oriani, V. B., Alvim, I. D., Consoli, L., Molina, G., Pastore, G. M., & Hubinger, M. D. (2016). Solid lipid microparticles produced by spray chilling technique to deliver ginger oleoresin: Structure and compound retention. *Food Research International*, *80*, 41–49.

Pan-utai, W., & Iamtham, S. (2019). Physical extraction and extrusion entrapment of C-phycocyanin from Arthrospira platensis. *Journal of King Saud University-Science*, *31*(4), 1535–1542.

Periasamy, R., Nayaki, S. K., Sivakumar, K., & Ramasamy, G. (2020). Synthesis and characterization of host-guest inclusion complex of β-cyclodextrin with 4,4′-methylenedianiline by diverse methodologies. *Journal of Molecular Liquids*, *316*, 113843.

Rashidinejad, A., Birch, E. J., Sun-Waterhouse, D., & Everett, D. W. (2014). Delivery of green tea catechin and epigallocatechin gallate in liposomes incorporated into low-fat hard cheese. *Food Chemistry*, *156*, 176–183.

Reddy, C. K., Jung, E. S., Son, S. Y., & Lee, C. H. (2020). Inclusion complexation of catechins-rich green tea extract by β-cyclodextrin: Preparation, physicochemical, thermal, and antioxidant properties. *LWT*, *131*, 109723.

Rehman, A., Ahmad, T., Aadil, R. M., Spotti, M. J., Bakry, A. M., Khan, I. M., ... Tong, Q. (2019). Pectin polymers as wall materials for the nano-encapsulation of bioactive compounds. *Trends in Food Science & Technology*, *90*, 35–46.

Risch, S. J. (1995). *Encapsulation: Overview of uses and techniques*. ACS Publications.

Sablania, V., Bosco, S. J. D., Rohilla, S., & Shah, M. A. (2018). Microencapsulation of *Murraya koenigii* L. leaf extract using spray drying. *Journal of Food Measurement and Characterisation*, *12*, 892–901.

Saifullah, M., Shishir, M. R. I., Ferdowsi, R., Rahman, M. R. T., & Van Vuong, Q. (2019). Micro and nano encapsulation, retention and controlled release of flavor and aroma compounds: A critical review. *Trends in Food Science & Technology*, *86*, 230–251.

Santhalakshmy, S., Bosco, S. J. D., Francis, S., & Sabeena, M. (2015). Effect of inlet temperature on physicochemical properties of spray-dried jamun fruit juice powder. *Powder Technology*, *274*, 37–43.

Sessa, M., Tsao, R., Liu, R., Ferrari, G., & Donsì, F. (2011). Evaluation of the stability and antioxidant activity of nanoencapsulated resveratrol during in vitro digestion. *Journal of Agricultural and Food Chemistry*, *59* (23), 12352–12360.

Shishir, M. R. I., Xie, L., Sun, C., Zheng, X., & Chen, W. (2018). Advances in micro and nano-encapsulation of bioactive compounds using biopolymer and lipid-based transporters. *Trends in Food Science & Technology*, *78*, 34–60.

Sillick, M., & Gregson, C. M. (2012). Spray chill encapsulation of flavors within anhydrous erythritol crystals. *LWT-Food Science and Technology*, *48*(1), 107–113.

Silva-Weiss, A., Quilaqueo, M., Venegas, O., Ahumada, M., Silva, W., Osorio, F., & Giménez, B. (2018). Design of dipalmitoyl lecithin liposomes loaded with quercetin and rutin and their release kinetics from carboxymethyl cellulose edible films. *Journal of Food Engineering*, *224*, 165–173.

Tampau, A., González-Martinez, C., & Chiralt, A. (2017). Carvacrol encapsulation in starch or PCL based matrices by electrospinning. *Journal of Food Engineering*, *214*, 245–256.

Tarone, A. G., Cazarin, C. B. B., & Junior, M. R. M. (2020). Anthocyanins: New techniques and challenges in microencapsulation. *Food Research International*, *133*, 109092.

Timilsena, Y. P., Adhikari, R., Barrow, C. J., & Adhikari, B. (2016). Microencapsulation of chia seed oil using chia seed protein isolate-chia seed gum complex coacervates. *International Journal of Biological Macromolecules*, *91*, 347–357.

Wandrey, C., Bartkowiak, A., & Harding, S. E. (2010). *Materials for encapsulation. Encapsulation technologies for active food ingredients and food processing* (pp. 31–100). Springer.

Wang, W., Waterhouse, G. I., & Sun-Waterhouse, D. (2013). Co-extrusion encapsulation of canola oil with alginate: Effect of quercetin addition to oil core and pectin addition to alginate shell on oil stability. *Food Research International*, *54*(1), 837–851.

Waterhouse, G. I., Wang, W., & Sun-Waterhouse, D. (2014). Stability of canola oil encapsulated by co-extrusion technology: Effect of quercetin addition to alginate shell or oil core. *Food Chemistry*, *142*, 27–38.

Xiao, Z., Li, W., & Zhu, G. (2015). Effect of wall materials and core oil on the formation and properties of styralyl acetate microcapsules prepared by complex coacervation. *Colloid and Polymer Science*, *293*(5), 1339–1348.

Ye, Q., Li, T., Li, J., Liu, L., Dou, X., & Zhang, X. (2020). Development and evaluation of tea polyphenols loaded water in oil emulsion with zein as stabilizer. *Journal of Drug Delivery Science and Technology*, *56*, 101528.

Zuidam, N. J., & Heinrich, E. (2010). *Encapsulation of aroma. Encapsulation technologies for active food ingredients and food processing* (pp. 127–160). Springer.

CHAPTER

Stability of plant extracts

5

Jyoti Nishad

Department of Food Technology, SRCASW, University of Delhi, New Delhi, India

5.1 Introduction

In recent years, change in eating habits and increased demand of healthy food have revolutionized the food market. Researchers are now emphasizing more on the functional properties of food along with their nutritional characteristics and esthetic attributes. The successful results of dietary interventions in treating chronic diseases, lifestyle diseases, and degenerative disorders have enforced food industries to focus more on improvising product development to meet the consumers' demand of nutritional and healthy food (Lorenzo et al., 2019). Numerous epidemiological studies revealing the potential of plant foods in reducing the risk of above-mentioned disease conditions are encouraging consumers to take fresh fruits and vegetables or their minimally processed products in their diets. Moreover, scientific developments have changed the concept of traditionally cooked food and explored new strategies of product preparation in combination with various plant parts or their extracts for providing a functional and quality product. The plant extracts constitute various polysaccharides, proteins, fatty acid esters, vitamin, minerals, metal ions, and polyphenolic compounds, accounting to their nutritional and functional properties (Mir, Shah, Ganai, Ahmad, & Gani, 2019). Incorporation of these phytochemicals in food or pharmaceutical formulations increases the exposure of plant extracts to various processing conditions or extraction processes, drastically affecting the stability of plant extracts through oxidation, condensation, polymerization, hydrolysis, or other degradation mechanisms (Thakur, Ghodasra, Patel, & Dabhi, 2011).

Among many factors influencing the plant extract stability, thermal processing is reported as the main cause of structural modifications and alterations in their bioactivity (Ioannou, Hafsa, Hamdi, Charbonnel, & Ghoul, 2012). Heat processing at domestic level (frying, boiling, baking, etc.) or industrial level (blanching, sterilization, roasting, drying, etc.) helps in extending the shelf life of food, bring significant changes in sensory attributes and contributing to their availability (Barba, Sant'Ana, Orlien, & Koubaa, 2018; Putnik et al., 2017). However, thermal treatment also induces the formation of free radicals and other compounds, causing oxidation of bioactives, nutrient loss, quality degradation, or increased toxicity (Boekel et al., 2010). Considering the adverse effects of thermal processing on foods, the nonthermal technologies are gaining more importance at commercial level (Gabrić et al., 2018). Application of high-pressure processing, sonication, irradiation, pulse electric field, electrical voltage, and cold plasma are some of the strategies used alone or in combination for processing (Barba et al., 2017, 2018). Investigations analyzing the effect of

Plant Extracts: ***Applications in the Food Industry*****. DOI: https://doi.org/10.1016/B978-0-12-822475-5.00007-7**

nonthermal processing on plant extracts have demonstrated significant variations in correlation between treatment conditions and food matrices (Fernández-Jalao, Sánchez-Moreno, & Ancos, 2017; He et al., 2016; Rodríguez-Roque et al., 2015). Furthermore, mechanical pretreatment and processing revealed similar trend with respect to retention and degradation kinetics of polyphenols (Pérez-Gregorio, García-Falcón, & Simal-Gándara, 2011; Radziejewska-Kubzdela & Olejnik, 2016; Makris & Rossiter, 2001). Many studies also conducted to evaluate the effect of storage, pH, light, and other environmental conditions on phytochemicals and other nutrients suggesting the optimized processing conditions to develop quality, and safe product with improved shelf life (Ali et al., 2018; Chaaban, Ioannou, Paris, Charbonnel, & Ghoul, 2017; Nagar et al., 2021). Researchers have also highlighted the contribution of individual phytochemical and other components of plant matrix (fibers, peptides, sugars, and resins) to the stability and bioaccessibility of the plant extracts (Chen et al., 2014; Zhu, 2018).

Different mechanisms inducing modifications in bioactive compounds have been explained and discussed, giving the scientific basis for the selection of various processing factors. These mechanisms include triggering of stress signal, alterations in enzyme activity, auto-oxidation of compounds, free radicals generation and the associated scavenging reactions, polymerization and condensation reactions, stimulation of polyphenol synthesis under different stress conditions, and the presence of other functional groups (Barba et al., 2017; Bayliak, Burdyliuk, & Lushchak, 2016; He et al., 2016; Ioannou et al., 2012; Ismaiel, El-Ayouty, & Piercey-Normore, 2016; Kotsiou & Tasioula-Margari, 2016; Oms-Oliu, Odriozola-Serrano, Soliva-Fortuny, Elez-Martínez, & Martín-Belloso, 2012; Wang, Chen, & Wang, 2009a,b). However, the existing knowledge on above reaction mechanisms is not consistent throughout the various plant products. Therefore it is imperative to develop statistical models to analyze and accurately predict the stability of bioactive compounds after processing or storage. Numerous scientific reports are available on the factors influencing the functional compounds during and after processing or storage, however, no study comprehensively elucidate the effect of all the thermal, nonthermal and storage factors, and presence of other plant components. This chapter describes all the possible parameters affecting the stability of phytochemicals along with their proposed mechanism for better understanding of the response of functional compounds and further process optimization for retaining their activity and stability.

5.2 Stability of plant extracts

The stability of plant extracts in terms of their amount, activity, bioaccessibility, and bioavailability is known to be a factor of the extraction conditions of temperature, pressure, sonication, radiation, cold plasma, electric field, pH, presence of oxygen, and light (Chaaban et al., 2017; Mehta, Sharma, Bansal, Sangwan, & Yadav, 2019; Nagar et al., 2021; Park & Lee, 2021; Rodríguez-Roque et al., 2015; Xia, Wang, Xu, Mei, & Li, 2017). Further incorporation of plant extracts in various food formulations and their processing, and storage also affect the phytochemical potential of the bioactive compounds. Processing parameters such as temperature, light, pressure, pH, and storage conditions of temperature, relative humidity, time, presence of oxygen and light, have been widely studied for their effects on the functional compounds. Besides, numerous researches on phytochemical potential of multicomponent food matrix have revealed the synergistic and antagonistic interaction of different polyphenolic compounds affecting their overall bioactivity. The bioaccessibility and bioavailability of these compounds in the body during digestion is also an area of interest for scientists nowadays where in vitro studies have been performed to predict the health benefits imparted by some bioactive. These studies further explain the effect of processing parameters on the polyphenolic compounds when these are present along with other dietary components (Lorenzo et al., 2019). The following sections explain the various factors viz. thermal, nonthermal and storage conditions on the plant extract stability.

5.2.1 Effect of processing

Subjecting the plant or food matrices to different stages of processing starting from the extraction of bioactives to incorporation into food and further processing of food using heat or mechanical energy significantly affects the stability of plant active compounds. Along with the type of processing factor, intensity of the treatment has also shown to effect the stability to varying degrees (Galaz et al., 2017; Li, Akram, Al-Zuhair, Elnajjar, & Munir, 2020; Mehta et al., 2019; Teixeira-Guedes, Oppolzer, Barros, & Pereira-Wilson, 2019). Therefore it is imperative to understand the effect of these thermal and nonthermal processing on bioactivity of plant components before selecting the suitable technologies for achieving maximum yield of phenolic compounds without degrading their activity.

5.2.1.1 Thermal processing

Heat processing is one of the most commonly used technologies having significance in food preservation (blanching, pasteurization, and sterilization), and making the food more esthetic for its consumption (boiling, steaming, roasting, and frying). The latter is generally practiced as domestic food processing operations and is very important in deciding the bioaccessibility of polyphenols. There are various factors of type of food matrix, intensity of heat treatment, type of phenolic compound and other related processing effects which are being investigated for their role in predicting nutritional value of food (Tables 5.1 and 5.2). The studies revealed a nonuniform trend in the polyphenolic yield and their antioxidant activities, where some cooking methods facilitate the release of bioactives and others lead to degradation of those compounds (Martini, Conte, Cattivelli, & Tagliazucchi, 2021). A comparative study on steaming, frying and boiling revealed a significant increase in the phenol content of cauliflower with a slow rate of degradation during storage after steaming. These findings were explained by deleterious effects of boiling and frying, where, softening of the food matrix facilitates lixiviation of polyphenols to the solvent after boiling, and reactive species formation and high temperature of frying degrade the active compounds (Girgin & El, 2015).

Besides, the same unit operation does also imply a dual effect on phenolics release and their stability during processing, suggesting an increase in the antioxidant activity to certain level followed by a significant decline of the same (Park & Lee, 2021). The increase in bioaccessibility of compounds in the commodities like cereals was suggested due to softening and disruption of lignocellulosic structure after domestic heat treatment of boiling and steaming. Also, some authors have discussed the formation of more soluble phenolic compounds (Lima et al., 2017). Another important aspect is difference in heat stability within phenolic groups when subjected to same processing conditions. In a study on processing of blackberry jam, the results revealed variation in processing

Table 5.1 Effect of extraction methods on plant extract stability.

S. No.	Extraction method	Food matrix	Processing conditions			Impact on extract stability			References
			Temperature	Pressure	Other parameters	TP	AOX	Others	
1.	Ultrasound assisted solvent extraction	Black chokeberry	20°C, 40°C, 60°C, 80°C	–	Time- 4 h, initial solid–solvent ratio- 1:10, 1:20, 1:40; continuous mode sonication at 30.8 kHz frequency; 100 W power	Threefold increase at 60°C than at 20°C.	Increased with temperature	–	d'Alessandro et al. (2012)
2.	Subcritical water extraction	Date pits	120°C – 180°C	35 bar	Extraction time- 10 –30 min; impeller speed 500 rpm; solid content- 2 –10% w/w	First increased up to 150°C, and then decreased	Decreased with increase in temperature	TEY increased up to 150°C, dietary fibers increased with temperature	Li et al. (2020)
3.	Pressurized hot water extraction	Thyme	50°C – 200°C	1500 psi	Contact time- 5 – 30 min	Twofold increase in TP with temperature increment from 50°C to 200°C; high temperatures with long extraction times produced small number of polyphenols (methylflavones and flavonols)	1. fivefold increase in AOX with temperature increment from 50°C to 200°C	>twofold increase in TEY with increase in temperature from 100°C to 200°C	Vergara-Salinas et al. (2012)
4.	Hydro-alcoholic extraction	Oregano	22°C, 40°C, 60°C	–	Solvent:Ethanol–water; ethanol conc.%-0%, 60%, 80%, 96%, solid-to-liquid ratio- 1:20, 1:40 g/mL; particle size of plant material- <315 μm, 315–600 μm, 600–800 μm, 800–100 μm, >1000 μm, and not ground	TP increased with increase in temperature, maximum yield (49.80 mg GAE/gdw) obtained at 40°C	–	Increase in solvent selectivity with increase in temperature up to 40°C	Oreopoulou, et al. (2020)

5.	Accelerated solvent extraction (ASE)	Olive fruit	60°C, 80°C, 100°C	100 atm	Extraction cycles- 2/run; static time between runs- 10 min; rinse volume- 20%; solvents- acetone, EtOH, water	High phenolic (gallic acid, quercetin, luteolin, rutin) yield obtained at 60°C	–	Water at 100°C showed maximum TEY and % recovery of 130 mg/g and 13%	Ahmad, Ahmad, Aljamea, Abuthayn, and Aqeel (2020)
6.	Subcritical water extraction	Onion skin	170°C –230°C	30 bar	Mixing at 400 rpm; extraction time-30 min	Decrease in TP after 165°C	Nearly 38% loss in AOX at 230°C	Highest TF yield observed at 230°C; quercetin yield was high at 170°C than at 230°C	Munir, Kheirkhah, Baroutian, Quek, and Young (2018)
7.	Hot pressurized liquid extraction	Grape Pomace	90°C, 120°C, and 150°C	10 MPa	Solvent: water-glycerol mixtures (15%, 32.5%, and 50%) solvent volume-50 mL, extraction cycle-1, 250 s nitrogen purge time, and static extraction time- 5 min	~34 times increase in TP at 150°C	–	~13 times and 18 times more gallic acid and flavanol yield at high temperature	Huamán-Castilla, Mariotti-Celis, Martínez-Cifuentes, and Pérez-Correa (2020)
8.	Microwave-assisted extraction	*Hibiscus sabdariffa*	50°C–150°C	–	Solvent- 15%–75% EtOH; 1500 W microwave power; extraction time- 5–20 min	Highest TP yield at 164°C	–	Highest TFC achieved at 158°C (14.43 mg QE per g), high TEY of ~47% obtained at 150°C	Pimentel-Moral, et al. (2018)

AOX, *Antioxidant activity;* TAC, *total anthocyanin content;* TF, *total flavonoid;* TEY, *total extraction yield;* TP, *total phenols content.*

Table 5.2 Effect of temperature on polyphenols during domestic and industrial processing of food.

S. No.	Food matrix	Processing technique	Processing conditions			Impact on plant extract			References
			Temperature	Pressure	Other parameters	TP	AOX	Others	
1.	Omija fruit	Roasting	120°C, 150°C, 180°C	–	Roasting time- 5, 10, 15 min.	Nearly twofold increase in TP at 180°C	–	5 to 15-fold increase in TF at 180°C; lignan content increased up to 150°C and decreased at 180°C, ester content decreased with increase in temperature	Park and Lee (2021)
2.	Pomegranate peel	Drum drying	Drying conditions- 422 s at 100°C, 400 s at 110°C, 257 s at 120°C	–	–	No significant effect on TP	An insignificant decrease of 7.5% observed at 100°C	Least change in color at 120°C for 257 s	Galaz et al. (2017)
3.	Mulberry leaves	Air drying	Air-drying conditions- 45 h at 40°C; 7 h at 60°C, 4.3 h at 70°C, 2.5 h at 80°C, 1.7 h at 110°C	–	–	TP increased up to 60°C and then decreased (three times) at >60°C	AOX increased up to 60°C and then decreased at >60°C	kaempferol; quercetin; chlorogenic acid; rutin; astragalin; isoquercitrin showed degradation after 60°C	Katsube, Tsurunaga, Sugiyama, Furuno, and Yamasaki (2009)
4.	Apple Pomace	Blanching and air drying	Blanching in stainless steel steam-jacketed kettle at 95°C to achieve a stable temperature of 86°C for 4 min; dehydration temperatures- 50°C, 60°C, 70°C, and 80°C until 0.05 kg H_2O/kg dry matter achieved	–	–	1.25-fold increase in TP after blanching; significant loss in TP at 70°C	–	Twofold increase in TF after blanching; significant loss in TF at 70°C; decreased browning in blanched samples	Heras-Ramírez et al. (2012)
5.	Plum extract	Thermal treatment	Anthocyanin solutions submerged in thermostatic bath at temperatures 25°C–110°C	–	Time: 10 min	Heating for 5 min at 70°C –90°C reduced TP between 4%–23%, and decreased the content to 43%–72% at >90°C	61% AOX loss occurred after 20 min. of heating at 110°C	Maximum reduction of 71% and 91% in TFC and TAC was recorded at 110°C, after 20 min of holding	Turturică et al., 2016

6.	Onions	Frying Boiling Microwaving Sautéing	Frying. onions were fried for 2 min, submerged in 350 mL of 100% soybean oil heated to 150°C Boiling. onions were boiled for 5 min in 100 mL of distilled water containing 1 or 3% NaCl Microwaving. at high heat for 1 min in a microwave oven SautéIng. Chopped onions were cooked on a stovetop for 3 min in soybean oil	–	–	–	–	Frying. 32.8% loss in TF Boiling. 13.7% loss in TF Microwaving. 4.4% loss in TF SautéIng. 20.6% loss in TF	Lee et al. (2008)
7.	Kale	Steaming, boiling, frying	Steaming (over boiling water, 5 min–60 min); Boiling (5–60 min); Frying (140°C, 5 min–20 min)	–	–	Steaming caused a TP decrease of 5%–30%; stir frying showed 35%–41% decrease in TP	Steaming showed an initial increase of AOX, followed by 15% decrease for >20 min. cooking time, frying decrease the AOX significantly	Total carotenes, β-carotene and lutein content increased after boiling, 54%–42% decrease in lutein observed after steaming, stir frying caused a 10%–23% decrease in β-carotene	Giambanelli, Verkerk, D'Antuono, & Oliviero, 2016)
8.	Pepper, squash, green beans	Boiling, steaming, microwaving	–	–	Boiling (5 min); Steaming (over boiling water, 7.5 min, atmospheric pressure); Microwaving (1000 W, 1 min for squash, 1.5 min for pepper & beans)	Increase in TP of pepper & green beans, decrease in TP of squash	Increase in AOX of pepper & green beans, no change in AOX of squash	–	Turkmen, Sari, and Velioglu (2005)

(Continued)

Table 5.2 Effect of temperature on polyphenols during domestic and industrial processing of food. *Continued*

S. No.	Food matrix	Processing technique	Processing conditions			Impact on plant extract			References
			Temperature	Pressure	Other parameters	TP	AOX	Others	
9.	Cauliflower	Boiling, steaming	Steaming (over boiling water, 10 min); Boiling (10 min)	–	–	Increase in TP by 14.83% after steaming, decrease in TP by 1.8% after boiling	AOX increased by 47% after steaming, and decrease 8% after boiling	–	Girgin and El (2015)
10.	Eggplant	Baking, boiling, frying, grilling	Boiling (20 min at 100°C); grilling (cooking at 120°C for 10 min on a grilling plate); baking (30 min at 180°C in electric oven); frying (170°C for 10 min)	–	–	34.5% decrease after grilling; increased 42%, 67% & >300% after boiling, baking and frying, respectively	Frying increased (~3.5 times) the AOX; grilling induced 20% decrease; baking and boiling had no effect on activity	74.2% of increase in hydroxycinnamic acids after frying; 27%, 51%, and 60% decrease in total hydroxycinnamic acids was observed after boiling, grilling, and baking, respectively	Martini et al. (2021)
11.	Kidney bean, pinto bean, black bean, soybean	Boiling and pressure cooking	Soaking (12 h at room temp.), followed by boiling (100°C, for 50 min) or pressure cooking (115°C, for 20 min)	–	–	175%, 80%, 72%, 56% increase in TP of kidney, pinto, and borlotti beans, respectively, after boiling	109%, 74%, 42%, 64% increase in AOX of kidney, pinto, and borlotti beans, respectively, after boiling	TEY increased for all beans after boiling and pressure cooking (94%, 50%, 30%, 5% increase after boiling; 130%, 20%, 5%, 10% increase after pressure cooking of kidney, pinto, black and soy beans, respectively)	Teixeira-Guedes et al. (2019)
12.	Wild rocket leaves	Hot water pretreatment and steaming	Hot water pretreatment (90°C for 5 min)	–	Steaming (10 min)	Pretreatment with hot water yielded highest TP (45.4 mg/100 g); steaming of leaves reduced TP by 20%	Pretreatment with hot water gave highest AOX (5.8 μmol Trolox/1 g)	Steaming and pretreatment with hot water reduced the glucosinolate content by 21%, and 37%, respectively	Radziejewska-Kubzdela, Olejnik, and Biegańska-Marecik (2019)

AOX, *antioxidant activity;* TAC, *total anthocyanin content;* TEY, *total extraction yield;* TF, *total flavonoid;* TP, *total phenols content.*

effect on different bioactive compounds where bioaccessibility of total phenolics increased with a significant decrease in the bioaccessibility of flavonoids. The decrease in activity was ascribed to either degradation of the polyphenolic (anthocyanin) compound caused by oxidation and cleavage of intramolecular bonds after exposure to heat or formation of complex with reduce activity (procyanidins) in association with other polyphenolics. On the contrary, release of bound phenolic compounds have an additive effect on the accessibility (Tomas et al., 2017).

Application of temperature during the extraction of plant extracts increases the extraction efficiency; however, after a certain range of temperature, the adverse effect of temperature on the yield of bioactive compounds is more prominent. The effect of different extraction methods (ultrasound assisted, hydroalcoholic, solvent, subcritical-water extraction etc.) and food processing techniques (roasting, drying, blanching, etc.) involving heat on the functional bioactive of plant extracts has been summarized in Table 5.1. In the process of extraction the temperature generally facilitate the extraction of polyphenols. However, after certain maxima the effect of temperature negatively affect the extraction yield and bioactivity of the compounds (d'Alessandro, Kriaa, Nikov, & Dimitrov, 2012; Li et al., 2020). The temperature aided high polyphenol yield is observed to be due to improved solvation power of the solvent, high solubility of polyphenolics at high temperatures along with increased diffusivities of the extracted compounds and improved mass transfer between the food matrix and the solvent. Furthermore, temperature-induced disruption of plant cell membrane and hydrolysis of ester or ether bonds between different complex bioactive compounds leads to the release of bound phytochemicals and therefore enhances the extraction yield (Gonzales et al., 2015; Prasad, Yang, Yi, Zhao, & Jiang, 2009). However, at very high temperature the content of some active compounds tend to decrease which is attributed to their thermal sensitivity and probable structural changes (Oreopoulou, Goussias, Tsimogiannis, & Oreopoulou, 2020; Park & Lee, 2021). Various studies on polyphenol extraction suggested the synergistic and antagonistic interaction between temperature and other extraction parameters such as type of food matrix, particle size, pretreatment of food, type of solvent, solvent concentration, solid and solvent ratio, time of extraction, pressure, presence of enzyme, microwave energy, ultrasound energy, etc. (Li et al., 2017; Nishad, Saha, & Kaur, 2019; Oreopoulou, et al., 2020; Vergara-Salinas, Pérez-Jiménez, Torres, Agosin, & Pérez-Correa, 2012). Few studies have also depicted an increase in total antioxidant activity whereas the concentration of individual polyphenolic compound decreased after extraction. Nishad et al. (2019) in their study on *Citrus sinensis* peel revealed similar findings where extraction resulted increase in antioxidant activity and a decrease in naringin and other phenolics attributed to the conversion of galloylated form of polyphenolic compounds to agalloylated form having higher reducing activities.

Moreover, heat processing of food has some effects on the polyphenolics yield. Treatment like pasteurization, blanching, drying, roasting, steam heating etc. plays an important role in signifying the phytochemical potential of the foods (Fuleki & Ricardo-Da-Silva, 2003; Galaz et al., 2017; Heras-Ramírez et al., 2012; Park & Lee, 2021; Zhang, Chen, Li, Pei, & Liang, 2010; Zhang, Cardon, Cabrera, & Laursen, 2010). Most of these studies depicted a significant loss in total phenolic content, flavonoids and antioxidant activity. Where the degree of loss is a factor of process method used for for example, effect of hot air drying on flavonoids is more intense than the freeze drying (Zainol, Abdul-Hamid, Bakar, & Dek, 2009). However, some studies showed positive impact of temperature on prevention of polyphenols by inactivating different hydrolytic and oxidative enzymes (Dewanto et al., 2002; Galaz et al., 2017). The effect of thermal treatment on water-soluble plant pigments having antioxidant potential is more prominent. Study on monomeric anthocyanins revealed highest degradation rate of anthocyanins to anthocyanidins after thermal processing attributed to oxidation and disruption of covalent bonds (Turturică, Stănciuc, Bahrim, & Râpeanu, 2016). On the contrary the water insoluble plant pigments like carotenes are more stable to heat treatment. A study on thermal processing of tomatoes depicted stability of lycopene at high temperatures of 130°C, beyond which the pigment start degrading but to a lower extent (Colle, Lemmens, Van Buggenhout, Van Loey, & Hendrickx, 2010).

Furthermore, individual polyphenolic compounds behave differently at similar thermal processing conditions. A study on buckwheat groats revealed higher thermal stability of rutin than other flavonoids (vitexin, isovitexin, homoorientin and orientin) (Zielinski, Michalska, Amigo-Benavent, del Castillo, & Piskula, 2009). Similarly, pasteurization of strawberry juices at 90°C for 60 s did not affect the quercetin and kaempferol contents but significantly reduced the naringin, quercetin, naringenin content in processed grapefruit juices (Igual, García-Martínez, Camacho, & Martínez-Navarrete, 2011; Odriozola-Serrano, Soliva-Fortuny, & Martin-Belloso, 2008). These findings are attributed to the structural differences in the polyphenolic compounds (Buchner, Krumbein, Rhon, & Kroh, 2006). The variation in thermal stability, however, could also be due to structural difference of food matrices where it act as barrier to the heat applied. More promising and uniform results for bioactive stability are reported in case of novel technologies viz. high-intensity pulsed electric fields, microwave, infra-red heating, high-pressure processing (Odriozola-Serrano et al., 2008; Srinivas, King, Monrad, Howard, & Zhang, 2011; Xi & Shouqin, 2007). The degradation of bioactives is not only a function of temperature but also depends on the pH, oxygen, and other phytochemicals (Buchner et al., 2006; Murakami, Yamaguchi, Takamura, & Matoba, 2004).

5.2.1.2 Nonthermal processing

Preparation and processing of foods include primary and secondary unit operations where mechanical processes of cutting, peeling, trimming, chopping, pressing, and filtration significantly affect the functional properties of food (Pap et al., 2012; Pérez-Gregorio et al., 2011; Renard et al., 2011). Further, with increasing concern of consumers for wholesome and safe food, various nonthermal processing have occupied significant place in food industries. Pressure, pulse electric field, electrical voltage, ultrasonication, microwave, ultraviolet radiation, cold plasma are few important components of these new processing and preservation technologies. Numerous studies are based on these technologies where researchers have focused on analyzing the short- and long- term effects on physicochemical and functional properties of different food components (Table 5.3).

5.2.1.2.1 Mechanical processing

Different stages of preparation of raw material viz. peeling, trimming, chopping, slicing, to further processing of food matrix through crushing, pressing, and filtration involve the use of mechanical energy. This processing has significant effect on phytochemicals with respect to their extraction yield, bioactivity, and bioavailability (Nicoli, Anese, & Parpinel, 1999). The major loss of polyphenolics was observed in initial stages of processing or preprocessing steps. The reduction is primarily attributed to the removal of parts during peeling and trimming of the fruits or vegetables, having high content of bioactives. Slicing also revealed a negative effect on the polyphenols (Makris & Rossiter, 2001). However, the cutting has shown contrary results in many studies where the step was responsible for a significant increase of flavonol content (Pérez-Gregorio et al., 2011; Tudela,

Table 5.3 Effect of nonthermal processing on plant extract stability.

S. No.	Treatment	Food matrix	Processing conditions			Impact on functional quality and stability of plant extract			References
			Pressure	Time	Others	TP	AOX	Others	
1.	HPH	Apple, grape and orange juice	250 MPa	10 min	–	Apple juice. 28.5% decrease in TP Grape juice. 14.6% increase in TP Orange juice. 29% increase in TP	Apple juice. 39% reduction in AOX Grape juice. 16% increase in AOX Orange juice. 29% increase in AOX	29% reduction of phenolic bioaccessibility in apple, and preservation in grape and orange juice	He et al. (2016)
2.	HPP	Onion	400 MPa	5 min	Temperature: 25°C; compression rate: 500 MPa/min	TP increased by 6%	9.6% increase in DPPH value	Bioaccessibility of total flavonols was preserved (~17.63%)	Fernández-Jalao et al. (2017)
3.	HVED, PEF and US	Exotic fruit juice with *Stevia rebaudiana*	–	–	Energy level. 32 kJ/kg, 256 kJ/kg; *PEF*. 50 to 400 pulses; electric field strength – 25 kV/cm; temperature- 35°C *HVED*. 50 to 400 discharge number, initial voltage peak amplitude- 40 kV *US*. 400 W, 24 kHz frequency; 100% amplitude, cycle- 1.	TP after high energy HVED and PEF were increased (2%–4%)	ORAC values depicted an increase (16%–22%) in AOX after PEF and US	Increase in bioaccessibility of TP (34.2%) after HVED at low energy level, further decrease to 16.7% at 256 kJ/kg; in case of PEF and US increase in TP bioaccessibility at both energy levels; Total carotenoids increased (18%) after low energy PEF treatment, while high energy HVED and USN caused 28%–45% decrease in total carotenoids	Buniowska, et al. (2017)

(*Continued*)

Table 5.3 Effect of nonthermal processing on plant extract stability. ***Continued***

S. No.	Treatment	Food matrix	Processing conditions			Impact on functional quality and stability of plant extract			References
			Pressure	Time	Others	TP	AOX	Others	
4.	HIPEF, HHP	Water-(WB), Milk-(MB) and soymilk-fruit (SB) juice beverage	HHP. 400 MPa	5 min	HIPEF. 35 kV/cm, 1800 μs	10% and 20% decrease in WB after HPP and HIPEF; whereas HPP and HIPEF of MB and SB reported ~20% increase	Around 12% and 16% decrease of AOX in WB; 30% and 27% increase in MB; 18% and 10% decrease in SB was observed after HIPEF and HPP, respectively	HIPEF reduced the vitamin C in the range of 8%–15% in all beverages; HPP revealed a decrease of 10.5% in vitamin C in SB	Rodríguez-Roque et al. (2015)
5.	HHP	Germinated brown rice	100, 300, and 500 MPa	10 min	Temperature-18°C	–	At 500 MPa, AOX increased by 12.72% as depicted by DPPH value	The *in vitro* bioaccessibility of calcium and copper was increased by 12.59%–52.17% and 2.87%–23.06%, respectively, after HHP	Xia et al. (2017)
6.	Thermal treatment, US, UV-C	Mango juice	–	–	*Thermal treatment*:90°C for 30 and 60 s *US*: 15, 30 and 60 min at 25°C, 40 kHz frequency *UV-C light treatment:* 15 min, 30 min and 60 min at 25°C	Significant improvement in extraction of quinic acid, ellagic acid, quercetin, gallic acid, kaempferol, mangiferin, and tannic acid after US and UV-C	Nearly 3% increase in AOX in non-thermally treated samples	Thermally and non-thermally treated samples depicted extended shelf life of 4–5 weeks at 4°C	Santhirasegaram et al. (2015)

7.	UV, US, cold plasma treatment, thermal processing	Tomato based beverage	–	–	*UV processing:* 10 min and 15 min in UV cabinet at 254 nm *Cold plasma treatment:* 260 V, 60 kV at 50 Hz for 10 min and 15 min *Ultrasonication:* 240 V, 37 kHz for 10 and 15 min *Thermal processing:* 80°C for 2 min	Significant increase in TP and individual phenolic compound after non-thermal treatment, highest increment was observed after cold plasma treatment	–	1 log reduction in bacterial, yeast and mold count; no effect on glucose and fructose content	Mehta et al. (2019)
8.	UV radiation	Starfruit juice	UV radiation dose of 2.158 J/m^2	30 and 60 min	Temperature: 25°C ± 1°C.	3.1% and 6.2% increase in TP at 30 min and 60 min, respectively	Percentage increase of 1.9 and 3.4 in AOX at 30 min and 60 min of UV exposure time, respectively	% increase in flavonoids was 17.4 at 30 min and 60 min; ascorbic acid was decreased by 10% and 20% after 30 min and 60 min, respectively	Bhat, Ameran, Voon, Karim, and Tze (2011)
9.	Pressing	Apple juice	20 kg apple/ pressing	–	Temperature 4°C, 11°C, 18 and 25°C	Increase in TP content after thermal pressing	–	Increase of the proanthocyanidins (> 50%)	Renard et al. (2011)
10.	Ultrafiltration	Blackcurrant juice	1 bar to 2.75 bars	–	recirculation flow rate of 220 L/h at feed temperatures of 25°C and 45°C	–	–	50% and 46% decrease in anthocyanin and flavonol content after filtration	Pap et al. (2012)
11.	HHP	Orange juice	600 MPa	4 min	–	~6% reduction in TP after treatment	25% loss in AOX as measured by FRAP assay	Preservation of bioaccessibility of TP	Mennah-Govela and Bornhorst (2017)

(Continued)

Table 5.3 Effect of nonthermal processing on plant extract stability. ***Continued***

S. No.	Treatment	Food matrix	Processing conditions			Impact on functional quality and stability of plant extract			References
			Pressure	Time	Others	TP	AOX	Others	
12.	Soaking	Coleslaw mix	–	5 min	soaking in 5 g/L ascorbic acid and 5 g/L citric acid solution	–	–	A reduction of 20% in glucosinolate, 26% in glucoiberin, and 14% in sinigrin levels after pretreatment	Radziejewska-Kubzdela and Olejnik (2016)
13.	Chopping	Asparagus	–	–	Chopped into small pieces (0.5-cm length), kept at room temperature, under open air, for 60 min	–	–	18.5% decrease of rutin content	Makris and Rossiter (2001)
14.	HVED	Cocoa	–	15, 30, and 45 min	40 and 80 Hz	–	–	11% and 70% loss in epicatechin gallate and catechin concentration, respectively	Barišić et al. (2020)
15.	UV-light	Elderberry fruit	–	Pulsed UV duration: 5, 10, 20, 30 s	3 energy dosages:4500, 6000, 11,000 J/m^2/pulse	Highest increase in TP ~50% was found with 11,000 J/m2/pulse for 10 s	–	–	Ramesh, Valérie, and Mark (2012)

AOX, *antioxidant activity;* HHP, *high hydrostatic pressure;* HIPEF, *high-intensity pulsed electric fields;* HPH, *high pressure homogenization;* HPP, *high pressure processing;* HVED, *high voltage electrical discharges;* MWH, *microwave heating;* PEF, *pulsed electric fields;* TAC, *total anthocyanin content;* TEY, *total extraction yield;* TF, *total flavonoid;* TP, *total phenols content;* US, *ultrasonication;* UV, *ultraviolet radiation.*

Cantos, Espin, Tomás-Barberán, & Gil, 2002). The results were supported by the fact that wounding induces activation of plant defense system which further activate enzymes like phenylalanine ammonia-lyase, enhancing the biosynthesis of polyphenols (Tudela et al., 2002).

Further studies on pressing and filtration have a positive effect on flavonoid content in association with temperature, solvent, and enzyme; however, the single step processing yielded a significant decrease in the values (Oszmianski, Wojdylo, & Kolniak, 2009; Renard et al., 2011; Van Der Sluis, Dekker, Skrede, & Jongen, 2004). The reduction was associated with the compounds left behind in the pomace or residues. Application of other factors (temperature, solvent, and enzyme) lead to cell lysis, disruption of bonds, breakdown of complex food structure, solubilization of active compounds, and their release into the solvent, thereby improving the yield of phytochemicals with simultaneously reducing their loss in pomace or residues (Korus, Słupski, Gębczyński, & Banaś, 2014; Pap et al., 2012; Van Der Sluis et al., 2004).

5.2.1.2.2 Novel processing technologies

Application of hydrostatic pressure, pulse electric field, ultrasonication, pressure-induced homogenization, ultraviolet radiation, irradiation, and cold plasma are extensively used in food industries. The commercialization of these technologies for food processing and preservation has been attracting researchers to investigate their interaction with different food components responsible for bringing nutritional and sensory changes in the products. Largely, these processing methods have reported to impart positive effect on phytochemical bioavailability and bioaccessibilty (Table 5.3).

The pressure treatment in the form of high hydrostatic pressure (HPP) processing or high pressure homogenization (HPH) reduces the negative effects of processing on phytochemicals, and even increase their content (Lorenzo et al., 2019; Rodríguez-Roque et al., 2015). However, depending upon the food matrix the treatment may reduce the bioaccessibility (He et al., 2016). This variation in results was explained by the difference in composition of the food. The increase in after treatment phenolic content is attributed to structural changes in tissues, facilitating the release of bioactives. High pressure cause alterations in physical structure, induces chemical reactions, and decrease the volume under the effect of molecular volume change. The compression in volume affects the integrity of cell structure and bring modifications in its permeability, thus facilitating the release of bioactives from the cell and further improve the solubilization of the compounds (Patterson, 2014). Further this can also affect the in-vivo accessibility and availability of the compound, though no correlation has yet been found between the structure disruption and bioavailability of phytochemicals (Barba, Terefe, Buckow, Knorr, & Orlien, 2015). Conversely, the reduction in bioactive content could be due to release of polyphenol oxidase along with other cytoplasmic compounds which are known to trigger oxidation reactions, epimerization, and finally the degradation (He et al., 2016).

Other processing treatments also revealed on par results with pressure treatment where the stability of plant extracts were manifested by the polyphenol content and antioxidant activity (Fig. 5.1) (Buniowska, Carbonell-Capella, Frigola, & Esteve, 2017; Rodríguez-Roque et al., 2015). Treatment of food using pulse electric field (PEF) revealed a significant interaction of food matrix and individual phenolic compound and their effect on the yield and bioaccessibility of phytochemicals (Rodríguez-Roque et al., 2015).

The optimization of PEF process involves tuning of several factors of electric field intensity, properties of pulse, treatment time, electrode configuration, and temperature. In addition the

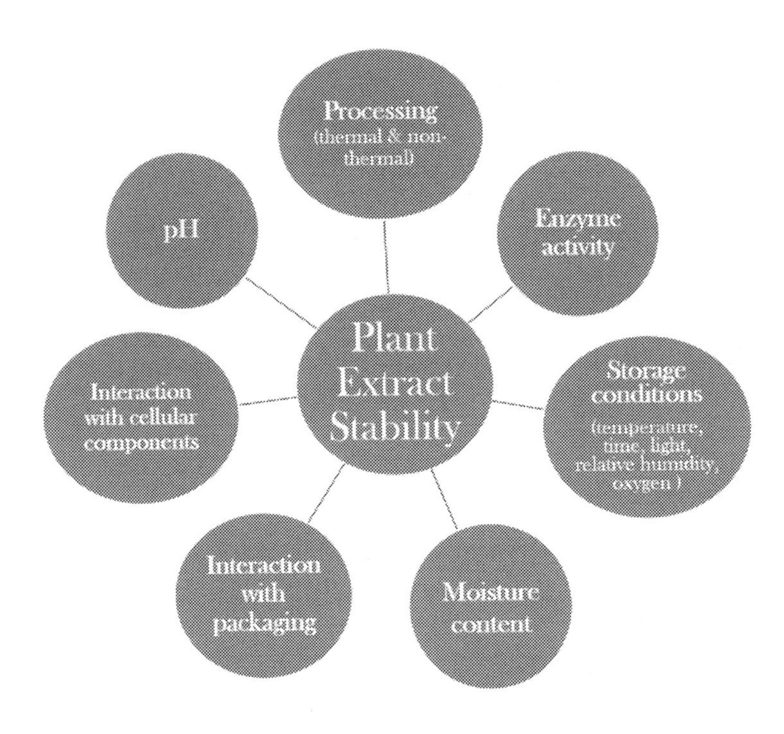

FIGURE 5.1

Factors affecting stability of plant extracts.

properties of food including type of matrix, pH, the ionic strength, and conductivity also become crucial during the PEF treatment of food with the objective of maximizing the phytochemical yield (Knorr et al., 2011; Li & Farid, 2016). The increase in phenolics content and their activity is suggested to be an effect of electroporation. This phenomenon causes perforation and compression of the cell membrane and leakage of intracellular content, breaking of bonds and complex compounds, thereby increasing their solubility and extractability (Barba et al., 2017).

Another important novel technology is based on sound waves ultrasonication. The positive effects of this treatment on bioactive release is associated with acoustic cavitation. Sonication generates microbubbles in the system which continuously grow in the medium, and collapse after certain time, giving rise to shock waves. This is these waves which generate high temperatures and pressures, resulting in the cavitation mechanism (Cravotto & Binello, 2016). The cavitation in food matrix is responsible for producing shear forces and thereby disrupt the cell structure, breaks down the polysaccharides or other polymers, and hence helps in releasing bound phytochemicals. Also disruption in cell structure leads to an increase in the contact between solvent and solute (bioactive compounds), enhancing penetration of solvent and improving extractability of phytochemicals. Ultrasonication under controlled parameters are widely reported to increase the phenolics yield and antioxidant activity. However, the treatment has also depicted negative effects on plant phenolics when the matrix was subjected to extreme conditions of sonication energy, this is attributed to the degradation of the compounds due to their scavenging actions on newly formed free radicals

(Nishad et al., 2019). Sonication as pretreatment is also reported to be very efficient in increasing the bioaccessibility of the compounds (Fonteles et al., 2016).

Ultraviolet radiation (UV) is also another important processing technology, which is widely investigated for its effects on the levels of phytochemicals and on the capability of plants to produce them at different levels. Various conditions of radiation exposure, exposure and storage temperatures, wounding of the plant matrix, sensitivity of the compounds, and effect of treatment on other constituents which are responsible for production or accumulation of phytochemicals, may increase the concentration of bioactive compounds. Enhanced antioxidant activity of a plant after treatment is mainly attributed to two mechanism: (1) increase in enzyme activity of phenylalanine ammonia-lyase and peroxidase, (2) increase in extractability of compounds from the cellular matrix. Increase in activity of these enzymes trigger the polyphenolic synthesis and hence add to their content. On the contrary, inactivation of some enzymes like polyphenol oxidase tends to retain the antioxidant power of the system (Alothman et al., 2009; Oms-Oliu et al., 2012). Also, disruption in chemical bonds of polyphenols under the effect of treatment facilitates the release of more soluble phenols of low molecular weight, enhancing the phenolic yield and antioxidant activity. Numerous studies have also reported formation of free radicals after radiation treatment which further results in a decrease of phytochemical content (Sajilata & Singhal, 2006).

Similarly, other novel technologies like high-voltage electrical discharges (HVED) showed promising results in increasing the yield of phenolics and their bioavailability (Buniowska et al., 2017). These findings highlight the importance of nonthermal processing techniques to stabilize the plant extracts for their further utilization in functionalization.

5.2.2 Effect of pH

pH is another very important factor widely investigated for its effects on extraction yield and stability of phytochemicals (Table 5.4). The plant matrix, their extracts, and foods are generally exposed to a wide pH range during their processing and storage, affecting their functionality. Therefore it becomes imperative to know about the antioxidant systems and their sensitivity toward acidity or alkalinity for the optimization of technological and processing conditions. Many researchers have confirmed a significant interaction between the type of phenolic compound and the pH. Many polyphenolic compounds like anthocyanin are more stable at acidic pH and showed a decrease in bioaccessibility at high pH, on the contrary phenolic compounds like procyanidin B, kaempferol, chlorophyll revealed opposite trend with respect to their stability (Ismaiel et al., 2016; Nagar et al., 2021; Roy & Urooj, 2013). There are many other factors associated with the effect of pH where pretreatment, heat processing, exposure to high pressure, presence of other food components or polysaccharides have depicted the change in trend of availability, extraction and overall stability of phytochemicals (Nagar et al., 2021; Roy & Urooj, 2013). In a study on blue green algae, the spirulina was found to be stable at wider pH range but showed a decreased growth rate at pH above 10 and depicted a significant color change with the inhibition of chlorophyll and carotenoids synthesis. This degradation in growth rate was explained by correlation of photosynthetic activity and pH, high pH inhibit the photosynthesis due to inaccessibility of carbon dioxide. The bicarbonate if available as an only source of carbon then it cause a rise in pH to 8 and eventually limiting the content of free CO_2. This deficiency further brings cells under stress and induce formation of free radicals or ROS or oxidative stress. The variation in the content of phenolics at high pH is further

Table 5.4 Effect of pH on plant extract stability.

S. no.	pH range	Other processing conditions	Food matrix	Impact on plant extract			References
				TP	AOX	Others	
1.	3.0, 5.0, 7.0 and 9.0	Storage: at 4°C in closed containers (0.5-l flasks) in the dark for up to 400 days	Grape marc extract	No significant change in TP for all pH	Loss of AOX at pH 7 and 9 over the first 100 days	Clouding of extracts was observed in all the pH solutions during stoarge	Amendola, De Faveri, and Spigno (2010)
2.	4.0, 7.0 and 9.0	Preincubation at various pH values for 24 h	Methanolic antioxidant extract of leaves of: Pomegranate (PM), sweet potato (SPL), carrot (CL), kilkeerae (KL), shepu (SH), beet greens (BL)	–	Highest AOX activity of PM, CL and KL at pH 4, for SPL and SH at pH 9, for BL at pH 7	–	Roy and Urooj (2013)
3.	3.0, 4.0, 5.0, 6.0 and 7.0	At different temperatures (4°C, 25°C, and 100°C) for 24 h	Tea	Tea polyphenols were more stable at low pH; Catechins degraded (21%) at pH 7 after 24 h	–	~50% decrease in clarity of the extracts at high pH of 7.0°C and 100°C	Zeng, Ma, Li, & Luo, (2017)
4.	3.0, 5.0, 7.0 and 8.0	–	Sweet potato leaf	TPC was higher in neutral and weak acid solvent; optimum pH- 5.0–7.0	AOX were higher (2.71 mg TE/mL) at pH 7	–	Sun, Mu, and Xi (2017)
5.	7.5, 8.0, 8.5, 9.0, 9.5, 10.0, 10.5 and 11.0	Incubation: at 31°C with continuous cooling white fluorescent lights (60 μmol/photons m^2 s)	Blue green algae	Highest TP (12.1 mg GAE/g DW) observed at pH 9.5	Highest AOX at pH 9.0 was reported with a percent increase of 567%	Optimum pH for growth was 9.0; highest chlorophyll a (10.6 mg/g DW), carotenoids (2.4 mg/g DW) at alkaline pH of 8.5	Ismaiel et al. (2016)
6.	6.0 and 7.8	Plant extracts were mixed with 50 mM KPi, pH 6.0 or 7.8, at the ratio 1:20 (v/v) and incubated for 30 min in	*Rosa canina* L., *Rhodiola rosea* L., *Hypericum perforatum* L.,	-	AOX at pH 6.0 was 1.4-fold higher for *R. rosea* and *G. lutea* extracts and 1.2-fold higher for	–	Bayliak et al. (2016)

		the presence of 10 mM H_2O_2	and *Gentiana lutea*		*R. canina* and *H. perforatum* extracts		
7.	2.5 and 6.5	*HPP:* samples were treated for 10 min at 600 MPa (initial temperature 25°C) *Pasteurization*: 90°C for 30 s	Strawberries	Decrease in polyphenols at high pH except procyanidin B, kaempferol-3-O-glucuronide and kaempferol-3-O-malonyl-glucoside which showed opposite trend	–	Decrease in bioaccessibility of anthocyanins by 50% at neutral and high pH	Nagar et al. (2021)
8.	4.0 and 9.0	Storage: (a) in dark under refrigeration (5°C) (b) in dark at room temperature (25°C)	Drumstick leaves, mint leaves and carrot tuber	–	AOX of mint leaves and carrot tuber extracts was higher at pH 9 than at pH 4, for drumstick leaves it was unchanged in both alkaline and acid pH	–	Arabshahi-D, Devi, and Urooj (2007)
9.	3.3, 6.3 and 8.3	Extraction: 50% ethanol with a solid/liquid ratio of 1/15; 40°C for 40 min	Blueberry pomace	TPC significantly increased (9%) with increase in pH above 6.3	AOX increased by 9% at pH 8.3	TAC significantly decreased (5.5%) above pH 6.3	Bamba et al. (2018)
10.	4.0 to 9.0	–	Lettuce extract (LE) with quercetin (QC), green tea extract (GTE) or grape seed extract (GSE)	–	14%–40% increase in AOX at pH 9.0 in LE with different phenolics	–	Altunkaya, Gökmen, and Skibsted (2016)

AOX, *antioxidant activity;* HPP, *high pressure processing;* TEY, *total extraction yield;* TP, *total phenols content;* TF, *total flavonoid;* TAC, *total anthocyanin content.*

explained by the generation of oxidative stress. The increase in the concentration corresponds to plant reaction toward the oxidation for alleviation of free radicals. The further increase in pH demonstrated a reduction of phenolics, attributing to inability of cells to function under very high pH conditions (10.5–11.0). Impairment in cell functioning results in cessation of phenolics production, and growth rate, depicting low biomass, and pigment production. The hydrogen peroxide model have been successfully used to evaluate the prooxidant/oxidant properties of the plant extracts at varying pH range (Bayliak et al., 2016). Thus the pH showed effects on auto-oxidation, nonenzymatic antioxidants (phycocyanin and phenolics), and antioxidant enzymes activities, which are responsible for the oxidative stability of plant extracts (Chu, Lim, Radhakrishnan, & Lim, 2010; Ismaiel et al., 2016).

5.2.3 Effect of storage

Storage of food products is an integral part of food supply chain and has been widely studied for its impact on nutritional and functional profiles of foods. This is a major concern for food industries to control and monitor the quality loss during storage. Storage conditions of time, temperature, relative humidity, presence of oxygen and light are the major factors deciding the degree of loss in foods. Mostly researchers have investigated combinations of different factors for their effect on fresh produce and processed products (Table 5.5). Significant variability in the responses of plants was observed on the basis of their processing state i.e. fresh or processed, where fresh commodities or juices revealed no effect on the polyphenolics, on the contrary processing of matrices induced the degradation of active compounds (Ali et al., 2018; Bazinet, Araya-Farias, Doyen, Trudel, & Têtu, 2010; Radziejewska-Kubzdela & Olejnik, 2016; Santhirasegaram, Razali, George, & Somasundram, 2015). However, the effect of processing is not consistent and did depict a different trend in other products (Odriozola-Serrano, Soliva-Fortuny, & Martín-Belloso, 2009; Zafrilla, Ferreres, & Tomás-Barberán, 2001). This discrepancy could be explained by the fact that the parameters of storage have synergistic or antagonistic effects on bioactive stability and shelf life (Ioannou et al., 2012). Photostability of plant and their extracts is another important deciding factor of its antioxidant activity and stability. Presence of light act as stress signal and expedite the process of phenol degradation by keeping up the mechanism of their synthesis active even at low temperature (Tudela et al., 2002; Wang et al., 2009a,b). Light is known to effect phytochemicals during different stages of plant growth, its processing, and storage. Further, the efficiency of light in affecting the plant extracts is mainly a factor of wavelength of light, duration of exposure, pH of matrix, physicochemical properties, concentration of the compound, and the structure (Ioannou et al., 2012). Exposure to light with low wavelength revealed photo-induced molecular rearrangement when compared with high wavelength light which triggered photooxidation (Tommasini et al., 2004). Polyphenolic compounds can either increase or degrade in presence of light depending on the processing state of the food. In fresh foods, light induce a stress signal and triggers the mechanism of synthesis of active compounds (Pérez-Gregorio et al., 2011). Conversely, in processed food products photo degradation of the functional compounds was observed by many researchers. Among different polyphenolic compounds some are more prone to get affected by light such as anthocyanins, chlorophylls, carotenoids etc. (Boon, McClements, Weiss, & Decker, 2010; Lee, Ahn, & Choe, 2014; Março, Poppi, Scarmino, & Tauler, 2011). On the contrary some studies yielded insignificant or no effect on anthocyanins after exposure to light (Dyrby,

Table 5.5 Effect of storage conditions on plant extract stability.

S. no.	Food matrix	Storage conditions				Impact on plant extract			References
		Temperature	Duration	Light/gaseous atmosphere	Other conditions	TP	AOX	Others	
1.	*Piper betle* extracts	5°C and 25°C	6 months	With and without light	–	TP extract stored at 5°C with and without exposure to light showed >99% retention; storage at 25°C lead to retention of 97% (dark) and 93% (light)	High AOX stability was observed at 5°C in dark which retained 99.98% activity after 180 days; extracts at 25°C with light were least stable (90% of activity)	Complete loss of isoeugenol at 25°C with or without the presence of light after 30 days of storage	Ali et al. (2018)
2.	Sweet potato leaf	55°C, 65°C, 80°C, and 100°C	0, 10, 30, 60, and 90 min	Light treatments: direct sunlight from 10:00 am to 3:00 pm	–	TPC of samples after thermal treatment was higher than 91% and after light treatment for 5 h was 98%	AOX retention at 80°C and 100°C, decreased significantlyafter 90 min, and remained at 62.14% and 61.86%, respectively; light treatment for 5 h retained the AOX up to 92.5%	–	Sun et al. (2017)
3.	Rutin, naringin, mesquitol, eriodictyol, luteolin, luteolin-7-O-glucoside	25°C	15 days	Model solutions exposed to 0% light (darkness) and 100% light (equivalent to exposure at 16.5 klux) 2 models: nitrogen bubbling of 2 min (O_2 conc. of 15%); without bubbling (O_2 conc. of 85%)	–	–	Antioxidant activity of rutin and mesquitol increased in the presence of O_2 and light	Decrease in flavonoid content as a function of light intensity, O_2 conc. and structure of compound	Chaaban et al. (2017)

(*Continued*)

Table 5.5 Effect of storage conditions on plant extract stability. ***Continued***

S. no.	Food matrix	Storage conditions				Impact on plant extract			References
		Temperature	Duration	Light/gaseous atmosphere	Other conditions	TP	AOX	Others	
4.	Coleslaw mix	4°C	12 days	Packaged under modified atmosphere consisting of 5/10/85, 20/25/55, 50/30/20, 70/30/0% of O_2/CO_2/N_2, as well as air	Pretreatment-soaking in an ascorbic acid (5 g/L) and citric acid (5 g/L) solution	–	–	Total glucosinolates was highest with 38% increase in the samples packaged under the modified atmospheres with 5/10/85%O_2/CO_2/N_2	Radziejewska-Kubzdela and Olejnik (2016)
5.	Garlic clove	20°C ± 2°C	12 weeks	1000 lx light/dark (12 h/12 h) cycle	45% RH	TP reached maximum values (839.96 μg/g DW) after 6 weeks	AOX reached maximum values at 8 weeks (26% incraese), then decreased significantly	Levels of 18 organosulfur compounds increased from 2 to 6 weeks, reached maximum level 41.36 ± 2.34% at 8 weeks, then decreased significantly	Fei, Tong, Wei, and De Yang (2015)
6.	Mango juice	4°C ± 1°C	5 weeks	–	–	–	–	Aerobic plate counts increased from 2.74 to 8.32 log CFU/mL and yeast and mold counts increased from 2.42 to 6.10 log CFU/mL after 5 weeks storage	Santhirasegaram et al. (2015)
7.	*Anemopsis californica*	50°C, 25°C, 4°C, and −20°C	180 days (stability was measured every 30 days)	Under ambient light and dark conditions (using amber bottle)	–	79% of total phenols was conserved at the end of storage, at −20°C in dark	Best conditions for AOX stability were 4°C (95%) and −20°C (98%) under dark	Retention of 73% of total flavonoid after 180 days at −20°C in dark	Del-Toro-Sánchez et al. (2015)
8.	*Hypericum perforatum*	–	6 months	Storage conditions: 1. 25°C with uncontrolled humidity with daylight 2. 25°C with uncontrolled humidity without daylight, dark	–	–	–	Chlorogenic acid was the most stable, decay of phenolics was lowest at −20°C and highest at 40° C and 75% RH; dark condition decreases breakdown within 4 months	Koyu and Haznedaroglu (2015)

				3. 25°C, 65% relative humidity 4. 40°C, 75% RH 5. −20°C 6. 4°C					
9.	Seeds from four *Brassica oleracea* varieties: Broccoli, kale, Penca cabbage, and red cabbage	–	Harvesting of sprouts: green sprouts (GS) after 7, 9, 12 and 15 days of germination and white sprouts (WS) after 5, 6, 7, 9 and 12 days	Photoperiod regimes: for GS production a cycle of 16 h of light and 8 h of darkness; for WS under dark	Sprouting: 12 h in darkness, at room temperature, light agitation; 25°C temperature, and different photoperiod regimes	TP of red cabbage and Penca cabbage WS decreased about 10% along all the experiment (from day 5 to day 12). WS of kale and broccoli showed an increase in TP from day 5 to 7, followed by a decrease (day 9 to day 12 of 20% and 10%, respectively)	Red and Penca cabbage sprouts produced under light cycles showed highest AOX (57.11 μg/mL) than kale	Seeds revealed TFC maxima after different germination days; Broccoli GS showed after 12 days of germination (24.0 mg QE/g), red cabbage and Penca cabbage after 7 days (41.2 mg and 24.7 mg QE/g) and Galega kale after 9 days of germination (25.4 mg QE/g)	Vale, Cidade, Pinto, and Oliveira (2014)
10.	Hazelnut	In-shell hazelnuts storage-ambient temperatures (10°C–26°C) Shelled hazelnuts: 4°C	1 year	Shelled hazelnuts: with or without modified atmosphere (1% O_2, 99% N_2) for	In-shell hazelnuts: 60%–80% RH Shelled hazelnuts: 55% RH	TP decreased after 8th month for shelled and in-shell hazelnuts, then showed no change, with a slight increase (13%) in refrigerated kernels	Highest AOX (6.29 TE mmol/kg) in refrigerated kernels after 12 months	Refrigerated storage reduced the lipid oxidation, with best result in modified atmosphere (0.057 O_2 mmol/kg) after 12 months	Ghirardello et al. (2013)
11.	Pomegranate peel		0, 1, 5, 10, 30, 60, 90, and 180 days	2 packaging methods (no light and exposure to light)	pH values (3.5, 5.0, and 7.0)	67% retention of TP at low pH (3.5) in dark packaging	AOX was retained to 58% at low pH (3.5) in dark packaging	Extracts stored at high pH were more opaque and looked darker and chromaticity deeper in color than that at the low pH	Qu, Breksa, Pan, Ma, and Mchugh (2012)

(Continued)

Table 5.5 Effect of storage conditions on plant extract stability. ***Continued***

S. no.	Food matrix	Storage conditions				Impact on plant extract			References
		Temperature	Duration	Light/gaseous atmosphere	Other conditions	TP	AOX	Others	
12.	Fresh cut onions	–	0, 1, 3, 8 and 16 days	Treatment: white visible light at a 45-cm distance (fluorescent tube-lamp at 14 W 230–240 V, 50–60 Hz)	Packaging 1: under vacuum and refrigerated storage at 1–2°C in the absence of light in PA/PE 20/70 (90-lm thickness) bags Packaging 2 and 3: in closed cups of PS or PET cups (12.5X 9 cm, 4 cm depth) stored under refrigeration (1°C–2°C) in the absence of light Packaging 4: closed cups of PS stored under refrigeration in the presence of visible light	–	–	A 12%–30% reduction in TAC after 16 days in different packaging; increase of total flavonols by 28% under PS packaging, increase of total flavonols by 58% and total anthocyanins by 39% in presence of light	Pérez-Gregorio et al. (2011)
13.	Litchi	–	–	Continuous flow (30 mL/min) of humidified air (control) and 100% O_2	Dipping treatment: dipped for 3 min in 0.1% Sportak fungicide solution and air-dried for 2 h at 28°C	After 4 and 6 days of storage, TP in pure oxygen–exposed fruits was maximum	Exposure to oxygen reduces the reduction rate of AOX	Exposure to pure oxygen enhanced the activity of superoxide dismutase by 20%, and catalase by 40%, compared with that of 0 days	Duan et al. (2011)

14.	Sprouted seeds (wheat, radish and lentils)	27°C	–	Light wavelengths: 385, 445, 510, 595, 638, 669, and 731 nm; photosynthetic photon flux density of about 100 μmol/m^2 s and a 12 h photoperiod were maintained during treatment	Sprouting: for 24 h in germination plates at 18°C; under light conditions	Accumulation of TP was greater in red light radiated lentil, radish, and wheat seeds	After 3 days germination, the AOX was increased by about 12% in wheat seeds in green light (510 nm), whereas decreased by 50% in radish	Positive effect of green light on vitamin C was observed in all treated seeds, whereas red radiation led to a significant decrease of vitamin C	Samuolienė et al. (2011)
15.	Enriched tea drink	4°C and 25°C	6 months	-	-	-	-	No effect on catechin was observed at 4°C however at 25°C reached to 0 after 30 days	Bazinet et al. (2010)
16.	Ginger	-	-	The plants were grown under four level of glasshouse shade (0%, 20%, 40% and 60% shade) corresponding to 790, 630, 460 and 310 μmol/m^2s of photosynthetically active radiation (PAR)	Harvesting- 16 weeks	790 μmol/m^2s was best for maximum TP production	AOX were higher in the leaves under 310 μmol/m^2s	Higher TF (5.95 mg/g DW and 8.45 mg/g DW) under 310 μmol m^2s of light intensity	Ghasemzadeh, Jaafar, Rahmat, Wahab, and Halim (2010)
17.	Mulberry fruits	40°C, 50°C and 70°C	0, 1, 2, 3, 6, and 10 h	Fruit extracts placed under normal fluorescent lights (220 V, 50 Hz and 0.37 A) at about 18 in. distance at room temperature for 10 h	–	–	After thermal and light treatment for 10 h, the AOX decreased significantly	Around 25% loss in TAC after 10 h at 70°C; 18% decrease in TAC after light treatment for 10 h at room temperature	Aramwit, Bang and Srichana (2010)

(Continued)

Table 5.5 Effect of storage conditions on plant extract stability. ***Continued***

S. no.	Food matrix	Storage conditions				Impact on plant extract			References
		Temperature	Duration	Light/gaseous atmosphere	Other conditions	TP	AOX	Others	
18.	Raspberries	24°C during the day (0700–1900) and 16°C at night (1900–0700)	1, 2, 3, and 4 days	RH at 75% and under three different light intensities (fluorescent lamps for 12 h/day (0700–1900)); Light treatment: photosynthetically active radiation (PAR) of 56 ± 0.5 μmol/m^2s (H), 31 ± 0.2 μmol/m^2s (L), and in the dark (D)	Harvesting at five different stages based on surface red color: (1) 0%–5% red, (2) 20% red, (3) 50% red, (4) 80% red, (5) 100% red	Fruits of greener stages showed highest TP; fruit exposed to higher light intensities had higher TP especially during the first 2 days of storage	Fruit harvested at greener stages (5% and 20%) consistently yielded higher AOX	TAC increased with fruit maturity and during storage	Wang et al. (2009a, 2009b)
19.	Blueberries	–	–	Light treatment: UV-C lamp, 254 nm, time- 1, 5, 10, and 15 min equal to the dosages of 0.43, 2.15, 4.30, and 6.45 kJ/m^2	–	2.15 and 4.30 kJ/m^2 yielded highest TP (~60%)	4.30 kJ/m^2 yielded highest AOX (44%)	54% increase in anthocyanin at 4.30 kJ/m^2	Wang et al. (2009a, 2009b)
20.	Longan fruit	28°C	6 days	Atmosphere of 5%, 21% (control) or 60% O_2 (balance N_2) at 28°C and 90%–95% relative humidity	–	Highest TP (12 mg/g FW) was observed in fruits stored at 5% O_2 after 4 days of storage	Fruit exposed to 5% O_2 exhibited highest AOX after 4 and 6 days of storage	Exposure of 5 or 60% O_2 resulted in a higher level of total soluble solids and lower ascorbic acid	Cheng et al. (2009)

AOX, *antioxidant activity;* TAC, *total anthocyanin content;* TEY, *total extraction yield;* TF, *total flavonoid;* TP, *total phenols content.*

Westergaard, & Stapelfeldt, 2001; López-Rubira, Conesa, Allende, & Artés, 2005). Additionally, position of hydroxyl group in benzene ring decides the stability of compound against light, where glycosylation in position 3 gives more stability (Zhang, Cardon, et al., 2010; Zhang, Chen, et al., 2010).

Other factors of pretreatment or mechanical processing of foods prior to storage also play a significant role in the activity and stability of phytochemicals by stimulating the process of oxidation. Storage under modified atmospheric conditions depicts an increase in phenolic content at elevated CO_2 content during certain period and shows a decline over long duration storage. This finding is supported by the stress-induced accumulation of phenolic compounds under high carbon dioxide and oxygen levels, and mechanical damage. Some researchers have also suggested inactivation of enzymes at high CO_2 level, leading to increased content of bioactives. Whereas, during long period storage high CO_2 and oxygen levels may cause damage in cell membrane and enzymatic degradation of polyphenolics. Higher levels of oxygen during high demand for energy can further augment the oxidation of respiratory substrates and make them incapable of transporting increased electrons. This trigger the formation of free radicals and depletion of phytochemicals in scavenging mechanism (Radziejewska-Kubzdela & Olejnik, 2016; Sørensen, 1990; Xu, Guo, Yuan, Yuan, & Wang, 2006). Storage at higher temperatures has negative impact on polyphenolics and ascorbic acid due to increase in oxidation and hydrolytic reactions, which has more deteriorative effects in presence of light (Del-Toro-Sánchez et al., 2015; Kotsiou & Tasioula-Margari, 2016).

Similar to other factors discussed earlier, the stability of plant extracts during storage does also get effected by individual bioactive compound due to presence of structural differences. The reactivity of the polyphenolic is dependent on the position of functional groups. Rice-Evans, Miller, and Paganga (1996) have reported susceptibility of position 3 and 4 in flavonoids toward dihydroxylation than others. Further, presence of hydroxyl group decreases the stability whereas methyl groups makes more stable compounds (Bkowska-Barczak, 2005).

Anthocyanin, in many studies, revealed instability during storage ascribed to residual activity of enzymes, susceptibility of monomeric anthocyanins toward polymerization, and condensation reactions with other phenolics (Brownmiller, Howard, & Prior, 2008; Ochoa, Kesseler, Vullioud, & Lozano, 1999; Reed, Krueger, & Vestling, 2005). Additionally, pH, temperature, oxygen, light, sugars, metal ions have also investigated as limiting factor of anthocyanin stability (Ioannou et al., 2012).

5.2.4 Miscellaneous factors

A food matrix constitute of many components viz. carbohydrates, protein, fats, vitamins, minerals, and polyphenolic compounds. The interaction of these constituents is very significant for the nutritional and functional values of the food, corresponding to their positive and negative synergies. Different bioactive compounds when present in a solution have also shown synergistic interaction with respect to their antioxidant activity and stability, whereas, a different combination of polyphenolics revealed antagonistic effect (Hidalgo, Sanchez-Moreno, & De Pascual-Teresa, 2010). Plant polyphenols are located within the cell prior to processing and do not interact with other cell wall material, however, on processing or extraction these compounds are released and come in contact with carbohydrates, minerals and metal ions and form complexes (Zhu, 2018). The binding of active compounds to these components can either increase or decrease their activity, bioaccessibility, and bioavailability. The interactions between polyphenols and polysaccharides could be covalent or non-covalent and their strength is affected by molecular size of compounds and the conformational flexibility, demonstrating a difference in the behavior of the polyphenols (Chirug, Okun, Ramon, & Shpigelman, 2018). A study on dietary fiber interaction suggested a negative effect on bioavailability of polyphenol glucoside, attributing to gelation, increased viscosity, or binding and entrapment of compounds (Bohn et al., 2015). Presence of ascorbic acid in conjunction with phenols has also reported to positively affect the stability of both the compounds, suggesting an antioxidation and cooperation effect between the two (Chen et al., 2014). Moreover, investigations have depicted protective effect of water-soluble antioxidants on lipids as compare to lipid-soluble ones, due to the "polar paradox" (Porter, Levasseur, & Henick, 1977).

The naturally occurring plant enzymes including catalases, polyphenol oxidases, peroxidases, amylases are also extensively studied for their impact on plant bioactives. These enzymes can break the complex structure of plant cells, chemically modifying the secondary metabolites, inducing oxidation, and significantly affecting the antioxidant activity and stability of the plant extracts (Ravimannan & Nisansala, 2017; Sachadyn-Król et al., 2016).

Further, the water present in the raw material is critical in plant extract stability. Presence of moisture allows the redox reaction to produce free radicals, having detrimental effects on primary and secondary plant metabolites. Water also facilitate the enzyme activity and degrade the functional quality of the extracts. High moisture further allows the microbial growth in the products and adversely affect the stability of bioactive components (Gafner & Bergeron, 2005).

5.3 Improving stability of plant extracts

Incorporation of plant extracts in food or other formulations subjects the phytochemicals to various processing conditions, which affect their activit and stability. The loss of active ingredients at different unit operations restrict the use of natural plant extracts in food and pharmaceutical industries. The bioavailability of the extracts is another important concern in utilizing the plant extracts in the food. The bioavailability is affected by the solubility of the active compound, stability during different digestion stages, and absorption of the nutrients in the body. Thus it has attracted researchers to find plausible solutions for utilizing the antioxidant potential and improving the solubility, sustainability, absorption, availability, and stability of bioactive compounds (Rahman et al., 2020).

There are many approaches studied for their effectiveness in improving the stability and shelf life of phytochemicals. Characterization of the plant extract is the first step in the process of stabilization where knowledge of physical, chemical, functional, and biological attributes direct the selection of appropriate technique. Nanotechnology is the widely accepted method in stabilization including the formation of nanocoating on the active component and nanoemulsion preparations (Zorzi, Carvalho, von Poser, & Teixeira, 2015). Besides, encapsulation of bioactives, stabilization using water-soluble chelating agents for example, polyvinylpyrrolidone (PVP), use of suspending agent and nonionic surfactant for sparingly soluble or insoluble or sparingly insoluble plant extracts revealed their effect on protecting phytochemicals (Armendáriz-Barragán et al., 2016; Bosch et al., 2004; Rijo et al., 2014; Thakur et al., 2011; Wolf et al., 2007). Controlled storage condition also play an important role in alleviating the adverse effect of temperature, oxygen, light, moisture, interactions with other ingredients, cross metal contamination from containers, and microbes (Thakur, Prasad, & Laddha, 2008). Furthermore, in a biological plant system inactivation of enzymes and chemical modification of the phytochemicals are very successful in stabilization of the plant bioactives especially pigments. Pretreatments using high temperatures such as steaming, and hot water blanching, and chemical soaking in acid or alkaline solution are the widely used methods for stabilization (Ngamwonglumlert, Devahastin, & Chiewchan, 2017). Biotechnology has also emerged as a novel method of providing stability to phytochemicals where plants are genetically transformed for robustness (Miller, Fatnon, & Webb, 2004).

Nanocarrier is another strategy to thwart the existing problem of pharmaceutical industries. Encapsulation of plant metabolites into biocompatible and biodegradable nanoparticles facilitates the targeted delivery of the compounds and improves their bioavailability (Bharali et al., 2011). Besides, nanonization has other advantages of improved solubility, reducing recommended doses and side effects, and increasing the absorption of pharmaceutical herbs over their crude counterparts. Nanocarrier in the form of solid lipid nanoparticle, nanostructured lipid carrier, nanoemulsion, nanocapsule, drug conjugates, liposome, transferosome, nanosphere, nanocrystals, nanofiber, metal nanoparticle, nanotube, and biological nanocarrier overcome the limitations of utilizing natural plant extracts (Loredo-Tovias et al., 2017; Luo et al., 2011; Rahman et al., 2020; Tully, Fakhrullin, & Lvov, 2015).

5.4 Conclusion

Plant extract is a concoction of numerous components and holds an important place in human nutrition and dietary interventions. Bioaccessibility, bioavailability, and stability thus become imperative to utilize the potential of these phytochemicals. This chapter comprehensively illustrates the factors of consideration for bioactive stability. There are many intrinsic and extrinsic factors influencing the concentration and antioxidant activity of the compounds. Processing of food using thermal or mechanical energy has revealed their effect on individual polyphenols differently. Similarly, other factors of pH, storage conditions (duration, temperature, relative humidity, gaseous environment, and light), and other food components also demonstrated positive and negative correlation with the phytochemical stability and availability. Besides, characteristic of phenolic compounds, their structure conformations, and molecular size decide their sensitivity toward processing factors. A wide variation has been observed between operating conditions and selected samples, which are used in different studies; therefore it is difficult to give a comparable explanation for the effects of processing and food matrix. Presently, there is no explicit scientific ground that can be used for predicting the plant extract stability after processing or during storage. Therefore development of statistical models to analyze the interaction between different polyphenolic compounds and with other food components and surroundings, and understanding of degradation mechanisms would facilitate the accurate prediction of phytochemical stability. The phytochemicals have wider applications in food industries, pharmaceuticals, and cosmetics, requiring more extensive research for developing advance food processing technologies and efficient delivery system. Nonconventional processing technologies like high pressure processing, pulse electric field, cold plasma, and irradiation along with encapsulation and nanoemulsification should be further exploited for increasing their effectiveness in protecting the potential of plant extracts.

References

Ahmad, R., Ahmad, N., Aljamea, A., Abuthayn, S., & Aqeel, M. (2020). Evaluation of solvent and temperature effect on green accelerated solvent extraction (ASE) and UHPLC quantification of phenolics in fresh olive fruit (Oleaeuropaea). *Food Chemistry*, 128248.

Ali, A., Chong, C. H., Mah, S. H., Abdullah, L. C., Choong, T. S. Y., & Chua, B. L. (2018). Impact of storage conditions on the stability of predominant phenolic constituents and antioxidant activity of dried Piper betle extracts. *Molecules (Basel, Switzerland)*, *23*(2), 484.

Alothman, M., Bhat, R., & Karim, A. A. (2009). UV radiation-induced changes of antioxidant capacity of fresh-cut tropical fruits. *Innovative Food Science & Emerging Technologies*, *10*(4), 512–516.

Altunkaya, A., Gökmen, V., & Skibsted, L. H. (2016). pH dependent antioxidant activity of lettuce (L. sativa) and synergism with added phenolic antioxidants. *Food Chemistry*, *190*, 25–32.

Amendola, D., De Faveri, D. M., & Spigno, G. (2010). Grape marc phenolics: Extraction kinetics, quality and stability of extracts. *Journal of Food Engineering*, *97*(3), 384–392.

Arabshahi-D, S., Devi, D. V., & Urooj, A. (2007). Evaluation of antioxidant activity of some plant extracts and their heat, pH and storage stability. *Food Chemistry*, *100*(3), 1100–1105.

Aramwit, P., Bang, N., & Srichana, T. (2010). The properties and stability of anthocyanins in mulberry fruits. *Food Research International*, *43*(4), 1093–1097.

Armendáriz-Barragán, B., Zafar, N., Badri, W., Galindo-Rodríguez, S. A., Kabbaj, D., Fessi, H., & Elaissari, A. (2016). Plant extracts: From encapsulation to application. *Expert Opinion on Drug Delivery*, *13*(8), 1165–1175.

Bamba, B. S. B., Shi, J., Tranchant, C. C., Xue, S. J., Forney, C. F., & Lim, L. T. (2018). Influence of extraction conditions on ultrasound-assisted recovery of bioactive phenolics from blueberry pomace and their antioxidant activity. *Molecules (Basel, Switzerland)*, *23*(7), 1685.

Barba, F. J., Mariutti, L. R., Bragagnolo, N., Mercadante, A. Z., Barbosa-Canovas, G. V., & Orlien, V. (2017). Bioaccessibility of bioactive compounds from fruits and vegetables after thermal and nonthermal processing. *Trends in Food Science and Technology*, *67*, 195–206.

Barba, F. J., Sant'Ana, A. S., Orlien, V., & Koubaa, M. (2018). *Innovative technologies for food preservation: Inactivation of spoilage and pathogenic microorganisms* (1st ed.). Elsevier: Academic Press.

Barba, F. J., Terefe, N. S., Buckow, R., Knorr, D., & Orlien, V. (2015). New opportunities and perspectives of high pressure treatment to improve health and safety attributes of foods. A review. *Food Research International*, *77*, 725–742.

Barišić, V., Flanjak, I., Križić, I., Jozinović, A., Šubarić, D., Babić, J., ... Ačkar, Đ. (2020). Impact of high-voltage electric discharge treatment on cocoa shell phenolic components and methylxanthines. *Journal of Food Process Engineering*, *43*(1), e13057.

Bayliak, M. M., Burdyliuk, N. I., & Lushchak, V. I. (2016). Effects of pH on antioxidant and prooxidant properties of common medicinal herbs. *Open Life Sciences*, *11*(1), 298–307.

Bazinet, L., Araya-Farias, M., Doyen, A., Trudel, D., & Têtu, B. (2010). Effect of process unit operations and long-term storage on catechin contents in ECG-enriched teadrink. *Food Research International*, *43*, 1692–1701.

Bharali, D. J., Siddiqui, I. A., Adhami, V. M., Chamcheu, J. C., Aldahmash, A. M., Mukhtar, H., & Mousa, S. A. (2011). Nanoparticle delivery of natural products in the prevention and treatment of cancers: Current status and future prospects. *Cancers*, *3*(4), 4024–4045.

Bhat, R., Ameran, S. B., Voon, H. C., Karim, A. A., & Tze, L. M. (2011). Quality attributes of starfruit (Averrhoa carambola L.) juice treated with ultraviolet radiation. *Food Chemistry*, *127*(2), 641–644.

Bkowska-Barczak, A. (2005). Acylatedanthocyanins as stable, natural food colorants—a review. *Polish Journal of Food and Nutrition Sciences*, *14*, 55.

Boekel, M., Fogliano, V., Pellegrini, N., Stanton, C., Scholz, G., Lalljie, S., et al. (2010). A review on the beneficial aspects of food processing. *Molecular Nutrition & Food Research*, *54*(9), 1215–1247.

Bohn, T., McDougall, G. J., Alegría, A., Alminger, M., Arrigoni, E., Aura, A. M., & Martínez- Cuesta, M. C. (2015). Mind the gap—deficits in our knowledge of aspects impacting the bioavailability of phytochemicals and their metabolites—a position paper focusing on carotenoids and polyphenols. *Molecular Nutrition & Food Research*, *59*(7), 1307–1323.

Boon, C. S., McClements, D. J., Weiss, J., & Decker, E. A. (2010). Factors influencing the chemical stability of carotenoids in foods. *Critical Reviews in Food Science and Nutrition*, *50*(6), 515–532.

Bosch, H., Hilborn, M., Hovey, D., Kline, L., Lee, R., Pruitt, J., . . . Xu, S. (2004). *U.S. Patent Application No. 10/619,539*.

Brownmiller, C., Howard, L. R., & Prior, R. L. (2008). Processing and storage effects on monomeric anthocyanins, percent polymeric colour, and antioxidant capacity of processed blueberry products. *Journal of Food Science*, *5*(73), H72eH79.

Buchner, N., Krumbein, A., Rhon, S., & Kroh, L. W. (2006). Effect of thermal processing on the flavonolsrutin and quercetin. *Rapid Communications in Mass Spectrometry*, *20*, 3229–3235.

Buniowska, M., Carbonell-Capella, J. M., Frigola, A., & Esteve, M. J. (2017). Bioaccessibility of bioactive compounds after non-thermal processing of an exotic fruit juice blend sweetened with Stevia rebaudiana. *Food Chemistry*, *221*, 1834–1842.

Chaaban, H., Ioannou, I., Paris, C., Charbonnel, C., & Ghoul, M. (2017). The photostability of flavanones, flavonols and flavones and evolution of their antioxidant activity. *Journal of Photochemistry and Photobiology A: Chemistry*, *336*, 131–139.

Chen, J., Sun, H., Wang, Y., Wang, S., Tao, X., & Sun, A. (2014). Stability of apple polyphenols as a function of temperature and pH. *International Journal of Food Properties*, *17*(8), 1742–1749.

Cheng, G., Jiang, Y., Duan, X., Macnish, A., You, Y., & Li, Y. (2009). Effect of oxygen concentration on the biochemical and chemical changes of stored longan fruit. *Journal of Food Quality*, *32*(1), 2–17.

Chirug, L., Okun, Z., Ramon, O., & Shpigelman, A. (2018). Iron ions as mediators in pectin-flavonols interactions. *Food Hydrocolloids*, *84*, 441–449.

Chu, W. L., Lim, Y. W., Radhakrishnan, A. K., & Lim, P. E. (2010). Protective effect of aqueous extract from Spirulinaplatensis against cell death induced by free radicals. *BMC Complementary and Alternative Medicine*, *10*(1), 53.

Colle, I., Lemmens, L., Van Buggenhout, S., Van Loey, A., & Hendrickx, M. (2010). Effect of thermal processing on the degradation, isomerization, and bioaccessibility of lycopene in tomato pulp. *Journal of Food Science*, *75*(9), C753–C759.

Cravotto, G., & Binello, A. (2016). Chapter 1 - Low-frequency, high-power ultrasonic assisted food component extraction. In K. Knoerzer, P. Juliano, & G. Smithers (Eds.), *Innovative Food Processing Techniques*. Cambridge: Woodhead Publishing Limited.

d'Alessandro, L. G., Kriaa, K., Nikov, I., & Dimitrov, K. (2012). Ultrasound assisted extraction of polyphenols from black chokeberry. *Separation and Purification Technology*, *93*, 42–47.

Del-Toro-Sánchez, C. L., Gutiérrez-Lomelí, M., Lugo-Cervantes, E., Zurita, F., Robles-García, M. A., Ruiz-Cruz, S., & Guerrero-Medina, P. J. (2015). Storage effect on phenols and on the antioxidant activity of extracts from *Anemopsis californica* and inhibition of elastase enzyme. *Journal of Chemistry*, *2015*.

Dewanto, V., Wu, X., Adom, K. K., & Liu, R. H. (2002). Thermal processing enhances the nutritional value of tomatoes by increasing total antioxidant activity. *Journal of Agricultural and Food Chemistry*, *50*(10), 3010–3014.

Duan, X., Liu, T., Zhang, D., Su, X., Lin, H., & Jiang, Y. (2011). Effect of pure oxygen atmosphere on antioxidant enzyme and antioxidant activity of harvested litchi fruit during storage. *Food Research International*, *44*(7), 1905–1911.

Dyrby, M., Westergaard, N., & Stapelfeldt, H. (2001). Light and heat sensitivity of redcabbage extract in soft drink model systems. *Food Chemistry*, *72*, 431–437.

Fei, M. L., Tong, L. I., Wei, L. I., & De Yang, L. (2015). Changes in antioxidant capacity, levels of soluble sugar, total polyphenol, organosulfur compound and constituents in garlic clove during storage. *Industrial Crops and Products*, *69*, 137–142.

Fernández-Jalao, I., Sánchez-Moreno, C., & Ancos, B. (2017). Influence of food matrix and high-pressure processing on onion flavonols and antioxidant activity during gastrointestinal digestion. *Journal of Food Engineering*, *213*, 60–68.

Fonteles, T. V., Leite, A. K. F., Silva, A. R. A., Carneiro, A. P. G., de Castro Miguel, E., Cavada, B. S., et al. (2016). Ultrasound processing to enhance drying of cashew apple bagasse puree: Influence on antioxidant properties and in vitro bioaccessibility of bioactive compounds. *Ultrasonics Sonochemistry*, *31*, 237–249.

Fuleki, T., & Ricardo-Da-Silva, J. M. (2003). Effects of cultivar and processing method on the contents of catechins and procyanidins in grape juice. *Journal of Agricultural and Food Chemistry*, *51*, 640–646.

Gabrić, D., Barba, F., Roohinejad, S., Gharibzahedi, S. M. T., Radojčin, M., Putnik, P., & Bursać Kovačević, D. (2018). Pulsed electric fields as an alternative to thermal processing for preservation of nutritive and physicochemical properties of beverages: A review. *Journal of Food Process Engineering*, *41*(1), e12638.

Gafner, S., & Bergeron, C. (2005). The challenges of chemical stability testing of herbal extracts in finished products using state-of-the-art analytical methodologies. *Current Pharmaceutical Analysis*, *1*(2), 203–215.

Galaz, P., Valdenegro, M., Ramírez, C., Nuñez, H., Almonacid, S., & Simpson, R. (2017). Effect of drum drying temperature on drying kinetic and polyphenol contents in pomegranate peel. *Journal of Food Engineering*, *208*, 19–27.

Ghasemzadeh, A., Jaafar, H. Z., Rahmat, A., Wahab, P. E. M., & Halim, M. R. A. (2010). Effect of different light intensities on total phenolics and flavonoids synthesis and anti-oxidant activities in young ginger varieties (*Zingiber officinale* Roscoe). *International Journal of Molecular Sciences*, *11*(10), 3885–3897.

Ghirardello, D., Contessa, C., Valentini, N., Zeppa, G., Rolle, L., Gerbi, V., & Botta, R. (2013). Effect of storage conditions on chemical and physical characteristics of hazelnut (Corylusavellana L.). *Postharvest Biology and Technology*, *81*, 37–43.

Giambanelli, E., Verkerk, R., D'Antuono, L. F., & Oliviero, T. (2016). The kinetic of key phytochemical compounds of non-heading and heading leafy Brassica oleracea landraces as affected by traditional cooking methods. *Journal of the Science of Food and Agriculture*, *96*(14), 4772–4784.

Girgin, N., & El, S. N. (2015). Effects of cooking on in vitro sinigrin bioaccessibility, total phenols, antioxidant and antimutagenic activity of cauliflower (Brassica oleraceae L. var. Botrytis). *Journal of Food Composition and Analysis*, *37*, 119–127.

Gonzales, G. B., Raes, K., Vanhoutte, H., Coelus, S., Smagghe, G., & Van Camp, J. (2015). Liquid chromatography–mass spectrometry coupled with multivariate analysis for the characterization and discrimination of extractable and nonextractable polyphenols and glucosinolates from red cabbage and Brussels sprout waste streams. *Journal of Chromatography. A*, *1402*, 60–70.

He, Z., Tao, Y., Zeng, M., Zhang, S., Tao, G., Qin, F., et al. (2016). High-pressure homogenization processing, thermal treatment, and milk matrix affect in vitro bioaccessibility of phenolicsin apple, grape and orange juice to different extents. *Food Chemistry*, *200*, 107–116.

Heras-Ramírez, M. E., Quintero-Ramos, A., Camacho-Dávila, A. A., Barnard, J., Talamás-Abbud, R., Torres-Muñoz, J. V., & Salas-Muñoz, E. (2012). Effect of blanching and drying temperature on polyphenolic compound stability and antioxidant capacity of apple pomace. *Food and Bioprocess Technology*, *5*(6), 2201–2210.

Hidalgo, M., Sanchez-Moreno, C., & De Pascual-Teresa, S. (2010). Flavonoid-flavonoidinteraction and its effect on their antioxidant activity. *Food Chemistry*, *121*, 691–696.

Huamán-Castilla, N. L., Mariotti-Celis, M. S., Martínez-Cifuentes, M., & Pérez-Correa, J. R. (2020). Glycerol as alternative co-Solvent for water extraction of polyphenols from Carménère pomace: Hot pressurized liquid extraction and computational chemistry calculations. *Biomolecules*, *10*(3), 474.

Igual, M., García-Martínez, E., Camacho, M. M., & Martínez-Navarrete, N. (2011). Changes in flavonoid content of grapefruit juice caused by thermal treatmentand storage. *Innovative Food Science and Emerging Technologies*, *12*, 153–162.

Ioannou, I., Hafsa, I., Hamdi, S., Charbonnel, C., & Ghoul, M. (2012). Review of the effects of food processing and formulation on flavonol and anthocyanin behaviour. *Journal of Food Engineering*, *111*(2), 208–217.

Ismaiel, M. M. S., El-Ayouty, Y. M., & Piercey-Normore, M. (2016). Role of pH on antioxidants production by Spirulina (Arthrospira) platensis. *Brazilian journal of microbiology*, *47*(2), 298–304.

Katsube, T., Tsurunaga, Y., Sugiyama, M., Furuno, T., & Yamasaki, Y. (2009). Effect of air-drying temperature on antioxidant capacity and stability of polyphenolic compounds in mulberry (Morusalba L.) leaves. *Food Chemistry*, *113*(4), 964–969.

Knorr, D., Froehling, A., Jaeger, H., Reineke, K., Schlueter, O., & Schoessler, K. (2011). Emerging technologies in food processing. *Annual Review of Food Science and Technology*, *2*, 203–235.

Korus, A., Słupski, J., Gębczyński, P., & Banaś, A. (2014). Effect of preliminary processing and method of preservation on the content of glucosinolates in kale (Brassica oleracea L. var. acephala) leaves. *LWT-Food Science and Technology*, *59*(2), 1003–1008.

Kotsiou, K., & Tasioula-Margari, M. (2016). Monitoring the phenolic compounds of Greek extra-virgin olive oils during storage. *Food Chemistry*, *200*, 255–262.

Koyu, H., & Haznedaroglu, M. Z. (2015). Investigation of impact of storage conditions on Hypericum perforatum L. dried total extract. *Journal of Food And Drug Analysis*, *23*(3), 545–551.

Lee, E., Ahn, H., & Choe, E. (2014). Effects of light and lipids on chlorophyll degradation. *Food Science and Biotechnology*, *23*(4), 1061–1065.

Lee, S. U., Lee, J. H., Choi, S. H., Lee, J. S., Ohnisi-Kameyama, M., Kozukue, N., ... Friedman, M. (2008). Flavonoid content in fresh, home-processed, and light exposed onions and in dehydrated commercial onion products. *Journal of Agricultural and Food Chemistry*, *56*, 8541–8548.

Li, B., Akram, M., Al-Zuhair, S., Elnajjar, E., & Munir, M. T. (2020). Subcritical water extraction of phenolics, antioxidants and dietary fibres from waste date pits. *Journal of Environmental Chemical Engineering*, 104490.

Li, X., & Farid, M. (2016). A review on recent development in non-conventional food sterilization technologies. *Journal of Food Engineering*, *182*, 33–45.

Li, Y., Li, S., Lin, S. J., Zhang, J. J., Zhao, C. N., & Li, H. B. (2017). Microwave-assisted extraction of natural antioxidants from the exotic Gordoniaaxillaris fruit: Optimization and identification of phenolic compounds. *Molecules (Basel, Switzerland)*, *22*(9), 1481.

Lima, A. C. S., da Rocha Viana, J. D., Sousa Sabino, L. B., Silva, L. M. R., Silva, N. K. V., & Sousa, P. H. M. (2017). Processing of three different cooking methods of cassava: Effects on in vitro bioaccessibility of phenolic compounds and antioxidant activity. *LWT- FoodScience and Technology*, *76*, 253–258.

López-Rubira, V., Conesa, A., Allende, A., & Artés, F. (2005). Shelf life and overall qualityof minimally processed pomegranate arils modified atmosphere packaged andtreated with UV-C. *Postharvest Biology and Technology*, *37*, 174–185.

Loredo-Tovias, M., Duran-Meza, A. L., Villagrana-Escareño, M. V., Vega-Acosta, R., Reynaga-Hernández, E., Flores-Tandy, L. M., ... Ruiz-Garcia, J. (2017). Encapsidated ultrasmall nanolipospheres as novel nanocarriers for highly hydrophobic anticancer drugs. *Nanoscale*, *9*(32), 11625–11631.

Lorenzo, J. M., Estévez, M., Barba, F. J., Thirumdas, R., Franco, D., & Munekata, P. E. S. (2019). *Polyphenols: Bioaccessibility and bioavailability of bioactive components. In Innovative thermal and non-thermal processing, bioaccessibility and bioavailability of nutrients and bioactive compounds* (pp. 309–332). Woodhead Publishing.

Luo, C. F., Yuan, M., Chen, M. S., Liu, S. M., Zhu, L., Huang, B. Y., … Xiong, W. (2011). Pharmacokinetics, tissue distribution and relative bioavailability of puerarin solid lipid nanoparticles following oral administration. *International Journal of Pharmaceutics*, *410*(1–2), 138–144.

Makris, D. P., & Rossiter, J. T. (2001). Domestic processing of onion bulbs (*Allium cepa*) and Asparagus spears (*Asparagus officinalis*): Effect of flavonol content and antioxidant status. *Journal of Agricultural and Food Chemistry*, *49*, 3216–3222.

Março, P. H., Poppi, R. J., Scarmino, I. S., & Tauler, R. (2011). Investigation of the pH effectand UV radiation on kinetic degradation of anthocyanin mixtures extracted from Hibiscus acetosella. *Food Chemistry*, *125*, 1020–1027.

Martini, S., Conte, A., Cattivelli, A., & Tagliazucchi, D. (2021). Domestic cooking methods affect the stability and bioaccessibility of dark purple eggplant (Solanummelongena) phenolic compounds. *Food Chemistry*, 128298.

Mehta, D., Sharma, N., Bansal, V., Sangwan, R. S., & Yadav, S. K. (2019). Impact of ultrasonication, ultraviolet and atmospheric cold plasma processing on quality parameters of tomato-based beverage in comparison with thermal processing. *Innovative Food Science and Emerging Technologies*, *52*, 343–349.

Mennah-Govela, Y. A., & Bornhorst, G. M. (2017). Fresh-squeezed orange juice properties beforeand during in vitro digestion as influenced by orange variety and processing method. *Journal of Food Science*, *82*(10), 2438–2447.

Miller, T. J., Fatnon, M. J., & Webb, S. R. (2004). Transgenic plants producing soluble, stable immunoprophylactic and therapeutic compositions which released by disrupting plant cell wall. *PCT International Application*, 102.

Mir, S. A., Shah, M. A., Ganai, S. A., Ahmad, T., & Gani, M. (2019). Understanding the role of active components from plant sources in obesity management. *Journal of the Saudi Society of Agricultural Sciences*, *18* (2), 168–176.

Munir, M. T., Kheirkhah, H., Baroutian, S., Quek, S. Y., & Young, B. R. (2018). Subcritical water extraction of bioactive compounds from waste onion skin. *Journal of Cleaner Production*, *183*, 487–494.

Murakami, M., Yamaguchi, T., Takamura, H., & Matoba, T. (2004). Effects of thermaltreatment on radical-scavenging activity of single and mixed polyphenolic compounds. *Food Chemistry and Toxicology*, *69*, FCT7–FCT10.

Nagar, E. E., Berenshtein, L., Katz, I. H., Lesmes, U., Okun, Z., & Shpigelman, A. (2021). The impact of chemical structure on polyphenol bioaccessibility, as a function of processing, cell wall material and pH: A model system. *Journal of Food Engineering*, *289*, 110304.

Ngamwonglumlert, L., Devahastin, S., & Chiewchan, N. (2017). Natural colorants: Pigment stability and extraction yield enhancement via utilization of appropriate pretreatment and extraction methods. *Critical Reviews in Food Science and Nutrition*, *57*(15), 3243–3259.

Nicoli, M. C., Anese, M., & Parpinel, M. (1999). Influence of processing on the antioxidantproperties of fruit and vegetables. *Trends in Food Science and Technology*, *10*(3), 94–100.

Nishad, J., Saha, S., & Kaur, C. (2019). Enzyme-and ultrasound-assisted extractions of polyphenols from Citrus sinensis (cv. Malta) peel: A comparative study. *Journal of Food Processing and Preservation*, *43* (8), e14046.

Ochoa, M. R., Kesseler, A. G., Vullioud, M. B., & Lozano, J. E. (1999). Physical and chemical characteristics of raspberry pulp: Storage effect on composition and color. *LWT- Food Science and Technology*, *149*, 149–153.

Odriozola-Serrano, I., Soliva-Fortuny, R., & Martin-Belloso, O. (2008). Phenolic acids, flavonoids, vitamin C and antioxidant capacity of strawberry juices processed by high-intensity pulsed electric fields or heat treatments. *European Food Research Technology*, *228*, 239–248.

Odriozola-Serrano, I., Soliva-Fortuny, R., & Martín-Belloso, O. (2009). Influence of storage temperature on the kinetics of the changes in anthocyanins, vitamin C, and antioxidant capacity in fresh-cut strawberries stored under high-oxygenatmospheres. *Journal of Food Science*, *74*(2), C184–C191.

Oms-Oliu, G., Odriozola-Serrano, I., Soliva-Fortuny, R., Elez-Martínez, P., & Martín-Belloso, O. (2012). Stability of health-related compounds in plant foods through the application of non thermal processes. *Trends in Food Science and Technology*, *23*(2), 111–123.

Oreopoulou, A., Goussias, G., Tsimogiannis, D., & Oreopoulou, V. (2020). Hydro-alcoholic extraction kinetics of phenolics from oregano: Optimization of the extraction parameters. *Food and Bioproducts Processing*, *123*, 378–389.

Oszmianski, J., Wojdylo, A., & Kolniak, J. (2009). Effect of enzymatic mash treatment and storage on phenolic composition, antioxidant activity, and turbidity ofcloudy apple juice. *Journal of Agricultural and Food Chemistry*, *57*, 7078–7085.

Pap, N., Mahosenaho, M., Pongrácz, E., Mikkonen, H., Jaakkola, M., Virtanen, V., … Keiski, R. L. (2012). Effect of ultrafiltration on anthocyanin and flavonol content of black currant juice (Ribes nigrum L. *Food and Bioprocess Technology*, *5*(3), 921–928.

Park, M., & Lee, K. G. (2021). Effect of roasting temperature and time on volatile compounds, total polyphenols, total flavonoids, and lignan of omija (SchisandrachinensisBaillon) fruit extract. *Food Chemistry*, *338*, 127836.

Patterson, M. F. (2014). Food technologies: High pressure processing. In Y. Motarjemi (Ed.), *Encyclopedia of food safety* (Vol. 3, pp. 196–201). Waltham: Academic Press, Foods, materials, technologies and risks.

Pérez-Gregorio, M. R., García-Falcón, M. S., & Simal-Gándara, J. (2011). Flavonoids changes in fresh-cut onions during storage in different packaging systems. *Food Chemistry*, *124*(2), 652–658.

Pimentel-Moral, S., Borrás-Linares, I., Lozano-Sánchez, J., Arráez-Román, D., Martínez-Férez, A., & Segura-Carretero, A. (2018). Microwave-assisted extraction for *Hibiscus sabdariffa* bioactive compounds. *Journal of Pharmaceutical and Biomedical Analysis*, *156*, 313–322.

Porter, L. W., Levasseur, L. A., & Henick, A. S. (1977). Evaluationof some natural and synthetic phenolic antioxidants in linoleic acid monolayers on silica. *Journal of Food Science*, *42*, 1533–1535.

Prasad, K. N., Yang, E., Yi, C., Zhao, M., & Jiang, Y. (2009). Effects of high pressure extraction on the extraction yield, total phenolic content and antioxidant activity of longan fruit pericarp. *Innovative Food Science and Emerging Technologies*, *10*(2), 155–159.

Putnik, P., Barba, F. J., Lorenzo, J. M., Gabrić, D., Shpigelman, A., Cravotto, G., et al. (2017). An integrated approach to mandarin processing: Food safety and nutritional quality, consumer preference, and nutrient bioaccessibility. *Comprehensive Reviews in Food Science and Food Safety*, *16*(6), 1345–1358.

Qu, W., Breksa, A. P., III, Pan, Z., Ma, H., & Mchugh, T. H. (2012). Storage stability of sterilized liquid extracts from pomegranate peel. *Journal of Food Science*, *77*(7), C765–C772.

Radziejewska-Kubzdela, E., & Olejnik, A. (2016). Effects of pretreatment and modified atmosphere packaging on glucosinolate levels in coleslaw mix. *LWT*, *70*, 192–198.

Radziejewska-Kubzdela, E., Olejnik, A., & Biegańska-Marecik, R. (2019). Effect of pretreatment on bioactive compounds in wild rocket juice. *Journal of Food Science and Technology*, *56*(12), 5234–5242.

Rahman, H. S., Othman, H. H., Hammadi, N. I., Yeap, S. K., Amin, K. M., Samad, N. A., & Alitheen, N. B. (2020). Novel drug delivery systems for loading of natural plant extracts and their biomedical applications. *International Journal of Nanomedicine*, *15*, 2439.

Ramesh, M., Valérie, O., & Mark, L. (2012). Effect of pulsed ultraviolet light on the total phenol content of elderberry (*Sambucus nigra*) fruit. *Food and Nutrition Sciences*, *3*, 774–783.

Ravimannan, N., & Nisansala, A. (2017). Study on antioxidant activity in fruits and vegetables-A Review. *International Journal of Advanced Research in Biological Sciences*, *4*(3), 93–101.

Reed, J. D., Krueger, C. G., & Vestling, M. M. (2005). MALDI-TOF mass spectrometry of oligomeric food polyphenols. *Phytochemistry*, *66*(18), 2248–2263.

Renard, C. M. G. C., Le Quéré, J.-M., Bauduin, R., Symoneaux, R., Le Bourvellec, C., & Baron, A. (2011). Modulating polyphenolic composition and organoleptic properties of apple juices by manipulating the pressing conditions. *Food Chemistry*, *124*, 117–125.

Rice-Evans, C. A., Miller, N. J., & Paganga, G. (1996). Structure-antioxidant activity relationships of flavonoids and phenolic acids. *Free Radical Biology and Medicine*, *20*(7), 933–956.

Rijo, P., Falé, P. L., Serralheiro, M. L., Simões, M. F., Gomes, A., & Reis, C. (2014). Optimization of medicinal plant extraction methods and their encapsulation through extrusion technology. *Measurement*, *58*, 249–255.

Rodríguez-Roque, M. J., Ancos, B., Sánchez-Moreno, C., Cano, M. P., Elez-Martínez, P., & Martín-Belloso, O. (2015). Impact of food matrix and processing on the in vitro bioaccessibility of vitamin C, phenolic compounds, and hydrophilic antioxidant activity from fruit juice-based beverages. *Journal of Functional Foods*, *14*, 33–43.

Roy, L. G., & Urooj, A. (2013). Antioxidant potency, pH and heat stability of selected plant extracts. *Journal of Food Biochemistry*, *37*(3), 336–342.

Sachadyn-Król, M., Materska, M., Chilczuk, B., Karaś, M., Jakubczyk, A., Perucka, I., & Jackowska, I. (2016). Ozone-induced changes in the content of bioactive compounds and enzyme activity during storage of pepper fruits. *Food Chemistry*, *211*, 59–67.

Sajilata, M. G., & Singhal, R. S. (2006). Effect of irradiation and storage on the antioxidative activity of cashew nuts. *Radiation Physics and Chemistry*, *75*, 297–300.

Samuolienė, G., Urbonavičiūtė, A., Brazaitytė, A., Šabajevienė, G., Sakalauskaitė, J., & Duchovskis, P. (2011). The impact of LED illumination on antioxidant properties of sprouted seeds. *Open Life Sciences*, *6* (1), 68–74.

Santhirasegaram, V., Razali, Z., George, D. S., & Somasundram, C. (2015). Effects of thermal and non-thermal processing on phenolic compounds, antioxidant activity and sensory attributes of chokanan mango (Mangiferaindica L.) juice. *Food and Bioprocess Technology*, *8*(11), 2256–2267.

Sørensen, H. (1990). Glucosinolates: Structure-properties-function. In F. Shahidi (Ed.), *Canola and rapeseed* (pp. 149–172). US: Springer.

Srinivas, K., King, J. W., Monrad, J. K., Howard, L. R., & Zhang, D. (2011). Pressurized solvent extraction of flavonoids from grape pomace utilizing organic acid additives. *Italian Journal of Food Science*, *23*(1), 90–105.

Sun, H. N., Mu, T. H., & Xi, L. S. (2017). Effect of pH, heat, and light treatments on the antioxidant activity of sweet potato leaf polyphenols. *International Journal of Food Properties*, *20*(2), 318–332.

Teixeira-Guedes, C. I., Oppolzer, D., Barros, A. I., & Pereira-Wilson, C. (2019). Impact of cooking method on phenolic composition and antioxidant potential of four varieties of Phaseolus vulgaris L. and Glycine max L. *LWT*, *103*, 238–246.

Thakur, A. K., Prasad, N. A., & Laddha, K. S. (2008). *Stability testing of herbal products. The pharma review*. KONGPOSH Publications.

Thakur, L., Ghodasra, U., Patel, N., & Dabhi, M. (2011). Novel approaches for stability improvement in natural medicines. *Pharmacognosy Reviews*, *5*(9), 48.

Tomas, M., Toydemir, G., Boyacioglu, D., Hall, R. D., Beekwilder, J., & Capanoglu, E. (2017). Processing black mulberry into jam: Effects on antioxidant potential and in vitro bioaccessibility. *Journal of the Science of Food and Agriculture*, *97*, 3106–3113.

Tommasini, S., Calabrò, M. L., Donato, P., Raneri, D., Guglielmo, G., Ficarra, P., & Ficarra, R. (2004). Comparative photodegradation studies on 3-hydroxyflavone: Influence of different media, pH and light sources. *Journal of Pharmaceutical and Biomedical Analysis*, *35*, 389–397.

Tudela, J. A., Cantos, E., Espin, J. C., Tomás-Barberán, F. A., & Gil, M. I. (2002). Induction ofantioxidant flavonol biosynthesis in fresh-cut potatoes. Effect of domesticcooking. *Journal of Agricultural and Food Chemistry*, *50*, 5925–5931.

Tully, J., Fakhrullin, R., & Lvov, Y. (2015). *Halloysite clay nanotube composites with sustained release of chemicals*. *In* Nanomaterials and Nanoarchitectures (pp. 87–118). Dordrecht: Springer.

Turkmen, N., Sari, F., & Velioglu, Y. S. (2005). The effect of cooking methods on total phenolics and antioxidant activity of selected green vegetables. *Food Chemistry*, *93*(4), 713–718.

Turturică, M., Stănciuc, N., Bahrim, G., & Râpeanu, G. (2016). Effect of thermal treatment on phenolic compounds from plum (*Prunusdomestica*) extracts—A kinetic study. *Journal of Food Engineering*, *171*, 200–207.

Vale, A. P., Cidade, H., Pinto, M., & Oliveira, M. B. P. (2014). Effect of sprouting and light cycle on antioxidant activity of Brassica oleracea varieties. *Food Chemistry*, *165*, 379–387.

Van Der Sluis, A. A., Dekker, M., Skrede, G., & Jongen, W. M. (2004). Activity and concentration of polyphenolic antioxidants in apple juice. 2-Effect of novel production methods. *Journal of Agricultural and Food Chemistry*, *52*, 2840–2848.

Vergara-Salinas, J. R., Pérez-Jiménez, J., Torres, J. L., Agosin, E., & Pérez-Correa, J. R. (2012). Effects of temperature and time on polyphenolic content and antioxidant activity in the pressurized hot water extraction of deodorized thyme (Thymus vulgaris). *Journal of Agricultural and Food Chemistry*, *60*(44), 10920–10929.

Wang, C. Y., Chen, C. T., & Wang, S. Y. (2009a). Changes of flavonoid content and antioxidant capacity in blueberries after illumination with UV-C. *Food Chemistry*, *117*(3), 426–431.

Wang, S. Y., Chen, C. T., & Wang, C. Y. (2009b). The influence of light and maturity on fruit quality and flavonoid content of red raspberries. *Food Chemistry*, *112*, 676–684.

Wolf, A., Pouny, Y., Marton, I., Dgany, O., Altman, A., & Shoseyov, O. (2007). Use of robust plant chaperonin-like Sp1 proteins to stabilize therapeutic proteins in pharmaceutical use. *PCT Int App*, 155.

Xi, J., & Shouqin, Z. (2007). Antioxidant activity of ethanolic extracts of propolis by highhydrostatic pressure extraction. *International Journal of Food Science and Technology*, *42*(11), 1350–1356.

Xia, Q., Wang, L., Xu, C., Mei, J., & Li, Y. (2017). Effects of germination and high hydrostatic pressure processing on mineral elements, amino acids and antioxidants in vitro bioaccessibility, as well as starch digestibility in brown rice (Oryza sativa L.). *Food Chemistry*, *214*, 533–542.

Xu, C. J., Guo, D. P., Yuan, J., Yuan, G. F., & Wang, Q. M. (2006). Changes in glucoraphanin content and quinine reductase activity in broccoli (*Brassica oleraceavar. italica*) florets during cooling and controlled atmosphere storage. *Post harvest Biology and Technology*, *42*(2), 176–184.

Zafrilla, P., Ferreres, F., & Tomás-Barberán, F. A. (2001). Effect of processing and storage on the antioxidant ellagic acid derivatives and flavonoids of red raspberry (Rubusidaeus) jams. *Journal of Agricultural and Food Chemistry*, *49*, 3651–3655.

Zainol, M. M., Abdul-Hamid, A., Bakar, F. A., & Dek, S. P. (2009). Effect of different drying methods on the degradation of selected flavonoids in Centellaasiatica. *International Food Research Journal*, *16*(4), 531–537.

Zeng, L., Ma, M., Li, C., & Luo, L. (2017). Stability of tea polyphenols solution with different pH at different temperatures. *International Journal of Food Properties*, *20*(1), 1–18.

Zhang, M., Chen, H., Li, J., Pei, Y., & Liang, Y. (2010). Antioxidant properties of tartary buck wheat extracts as affected by different thermal processing methods. *LWT Food Science and Technology*, *43*, 181–185.

Zhang, X., Cardon, D., Cabrera, J. L., & Laursen, R. (2010). The role of glycosides in thelight-stabilization of 3-hydroxyflavone (flavonol) dyes as revealed by HPLC. *Microchimica Acta*, *169*, 327–334.

Zhu, F. (2018). Interactions between cell wall polysaccharides and polyphenols. *Critical Reviews in Food Science and Nutrition*, *58*(11), 1808–1831.

Zielinski, H., Michalska, A., Amigo-Benavent, M., del Castillo, M. D., & Piskula, M. K. (2009). Changes in protein quality and antioxidant properties of buckwheat seeds and groats induced by roasting. *Journal of Agricultural and Food Chemistry*, *57*(11), 4771–4776.

Zorzi, G. K., Carvalho, E. L. S., von Poser, G. L., & Teixeira, H. F. (2015). On the use of nanotechnology-based strategies for association of complex matrices from plant extracts. *Revista Brasileira de Farmacognosia*, *25*(4), 426–436.

CHAPTER

Plant extracts as food preservatives

6

Manzoor Ahmad Shah[1] and Shabir Ahmad Mir[2]

[1]Department of Food Science & Technology, Govt. Degree College for Women, Anantnag, India [2]Department of Food Science & Technology, Govt. College for Women, Srinagar, India

6.1 Introduction

Food preservation is a process in which food quality and safety are increased. Food is handled and treated in such a way to prevent or delay spoilage and prevent foodborne illnesses. The spoilage may be due to the inherent food properties or due to the microbial spoilage while the pathogenic microorganisms are responsible for various foodborne diseases. Different methods of preservation have been employed to overcome these problems. Among these methods, food preservatives have been used for a long time. Food preservatives may be categorized into antimicrobials, antioxidants, and antibrowning agents.

Antimicrobials are used to inhibit or prevent microbial growth in the food products, and their use has increased in recent years due to food safety concerns arising from microbial contamination. The antioxidants are used to inhibit or prevent oxidation reactions and are used primarily to prevent autoxidation, which results in rancidity and off-flavor development in food products. Antibrowning agents are chemicals used to inhibit or prevent browning (enzymatic and nonenzymatic) reactions in food products (Branen, Davidson, Salminen, & Thorngate, 2002).

These preservatives may be synthetic or natural. Despite the efficiency to prevent food spoilage and foodborne diseases, the use of synthetic preservatives has led to the buildup of chemical residues in the food products and thus adversely affects human health (Nazir et al., 2017). Owing to the safety concerns of synthetic chemicals and growing concerns about food safety and natural ingredients, the use of synthetic chemical is in a decline, whereas the research on natural ingredients has increased. The natural ingredients used in food preservation are obtained from plants as they are abundant and provide a wide range of ingredients needed for preservation. Almost every plant has been used in the form of extracts and evaluated for preservation in many food products.

Plant materials are used for the preparation of extracts using different types of techniques. These involve different types of solvents, procedures, and equipment to extract the bioactive components from the plant materials. These extraction methods are aimed to provide the extract with maximum yield and the highest quality. These extracts from plant sources are rich in various bioactive components and thus can be used to replace the synthetic preservatives. The antimicrobial properties of these extracts can be assessed by various microbiological tests, while the antioxidant properties are evaluated by using several chemical tests.

Plant Extracts: ***Applications in the Food Industry*. DOI: https://doi.org/10.1016/B978-0-12-822475-5.00010-7**

6.2 Sources

Plants are a renewable source of materials used for the preparation of plant extracts (Tayel & El-Tras, 2012). Different plants or their parts/products are being assessed for their ability to use as natural preservatives. These natural preservatives are prepared from various plant sources such as fruits, vegetables, herbs, and spices. Table 6.1 shows the plant source along with the part used for the preparation of extracts.

6.3 Extraction

Extraction is an important step for the preparation of plant extracts from different types of plant materials. Different types of procedures and techniques are used for extraction and can be categorized into conventional and nonconventional methods. The conventional methods like maceration and soxhlet extraction method have been used since ancient times and is still in use. Nonconventional methods such as microwave assisted extraction, ultrasound assisted extraction, pressurized liquid extraction and supercritical fluid extraction gained the popularity for the extraction of plant extracts due to less use of chemicals and eco-friendly in nature.

The extraction process is aimed to increase the yield of a substance with superior quality. A general procedure to prepare a plant extract is that the plant material to be used for extraction, is cleaned, dried, and powdered and then extracted using a solvent (Fig. 6.1). Different solvents have been exploited for preparation of extracts. These solvents may be used either alone or in combination in different ratios. The main solvents are water, methanol, ethanol and acetone (Shah, Bosco, & Mir, 2014). Sometimes some additional processes like defatting may also be done depending on the plant material. The quality of the extract is greatly influenced by the solvent and the method of extraction employed for the preparation of plant extracts. Literature shows that the extraction method can affect the bioactivity of the extracts (Chan, Lim, & Omar, 2007; Ding et al., 2012; Kothari, Pathan, & Seshadri, 2010; Kothari, 2011; Sikora, Cieslik, Leszcznska, Filipiak-Florkiewicz, & Pisulewski, 2008; Upadhyay, Ramalakshmi, & Rao, 2012; Yeh et al., 2014).

There are several factors that affect the extraction efficiency such as the composition and structure of phytochemicals, the extraction technique employed, nature of the sample, the type of solvent used, and the interfering components present in the sample (Stalikas, 2007). The extract yield is affected by the nature of the solvent, pH, temperature, time of extraction and chemical composition and structure of the sample. When the time and temperature are kept constant, then solvent and sample composition were found to be the most significant factors affecting the extract yield (Turkmen, Sari, & Velioglu, 2006).

6.4 Plant extracts as antimicrobials

The contamination of foods with microorganisms results in spoilage of food and/or various food-borne diseases. These problems can be solved by the application of antimicrobial agents to the foods. The plant extracts are considered as natural antimicrobial agents. Many researchers have

Table 6.1 Plant sources and their parts used for the preparation of natural preservatives.

Source	Scientific name	Part used
Acanthopanax	*Acanthopanax sessiliflorum*	Leaf
Amla	*Emblica officianalis*	Fruit
Black seed	*Nigella sativa*	Seeds
Blackcurrant	*Ribes nigrum* L.	Leaf
Bok choy	*Brassica campestris* L.	Leaf
Broccoli	*Brassica oleracea*	Flowering head
Butterbur	*Petasites japonicus*	Leaf
Chamnamul	*Pimpinella brachycarpa*	Leaf
Chinese chives/Leek	*Allium tuberosum*	Leaf
Cherry	*Prunus cerasus* L.	Leaf
Cinnamon	*Cinnamomum verum*	Bark
Cinnamon stick	*Cinnamomum burmannii*	Cortex
Clove	*Eugenia caryophylata*	Bud
Crown daisy	*Chrysanthemum coronarium*	Leaf
Curry	*Murraya koenigii*	Leaf
Cyprus	*Citrus aurantium* L.	Flower
Date	*Phoenix dactylifera*	Pits
Drumstick	*Moringa oleifera*	Leaf
Eleutherine	*Eleutherine americana*	Bulb
Fatsia	*Aralia elata* Seem	Leaf
Fenugreek	*Trigonella foenum-graecum*	Seed
Garlic	*Allium sativum*	Aerial parts, bulb
Ginger	*Zingiber officinale*	Rhizome
Grape	*Vitis vinifera*	Seed, pomace
Green tea	*Camellia sinensis*	Leaf
Hyssop	*Hyssopus officinalis*	Leaf and secondary branches
Kinnow	*Citrus reticulate*	Peel
Lemon balm	*Melissa officinalis*	Leaf
Lemon grass	*Cymbopogon citratus*	Leaf
Licorice	*Glycyrrhiza glabra*	Root
Lotus	*Nelumbo nucifera*	Leaf, rhizome knot
Mint	*Mentha spicata*	Leaf
Mugwort extract	*Artemisia princeps*	Leaf
Myrtle	*Myrtus communis myrtillus*	Leaf
Nettle	*Urtica dioica*	Leaf, flower
Onion	*Allium cepa* L.	Bulb
Oregano	*Origanum vulgare*	Leaf
Peanut	*Arachis hypogaea*	Skin
Pomegranate	*Punica granatum*	Peel
Potato	*Solanum tuberosum*	Peel

(*Continued*)

Table 6.1 ***Continued***

Source	Scientific name	Part used
Pumpkin	*Curcubita moschata*	Leaf
Roselle	*Hibiscus sabdariffa*	Flower
Rosemary	*Rosmarinus officinalis*	Leaf and secondary branches
Sea buckthorn	*Hippophae rhamnoides*	Seeds
Soybean	*Glycine max*	Leaf
Stonecrop	*Sedum sarmentosum*	Leaf
Summer savory	*Satureja hortensis*	Leaf

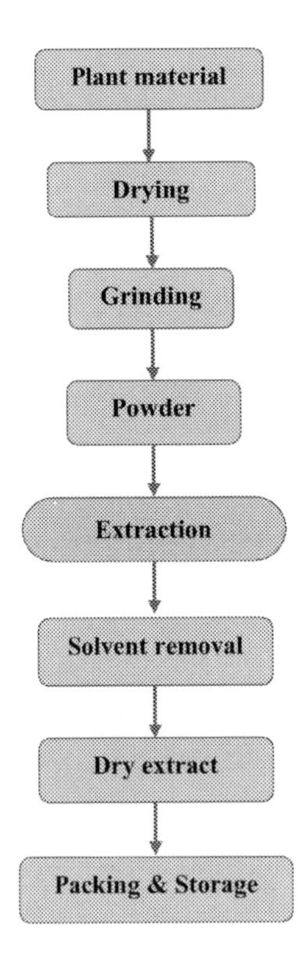

FIGURE 6.1

Flow sheet for the preparation of plant extract.

reported that plant extracts show antimicrobial properties against food spoilage as well as food-borne pathogenic microorganisms. Several methods have been used to assess the antimicrobial properties of plant extracts. The methods based on agar are pour plate disk diffusion, streak plate disk diffusion and agar-well diffusion method was assessed by using agar-well diffusion method.

The other method is the broth dilution assay (Othman et al., 2011). These methods are used to determine the minimum inhibitory concentration of the extracts.

Verma, Singh, Tiwari, Srivastava, and Verma (2012) studied the antibacterial properties of extracts (*Punica granatum*, *Citrus limon* and *Allium sativum*) against *Salmonella typhi*, *Escherichia coli*, *Bacillus cereus* and *Staphylococcus aureus*. They reported that these plant extracts were potentially effective against all of these microorganisms. Also, the extract prepared from *P. granatum* showed the highest effectiveness at a concentration of 500 mg/mL. Several other authors also revealed that the extracts prepared from the peels of *P. granatum* showed antimicrobial activity against *Micrococcus luteus*, *S. aureus*, *Bacillus megaterium*, *E. coli* and *Pseudomonas aeruginosa*. The concentration of the extract showing maximum effective ranged from 30 to 50 mg/mL (Dey et al., 2012; Duman, Ozgen, Dayisoylu, Eribl, & Durgac, 2009; Sadeghian, Ghorbani, Mohamadi_Nejad, & Rakhshandeh, 2011). The extracts from *P. granatum* exhibited antimicrobial activity against both the spoilage as well as pathogenic microorganisms. These extracts were found to be effective against *S. aureus*, *B. cereus*, *E. coli* and *S. typhi* (Alzoreky, 2009; Mahboubi, Asgarpanah, Sadaghiqani, & Faizi, 2015). Thus, these extracts can act as effective antimicrobial agents for food preservation and the prevention of foodborne illnesses.

Spices are rich sources of antimicrobial compounds, and several authors have their use against foodborne pathogens and food spoilage microorganisms. Cloves (*Syzygium aromaticum*) have a characteristic aroma and possess antimicrobial properties and thus can be used as a flavoring agent and as a preservative in foods. It has been reported that clove extracts show antimicrobial activity against *E. coli*, *S. aureus*, *B. cereus*, *L. monocytogenes*, *Listeria innocua* and *Salmonella enteric*. Cloves contain an active compound known as eugenol which inhibited the growth of *H. pylori* (Devi, Nisha, Sakthivel, & Pandian, 2010; Hill, Gomes, & Taylor, 2013; Sofia, Prasad, Vijay, & Srivastava, 2007). Mahfuzul_Hoque, Bari, Juneja, and Kawamoto (2007) reported that the clove extracts prepared using ethanol showed antimicrobial activity against *S. aureus*, *V. parahaemolyticus* and *P. aeruginosa* but were ineffective against *E. coli* and *Salmonella enteritidis*. In another study, Pandey and Singh (2011) reported that the MIC was between 0.1 and 2.31 mg/mL for methanolic extract from cloves and were effective against *S. aureus*, *P. aeruginosa* and *E. coli*.

Cumin (*Cuminum cyminum*) seeds extract showed antimicrobial activity against several types of bacteria involved in food poisoning. The cumin extracts with MIC range from 6.25 and 25 mg/mL, showed effectiveness against *E. coli*, *P. aeruginosa*, *S. aureus* and *B. pumilus* (Dua, Gaurav, Balkar, & Mahajan, 2013). Qader, Khalid, and Abdullah (2013) investigated the antimicrobial properties of several plant extracts prepared using ethanol and water against some pathogenic bacteria. They reported that the extract from *P. granatum* showed antimicrobial activity with MIC of 0.2 mg/mL. The extract from ginger (*Zingiber officinales*) also showed effectiveness against *P. aeruginosa* and *Klebsiella pneumonia* while thyme (*Thymus kotschyana*) extract showed effectiveness against *S. aureus* and *E. coli*. Soković, Glamočlija, Marin, Brkić, and van Griensven (2010) reported that thyme (*T. vulgaris*) extract exhibited a broad antibacterial activity against several microorganisms.

Peppermint (*Mentha piperita*) extracts exhibit broad spectrum antimicrobial properties and are very effective against several microorganisms including bacteria, yeasts, and molds (Carretto, Almeida, Furlan, Jorge, & Junqueira, 2010; Singh, Shushni, & Belkheir, 2015). It has been reported that the mint extracts showed activity against *S. aureus* and *S. pyogenes* (Singh et al., 2015). The ethanolic extract from mint exhibited antifungal activity against several *Candida spp.*, but its infusion lacked such properties (Carretto et al., 2010).

Cinnamon (*Cinnamomum zeylanicum* and *C. cassia*) extracts possess strong antimicrobial properties against a broad range of pathogenic and spoilage microorganisms. These extract show inhibit the growth of *B. cereus*, *B. coaguiaris*, *B. subtilis*, *P. aeruginosa*, *L. monocytogenes*, *Acinetobacter baumannii*, *E. cloacae*, *S. aureus*, *E. coli* and *A. lannensis* and *A. bogorensis* (Antolak, Czyzowska, & Kregiel, 2017; Bayoub, Baibai, Mountassif, Retmane, & Soukri, 2010; Hosseininejad et al., 2011; Khan et al., 2009; Ranasinghe et al., 2013; Tekwu et al., 2012)

Nettle (*Urtica dioica*) extract obtained using water as a solvent showed antimicrobial effect against *M. luteus*, *Proteus mirabilis*, *Citrobacter koseri*, *S. aureus*, *S. pyogenes*, *S. epidermidis*, *S. pneumoniae*, *E. aerogenes*, *E. coli*, and *C. albicans* (Gülçin, Küfrevioğlu, Oktay, & Büyükokuroğluc, 2004; Gülçin, Küfrevioglu, & Oktay, 2005; Modarresi-Chahardehi, Ibrahim, Fariza-Sulaiman, & Mousavi, 2012; Turker & Usta, 2008). Nettle extracts were effective against *Acinetobacter calcoaceticus*, *B. cereus*, *B. spizizenii*, *V. parahaemolyticus* and *K. pneumonia*. The antimicrobial properties of nettle extract were comparable to miconazole, amoxicillin and ofloxacin (Modarresi-Chahardehi et al., 2012). In another study, it was reported that the nettle extract showed inhibitory activity against *Asaia* spp. This bacterium is responsible for spoilage in several types of beverages (Antolak et al., 2017).

Sun et al. (2020) reported that the extract prepared from blueberry showed antimicrobial activity against *V. parahaemolyticus* present in salmon samples. This extract disrupts gene transcription, cell membrane structure, and energy transport in the microbial cells. Ahn, Grun, and Mustapha (2004) used extracts from grape seed and pine bark as antimicrobial agents in ground beef and reported that these extracts were effective against *E. coli* O157:H7, *S. typhimurium*, and *L. monocytogenes*. Aliakbarlu and Mohammadi (2015) reported that the sumac (*Rhus coraria* L.) and barberry (*Barberis vulgaris* L.) extracts reduced the microbial populations in ground sheep meat. Abdulla, Abdel Samie, and Zaki (2016) studied the effect of ziziphus leaf extract on microbial quality of sausages and reported that the extract reduced the microbial count in sausages during cold storage. Nowak, Czyzowska, Efenberger, and Krala (2016) used cherry and blackcurrant leaf extracts in vacuum-packed pork sausages and reported that these extracts were effective against *Pseudomonas*. Piskernik, Klancnik, Riedel, Brøndsted, and Smole Mozina (2011) reported that rosemary extract showed antimicrobial activity against *Campylobacter jejuni*-chicken meat.

Wei, Wolf-Hall, and Hall (2009) reported that raisin extracts showed inhibitory activity against the *Bacillus* species responsible for ropy-bread and also resulted in a decrease in the *B. licheniformis B. subtilis, A. flavus* and *P. chrysogenum* counts in liquid bread and bread systems. Degirmenci and Erkurt (2020) investigated the effect of cyprus (*Citrus aurantium* L.) flower extract on the pathogens in rice pudding and reported that the extract prepared in methanol exhibited the strongest antimicrobial effect against *E. coli* O157:H7, *S. typhimurium*, *L. monocytogenes*, and *B. cereus*. The use of this extract reduced the growth rate of all these pathogens and other natural flora in product.

6.5 Plant extracts as antioxidants

Lipid oxidation is one of the important types of chemical spoilage of foods. It leads to the production of several compounds that degrade the quality of foods with respect to sensory and nutritional aspects. These changes can be inhibited or minimized by the application of antioxidants.

Antioxidants prevent lipid oxidation in several ways. These compounds inhibit chain reaction by removing initiating radicals, break chain reaction, decompose peroxides, decrease localized oxygen concentrations, and bind the chain initiating catalysts during the lipid oxidation process (Dorman, Peltoketo, Hiltunen, & Tikkanen, 2003). A large number of substances have been found to act as antioxidants, but only a small number has been used in food. The application of antioxidants in food varies according to the regulatory laws of a country and international standards (Karre, Lopez, & Getty, 2013).

The antioxidants may be grouped into two categories: synthetic and natural. Synthetic antioxidants include butylated hydroxyanisole, butylated hydroxytoluene, tert-butylhydroquinone, and propyl gallate, and have been extensively used in different food products. These synthetic antioxidants have been associated with several negative impacts on health of consumers and thus consumers demand safer food and that too with an extended shelf life. This has forced the scientific community to evaluate the natural sources for the preservative potential, especially from plant sources (Shah et al., 2014).

The antioxidant potential of plant extracts can be evaluated by several assays such as diphenyl-1-picrylhydrazyl, superoxide anion scavenging assay, hydrogen radical scavenging assay, hydrogen peroxide scavenging assay, 2,2-azinobis-3-ethylbenzthiazoline-6-sulfonic acid radical scavenging activity and reducing power assay. The antioxidant capacity of the extract is influenced by the extraction procedure and the type of solvent used (Brewer, 2011; Shah et al., 2014).

Plant extracts have been applied to various food systems for the prevention of various deteriorative changes occurring due to the oxidation of various food components. Several authors have reviewed the use of plant extracts in meat and poultry products (Alirezalu et al., 2020; Karre et al., 2013; Nikmaram et al., 2018; Shah et al., 2014).

Mansour and Khalil (2000) used the extracts prepared from potato peels, ginger rhizomes, and fenugreek seeds in beef patties and reported that the extracts from ginger rhizome and fenugreek seed effectively reduced the lipid oxidation during cold storage. Fernandez-Lopez et al. (2003) studied the effect of extracts from rosemary and hyssop in cooked pork meat and reported that these extracts prevented the lipid oxidation as well as heme pigment degradation during cooking and storage. Ahn, Grun, and Mustapha (2007) applied Pycnogenol, ActiVin and Herbalox in cooked ground beef and reported that these antioxidants reduced lipid oxidation by 94%, 92% and 92% respectively compared to control. Also, these additives reduced the hexanal content of beef samples throughout the storage period.

Many researchers have used grape seed extract (GSE) in various types of meat products. Brannan and Mah (2007) evaluated the effect of GSE in raw and cooked ground meat and reported that GSE reduced the lipid oxidation more effectively than gallic acid. GSE inhibited the formation of primary and secondary oxidation compounds in different types of meat. Rojas and Brewer (2007) reported that GSE reduced the lipid oxidation and the off-odors associated with it in cooked beef and pork patties. Also, these authors reported a concentration dependent effect of antioxidant activity in GSE. In another study, Rojas and Brewer (2008) evaluated the effect GSE in vacuum packaged raw beef and pork patties kept under frozen conditions for 4 months and reported that the GSE prevented lipid oxidation in both types of meat. Similar reports on the application of GSE in various meat products have been reported by Shan, Cai, Brooks, and Corke (2009), Colindres and Brewer (2011), Ozvural and Vural (2012), and Reddy et al. (2013).

Kanatt, Chander, and Sharma (2007) evaluated the effect of mint leaf extract in lamb meat processed by irradiation and stored at chilled temperatures. They reported that the antioxidant properties of extract were equivalent to BHT. The extract lowered the TBARS values significantly in treated samples in comparison to the samples without any antioxidant.

Drumstick leaf extract was applied in cooked goat meat patties stored at refrigerated conditions by Das, Rajkumar, Verma, and Swarup (2012). They reported that the treated samples showed a reduction in lipid oxidation during refrigerated storage. The antioxidant properties of the extract were comparable to BHT. Furthermore, it had no adverse effect on the sensory parameters of patties. Similar results on the application of *Moringa* leaf extract were reported by Muthukumar, Naveena, Vaithiyanathan, Sen, and Sureshkumar (2014) for raw and cooked pork patties stored at refrigerated conditions. Shah, Bosco, and Mir (2015) reported that the *Moringa* leaf extract reduced the lipid oxidation in raw beef stored in high oxygen modified atmosphere packaging.

Akarpat, Turhan, and Ustun (2008) studied the effect of myrtle, rosemary, nettle, and lemon balm leaf extracts on beef patties kept at $-20°C \pm 2°C$ for 120 days and reported that these extracts had reduced the lipid oxidation in the patties. The highest antioxidant effect was shown by myrtle and rosemary extracts compared to nettle and lemon balm extracts. Also, the application of the extract from myrtle leaf (10% extract) to patties prevented the damage in lipids due to oxidation in beef patties during frozen storage. Alp and Aksu (2010) studied the effect of nettle (*U. dioica*) extract on the quality of ground beef kept under modified atmosphere and reported that the extract reduced the lipid oxidation in the beef samples.

Hayes, Stepanyan, Allen, O'Grady, and Kerry (2010a) evaluated the effect of extract from olive leaf in raw and cooked pork patties kept under aerobic and modified atmospheres for 8 and 12 days, respectively. Lipid oxidation was minimized by the olive leaf extract treatment. Also, it did not affect the sensory parameters but had changed the instrumental textural attributes. In another study, Hayes, Stepanyan, Allen, O'Grady, and Kerry (2010b) reported that the same leaf extract reduced the oxidation of lipid and oxymyoglobin in raw beef patties. Furthermore, Hayes, Stepanyan, Allen, O'Grady, and Kerry (2011) observed a reduction in lipid oxidation of fresh and cooked pork sausages packaged under aerobic and modified atmospheres by using olive leaf extract. Also, it did not affect the other physicochemical and sensory parameters of the sausages.

Vegetable oils are rich in several unsaturated compounds and thus are susceptible to various oxidative processes. Some studies have reported the use of plant extracts to prevent the oxidative damage in these oils. Potato peel extract was used to preserve vegetable oil from sunflower and soybean by Mohdaly, Sarhan, Mahmoud, Ramadan, and Smetanska (2010). The results showed that the peroxide value of these oils was reduced by the application of the potato peel extract. In another study, rosemary extract was used in preservation of soybean oil (Casarotti & Jorge, 2014). They reported that the extract improved the oxidative stability and other quality parameters of the oil. Wang, Fu, Wang, Yang, and Wu (2018) reported that the rosemary extracts used in omega-3 fatty-acid rich flaxseed oil reduced the lipid oxidation of the oil.

Milk and its products are rich source of lipids, and these lipids undergo several adverse changes during processing or storage. Literature shows that some studies have been carried out to preserve milk and milk products. Mango seed kernels extracts have been used to preserve pasteurized cow milk (Abdalla & El-Hamahmy, 2007) and ghee (Puravankara, Bohgra, & Sharma, 2000). Similarly, *Moringa* leaf extracts were used to preserve and increase the shelf life of ghee as it is rich in various natural antioxidant compounds (Siddhuraju & Becker, 2003).

Plant extracts have been investigated in several bakery products such as bread (Peng et al., 2010) and biscuits (Reddy, Urooj, & Kumar, 2005). GSE was used to enrich and preserve bread. It was reported that the antioxidant activity of the extract added was reduced due to the thermal degradation during baking process. However, the extract enriched the bread with antioxidant compounds (Peng et al., 2010). Amla (*Emblica officianalis*) and *Moringa* leaf extract were used as preservatives in cookies. The results revealed that these extracts showed the antioxidant effect comparable to BHT. The antioxidant potential of these extracts increased with the increase in extract concentration. The peroxide values also decreased in the treated samples (Reddy et al., 2005).

6.6 Plant extracts as antibrowning agents

Browning is a process, which results in color change of foods to brown or dark brown with the passage of time. It has a positive or negative impact on the quality of a food product (Martinez &Whitaker, 1995). In case of fruit and vegetable products, browning is not desirable but in case of certain foods it produces unique color and flavor for example, bread, coffee, cocoa, raisins, etc. Browning may proceed through enzymatic or nonenzymatic routes. The enzymatic browning requires enzymes to produce brown colored compounds while the nonenzymatic browning does not require any enzyme. Enzymatic browning occurs due to the polyphenol oxidase (PPO), a copper-containing enzyme present in food. Enzymatic browning in food products occurs during various processing procedures such as cutting, peeling, slicing, etc. The nonenzymatic browning occurs through a series of chemical reactions and can be categorized into the Maillard reaction, caramelization, and ascorbic acid oxidation. These changes affect the sensory and nutritional quality of the food products (Moon, Kwon, Lee, & Kim, 2020).

The enzymatic browning is responsible for economic losses in food industry and the main factor involved is the PPO. This enzyme catalyzes the oxidation of phenolic compounds in food products thus, leads to the antioxidant degradation and color alteration in them. Several synthetic compounds have been used to control the browning reactions in foods but the recent studies have shown that new antibrowning strategies are being developed to replace synthetic additives (sulfites) with natural extracts derived from plants (Tinello, Mihaylova, & Lante, 2020; Zocca, Lomolino, & Lante, 2011).

Plant extracts have been used to prevent the browning reactions in several food products as they are regarded as safe and fulfill the consumer demand of more natural additives in food. The extracts from onion have been used to prevent the browning in many food products. These extracts applied to potatoes cause inhibition of the PPO activity and thus suppressed the browning in them (Lee et al., 2002). Onion extract has been also used to reduce the browning in pear (Kim, Kim, & Park, 2005). In another study, this extract resulted in reduced browning in fresh apple juice (Lee, Seo, Rhee, & Kim, 2016).

Several fruits have shown promising results in the prevention of browning reactions in food products. Lozano-de-Gonzalez, Barrett, Wrolstad, and Durst (1993) used pineapple juice has been used to prevent the browning in fresh and dried apples. This extract has also been used as an antibrowning agent in banana slices by Chaisakdanugull, Theerakulkait, and Wrolstad (2007). Lemon and grape juice have been investigated on dough samples used for pastry production and the results revealed that these juices reduced the browning reactions in the dough (Brütsch et al., 2018).

Liu et al. (2019) investigated the effect of purslane (*Portulaca oleracea* L.) extract on potato slices and reported that the extract inhibited the activity of PPO in the potato slices and thus reduced the browning reaction during storage.

Tyrosinase is a copper-containing enzyme and is responsible for enzymatic browning in several foods such as button mushrooms. Fattahifar, Barzegar, Ahmadi Gavlighi, and Sahari (2018) investigated the effect of pistachio (*Pistacia vera* L.) green hull extract on the browning of button mushrooms and reported that this extract inhibited the activity of tyrosinase and thus reduced the browning reactions involving the role of tyrosinase.

6.7 Conclusion

Plants are rich sources of natural ingredients used in food preservation. Different plant parts have been used in the form of extracts and used as preservatives in many food products. As these preservatives are obtained from edible plant sources, they are generally regarded as safe and are preferred by consumers over the synthetic preservatives. These preservative are categorized into antimicrobials, antioxidants, and antibrowning agents according to their function in the food. These extracts show antimicrobial properties against food spoilage as well as foodborne pathogenic microorganisms. These extracts applied to various food systems help in preventing various deteriorative changes occurring due to the oxidation of various food components. Furthermore, these extracts effectively reduced the browning reactions in several food products.

References

Abdalla, A. E. M., & El-Hamahmy, S. M. (2007). Egyptian mango by-product 2: Antioxidant and antimicrobial activities of extract and oil from mango seed kernel. *Food Chemistry, 103*, 1141–1152.

Abdulla, G., Abdel Samie, M. A.-S., & Zaki, D. (2016). Evaluation of the antioxidant and antimicrobial effects of ziziphus leaves extract in sausage during cold storage. *Pakistan Journal of Food Science, 26*, 10–20.

Ahn, J., Grun, I. U., & Mustapha, A. (2004). Antimicrobial and antioxidant activities of natural extracts in vitro and in ground beef. *Journal of Food Protection, 67*, 148–155.

Ahn, J., Grun, I. U., & Mustapha, A. (2007). Effects of plant extracts on microbial growth, color change, and lipid oxidation in cooked beef. *Food Microbiology, 24*, 7–14.

Akarpat, A., Turhan, S., & Ustun, N. S. (2008). Effects of hot-water extracts from myrtle, rosemary, nettle and lemon balm leaves on lipid oxidation and color of beef patties during frozen storage. *Journal of Food Processing and Preservation, 32*, 117–132.

Aliakbarlu, J., & Mohammadi, S. (2015). Effect of sumac (*Rhus coraria* L.) and barberry (*Barberis vulgaris* L.) water extracts on microbial growth and chemical changes in ground sheep meat. *Journal of Food Processing and Preservation, 39*, 1859–1866.

Alirezalu, K., Pateiro, M., Yaghoubi, M., Alirezalu, A., Peighambardoust, S. H., & Lorenzo, J. M. (2020). Phytochemical constituents, advanced extraction technologies and techno-functional properties of selected Mediterranean plants for use in meat products. A comprehensive review. *Trends in Food Science & Technology, 100*, 292–306.

Alp, E., & Aksu, M. I. (2010). Effects of water extract of *Urtica dioica* L. and modified atmosphere packaging on the shelf life of ground beef. *Meat Science, 86*, 468–473.

Alzoreky, N. S. (2009). Antimicrobial activity of pomegranate (*Punica granatum* L.) fruit peels. *International Journal of Food Microbiology*, *134*, 244–248.

Antolak, H., Czyzowska, A., & Kregiel, D. (2017). Antibacterial and antiadhesive activities of extracts from edible plants against soft drink spoilage by *Asaia* spp. *Journal of Food Protection*, *80*(1), 25–34.

Bayoub, K., Baibai, T., Mountassif, D., Retmane, A., & Soukri, A. (2010). Antibacterial activities of the crude ethanol extracts of medicinal plants against *Listeria monocytogenes* and some other pathogenic strains. *African Journal of Biotechnology*, *9*, 4251–4258.

Branen, A. L., Davidson, P. M., Salminen, S., & Thorngate, J. H., III (2002). *Food additives* (2nd Edition). New York: Marcel and Dekker Inc.

Brannan, R. G., & Mah, E. (2007). Grape seed extract inhibits lipid oxidation in muscle from different species during refrigerated and frozen storage and oxidation catalyzed by peroxynitrite and iron/ascorbate in a pyrogallol red model system. *Meat Science*, *77*, 540–546.

Brewer, M. S. (2011). Natural antioxidants: Sources, compounds, mechanisms of action, and potential applications. *Comprehensive Reviews in Food Science and Food Safety*, *10*, 221–247.

Brütsch, L., Rugiero, S., Serrano, S., Städeli, C., Windhab, E., Fischer, P., & Kuster, S. (2018). Targeted inhibition of enzymatic browning in wheat pastry dough. *Journal of Agricultural and Food Chemistry*, *66*, 12353–12360.

Carretto, C. F. P., Almeida, R. B. A., Furlan, M. R., Jorge, A. O. C., & Junqueira, J. C. (2010). Antimicrobial activity of *Mentha piperita* L. against *Candida* spp. *Brazilian Dental Journal*, *13*(1), 4–9.

Casarotti, S., & Jorge, N. (2014). Antioxidant activity of rosemary extract in soybean oil under thermoxidation. *Journal of Food Processing and Preservation*, *38*(1), 136–145.

Chaisakdanugull, C., Theerakulkait, C., & Wrolstad, R. E. (2007). Pineapple juice and its fractions in enzymatic browning inhibition of banana [*Musa* (AAA Group) Gros Michel]. *Journal of Agricultural and Food Chemistry*, *55*, 4252–4257.

Chan, E. W. C., Lim, Y. Y., & Omar, M. (2007). Antioxidant and antibacterial activity of leaves of *Etlingera* species (Zingiberaceae) in Peninsular Malaysia. *Food Chemistry*, *104*, 1586–1593.

Colindres, P., & Brewer, M. S. (2011). Oxidative stability of cooked, frozen, reheated beef patties: Effect of antioxidants. *Journal of the Science of Food and Agriculture*, *91*, 963–968.

Das, A. K., Rajkumar, V., Verma, A. K., & Swarup, D. (2012). *Moringa oleifera* leaves extract: A natural antioxidant for retarding lipid peroxidation in cooked goat meat patties. *International Journal of Food Science and Technology*, *47*, 585–591.

Degirmenci, H., & Erkurt, H. (2020). Chemical profile and antioxidant potency of *Citrus aurantium* L. flower extracts with antibacterial effect against foodborne pathogens in rice pudding. *LWT - Food Science and Technology*, *126*, 109273. Available from https://doi.org/10.1016/j.lwt.2020.109273.

Devi, K. P., Nisha, S. A., Sakthivel, R., & Pandian, S. K. (2010). Eugenol (an essential oil of clove) acts as an antibacterial agent against *Salmonella typhi* by disrupting the cellular membrane. *Journal of Ethnopharmacology*, *130*(1), 107–115.

Dey, D., Debnath, S., Hazra, S., Ghosh, S., Ray, R., & Hazra, B. (2012). Pomegranate pericarp extract enhances the antibacterial activity of ciprofloxacin against extended spectrum b-lactamase (ESBL) and metallo-b-lactamase (MBL) producing Gram-negative bacilli. *Food and Chemical Toxicology: An International Journal Published for the British Industrial Biological Research Association*, *50*, 4302–4309.

Ding, S. H., An, K. J., Zhao, C. P., Li, Y., Guo, Y. H., & Wang, Z. F. (2012). Effect of drying methods on volatiles of Chinese ginger (*Zingiber officinale* Roscoe). *Food and Bioproducts Processing*, *90*, 515–524.

Dorman, H. J. D., Peltoketo, A., Hiltunen, R., & Tikkanen, M. J. (2003). Characterisation of the antioxidant properties of deodourised aqueous extracts from selected Lamiaceae herbs. *Food Chemistry*, *83*, 255–262.

Dua, A., Gaurav, G., Balkar, S., & Mahajan, R. (2013). Antimicrobial properties of methanolic extract of cumin (*Cuminum cyminum*). *seeds. International Journal of Research in Ayurveda and Pharmacy*, *4*(1), 104–107.

Duman, A. D., Ozgen, S. K. M., Dayisoylu, K. S., Eribl, N., & Durgac, C. (2009). Antimicrobial activity of six Pomegranate (*Punica granatum* L.) varieties and their relation to some of their pomological and phytonutrient characteristics. *Molecules (Basel, Switzerland)*, *14*, 1808–1817.

Fattahifar, E., Barzegar, M., Ahmadi Gavlighi, H., & Sahari, M. A. (2018). Evaluation of the inhibitory effect of pistachio (*Pistacia vera* L.) green hull aqueous extract on mushroom tyrosinase activity and its application as a button mushroom postharvest anti-browning agent. *Postharvest Biology and Technology*, *145*, 157–165.

Fernandez-Lopez, J., Sevilla, L., Sayas-Barbera, E., Navarro, C., Marin, F., & Perez-Alvarez, J. A. (2003). Evaluation of the antioxidant potential of hyssop (*Hyssopus officinalis* L.) and rosemary (*Rosmarinus officinalis* L.) extracts in cooked pork meat. *Journal of Food Science*, *68*, 660–664.

Gülçin, I., Küfrevioglu, Ö. I., & Oktay, M. (2005). Purification and characterization of polyphenol oxidase from nettle (*Uritca dioica* L.) and inhibitory effects of some chemicals on enzyme activity. *Journal of Enzyme Inhibition and Medicinal Chemistry*, *20*, 297–302.

Gülçin, I., Küfrevioğlu, Ö. I., Oktay, M., & Büyükokuroğluc, M. E. (2004).). Antioxidant, antimicrobial, antiulcer and analgesic activities of nettle (*Urtica dioica* L.). *Journal of Ethnopharmacology*, *90*, 205–215.

Hayes, J. E., Stepanyan, V., Allen, P., O'Grady, M. N., & Kerry, J. P. (2010a). Effect of lutein, sesamol, ellagic acid and olive leaf extract on the quality and shelf-life stability of packaged raw minced beef patties. *Meat Science*, *84*, 613–620.

Hayes, J. E., Stepanyan, V., Allen, P., O'Grady, M. N., & Kerry, J. P. (2010b). Evaluation of the effects of selected phytochemicals on quality indices and sensorial properties of raw and cooked pork stored in different packaging systems. *Meat Science*, *85*, 289–296.

Hayes, J. E., Stepanyan, V., Allen, P., O'Grady, M. N., & Kerry, J. P. (2011). Evaluation of the effects of selected plant-derived nutraceuticals on the quality and shelf-life stability of raw and cooked pork sausages. *LWT—Food Science and Technology*, *44*, 164–172.

Hill, L. E., Gomes, C., & Taylor, T. M. (2013). Characterization of beta-cyclodextrin inclusion complexes containing essential oils (trans-cinnamaldehyde, eugenol, cinnamon bark, and clove bud extracts) for antimicrobial delivery applications. *LWT—Food Science and Technology*, *51*(1), 86–93.

Hosseininejad, Z., Moghadam, S. D., Ebrahimi, F., Abdollahi, M., Zahedi, M. J., Nazari, M., ... Sharififar, F. (2011). In vitro screening of selected Iranian medicinal plants against *Helicobacter pylori*. *International Journal of Green Pharmacy*, *5*, 282–285.

Kanatt, S. R., Chander, R., & Sharma, A. (2007). Antioxidant potential of mint (*Mentha spicata* L.) in radiation-processed lamb meat. *Food Chemistry*, *100*, 451–458.

Karre, L., Lopez, K., & Getty, K. J. K. (2013). Natural antioxidants in meat and poultry products. *Meat Science*, *94*, 220–227.

Khan, R., Islam, B., Akram, M., Shakil, S., Ahmad, A., Ali, S. M., ... Khan, A. U. (2009). Antimicrobial activity of five herbal extracts against multi drug resistant (MDR) strains of bacteria and fungus of clinical origin. *Molecules (Basel, Switzerland)*, *14*, 586–597.

Kim, M. J., Kim, C. Y., & Park, I. (2005). Prevention of enzymatic browning of pear by onion extract. *Food Chemistry*, *89*, 181–184.

Kothari, V. (2011). In vitro antibacterial activity in Seed extracts of *Phoenix sylvestris* Roxb (Palmae), and *Tricosanthes dioica* L (Cucurbitaceae). *Current Trends in Biotechnology and Pharmacy*, *5*(1), 993–997.

Kothari, V., Pathan, S., & Seshadri, S. (2010). Antioxidant activity of *M. zapota* and *C. limon* seeds. *Journal of Natural Remedies*, *10*(2), 175–180.

Lee, B., Seo, J. D., Rhee, J. K., & Kim, C. Y. (2016). Heated apple juice supplemented with onion has greatly improved nutritional quality and browning index. *Food Chemistry*, *201*, 315–319.

Lee, M. K., Kim, Y. M., Kim, N. Y., Kim, G. N., Kim, S. H., Bang, K. S., & Park, I. (2002). Prevention of browning in potato with a heat-treated onion extract. *Bioscience, Biotechnology, and Biochemistry*, *66*, 856–858.

Liu, X., Yang, Q., Lu, Y., Li, Y., Li, T., Zhou, B., & Qiao, L. (2019). Effect of purslane (*Portulaca oleracea* L.) extract on anti-browning of freshcut potato slices during storage. *Food Chemistry*, *283*, 445–453.

Lozano-de-Gonzalez, P. G., Barrett, D. M., Wrolstad, R. E., & Durst, R. W. (1993). Enzymatic browning inhibited in fresh and dried apple rings by pineapple juice. *Journal of Food Science*, *58*, 399–404.

Mahboubi, A., Asgarpanah, J., Sadaghiqani, P. N., & Faizi, M. (2015). Total phenolic and flavonoid content and antibacterial activity of *Punica granatum* L. Var. pleniflora flower (Golnar) against bacterial strains causing food borne diseases. *BMC Complementary and Alternative Medicine*, *15*, 366–373.

Mahfuzul_Hoque, M., Bari, M. L., Juneja, V. K., & Kawamoto, S. (2007). Antimicrobial activity of cloves and cinnamon extracts against food borne pathogens and spoilage bacteria and inactivation of *Listeria monocytogenes* in ground chicken meat with their essential oils. *Journal of Food Science and Technology*, *72*, 9–21.

Mansour, E. H., & Khalil, A. H. (2000). Evaluation of antioxidant activity of some plant extracts and their application to ground beef patties. *Food Chemistry*, *69*, 135–141.

Martinez, M. V., & Whitaker, J. R. (1995). The biochemistry and control of enzymatic browning. *Trends Food Science and Technology*, *6*, 195–200.

Modarresi-Chahardehi, A., Ibrahim, D., Fariza-Sulaiman, S., & Mousavi, L. (2012). Screening antimicrobial activity of various extracts of *Urtica dioica*. *Revista de Biologia Tropical*, *60*(4), 1567–1576.

Mohdaly, A., Sarhan, M., Mahmoud, A., Ramadan, M., & Smetanska, I. (2010). Antioxidant efficacy of potato peels and sugar beet pulp extracts in vegetable oils protection. *Food Chemistry*, *123*(4), 1019–1026.

Moon, K. M., Kwon, E. B., Lee, B., & Kim, C. Y. (2020). Recent trends in controlling the enzymatic browning of fruit and vegetable products. *Molecules (Basel, Switzerland)*, *25*, 2754. Available from https://doi.org/10.3390/molecules25122754.

Muthukumar, M., Naveena, B. M., Vaithiyanathan, S., Sen, A. R., & Sureshkumar, K. (2014). Effect of incorporation of *Moringa oleifera* leaves extract on quality of ground pork patties. *Journal of Food Science and Technology*, *51*, 3172–3180.

Nazir, F., Salim, R., Yousf, N., Bashir, M., Naik, H. R., & Hussain, S. Z. (2017). Natural antimicrobials for food preservation. *Journal of Pharmacognosy and Phytochemistry*, *6*(6), 2078–2082.

Nikmaram, N., Budaraju, S., Barba, F. J., Lorenzo, J. M., Cox, R. B., Mallikarjunan, K., & Roohinejad, S. (2018). Application of plant extracts to improve the shelf-life, nutritional and health-related properties of ready-to-eat meat products. *Meat Science*, *145*, 245–255.

Nowak, A., Czyzowska, A., Efenberger, M., & Krala, L. (2016). Polyphenolic extracts of cherry (*Prunus cerasus* L.) and blackcurrant (*Ribes nigrum* L.) leaves as natural preservatives in meat products. *Food Microbiology*, *59*, 142–149.

Othman, M., Loh, H. S., Wiart, C., Khoo, T. J., Lim, K. H., & Ting, K. N. (2011). Optimal methods for evaluating antimicrobial activities from plant extracts. *Journal of Microbiological Methods*, *84*(2), 161–166.

Ozvural, E. B., & Vural, H. (2012). The effects of grape seed extract on quality characteristics of frankfurters. *Journal of Food Processing and Preservation*, *36*, 291–297.

Pandey, A., & Singh, P. (2011). Antibacterial activity of *Syzygium aromaticum* (Clove) with metal ion effect against food borne pathogens. *Asian Journal of Plant Science and Research*, *1*(2), 69–80.

Peng, X., Ma, J., Cheng, K. W., Jiang, Y., Chen, F., & Wang, M. (2010). The effects of grape seed extract fortification on the antioxidant activity and quality attributes of bread. *Food Chemistry*, *119* (1), 49–53.

Piskernik, S., Klancnik, A., Riedel, Ch. T., Brøndsted, L., & Smole Mozina, S. (2011). Reduction of *Campylobacter jejuni* by natural antimicrobials in chicken meat-related conditions. *Food Control*, *22*, 718–724.

Puravankara, D., Bohgra, V., & Sharma, R. S. (2000). Effect of antioxidant principles isolated from mango (*Mangifera indica* L.) seed kernels on oxidative stability of buffalo ghee (butter-fat). *Journal of the Science of Food and Agriculture*, *80*, 522–526.

Qader, M. K., Khalid, N. S., & Abdullah, A. M. (2013). Antibacterial activity of some plant extracts against clinical pathogens. *International Journal of Microbiology and Immunology Research.*, *1*(5), 53–56.

Ranasinghe, P., Pigera, S., Premakumara, G. A. S., Galappaththy, P., Constantine, G. R., & Katulanda, P. (2013). Medicinal properties of 'true' cinnamon (*Cinnamomum zeylanicum*): A systematic review. *BMC Complementary and Alternative Medicine*, *13*, 275. Available from https://doi.org/10.1186/1472-688213-275.

Reddy, G. V. B., Sen, A. R., Nair, P. N., Reddy, K. S., Reddy, K. K., & Kondaiah, N. (2013). Effects of grape seed extract on the oxidative and microbial stability of restructured mutton slices. *Meat Science*, *95*, 288–294.

Reddy, V., Urooj, A., & Kumar, A. (2005). Evaluation of antioxidant activity of some plant extracts and their application in biscuits. *Food Chemistry*, *90*(1), 317–321.

Rojas, M. C., & Brewer, M. S. (2007). Effect of natural antioxidants on oxidative stability of cooked, refrigerated beef and pork. *Journal of Food Science*, *72*, S282–S288.

Rojas, M. C., & Brewer, M. S. (2008). Effect of natural antioxidants on oxidative stability of frozen, vacuum-packaged beef and pork. *Journal of Food Quality*, *31*, 173–188.

Sadeghian, A., Ghorbani, A., Mohamadi_Nejad, A., & Rakhshandeh, H. (2011). Antimicrobial activity of aqueous and methanolic extracts of pomegranate fruit skin. *Avicenna Journal of Phytomedicine.*, *1*(2), 67–73.

Shah, M. A., Bosco, S. J. D., & Mir, S. A. (2014). Plant extracts as natural antioxidants in meat and meat products. *Meat Science*, *98*, 21–33.

Shah, M. A., Bosco, S. J. D., & Mir, S. A. (2015). Effect of *Moringa oleifera* leaf extract on the physicochemical properties of modified atmosphere packaged raw beef. *Food Packaging and Shelf Life*, *3*, 31–38.

Shan, B., Cai, Y. Z., Brooks, J. D., & Corke, H. (2009). Antibacterial and antioxidant effects of five spice and herb extracts as natural preservatives of raw pork. *Journal of the Science of Food and Agriculture*, *89*, 1879–1885.

Siddhuraju, P., & Becker, K. (2003). Antioxidant properties of various solvent extracts of total phenolic constituents from three different agroclimatic origins of drumstick tree (*Moringa oleifera* Lam.) leaves. *Journal of Agricultural and Food Chemistry*, *51*, 2144–2155.

Sikora, E., Cieslik, E., Leszcznska, T., Filipiak-Florkiewicz, A., & Pisulewski, P. M. (2008). The antioxidant activity of selected cruciferous vegetables subjected to aquathermal processing. *Food Chemistry*, *107*, 55–59.

Singh, R., Shushni, M. A. M., & Belkheir, A. (2015). Antibacterial and antioxidant activities of *Mentha piperita* L. *Arabian Journal of Chemistry*, *8*(3), 322–328.

Sofia, K., Prasad, R., Vijay, V. K., & Srivastava, A. K. (2007). Evaluation of antibacterial activity of Indian spices against common foodborne pathogens. *International Journal of Food Science and Technology*, *42* (8), 910–915.

Soković, M., Glamočlija, J., Marin, P. D., Brkić, D., & van Griensven, L. J. L. D. (2010). Antibacterial effects of the essential oils of commonly consumed medicinal herbs using an in vitro model. *Molecules (Basel, Switzerland)*, *15*, 7532–7546.

Stalikas, C. D. (2007). Extraction, separation, and detection methods for phenolic acids and flavonoids. *Journal of Separation Science*, *30*, 3268–3295.

Sun, X. H., Hao, L. R., Xie, Q. C., Lan, W. Q., Zhao, Y., Pan, Y. J., & Wu, V. C. H. (2020). Antimicrobial effects and membrane damage mechanism of blueberry (*Vaccinium corymbosum* L.) extract against *Vibrio parahaemolyticus*. *Food Control*, *111*, 107020. Available from https://doi.org/10.1016/j.foodcont.2019.107020.

Tayel, A. A., & El-Tras, W. F. (2012). Plant extracts as potent biopreservatives for *Salmonella typhimurium* control and quality enhancement in ground beef. *Journal of Food Safety*, *32*, 115–121.

Tekwu, E. M., Askun, T., Kuete, V., Nkengfack, A. E., Nyasse, B., Etoa, F. X., & Beng, V. P. (2012). Antibacterial activity of selected Cameroonian dietary spices ethno-medically used against strains of mycobacterium tuberculosis. *Journal of Ethnopharmacology*, *142*, 374–382.

Tinello, F., Mihaylova, D., & Lante, A. (2020). Valorization of onion extracts as anti-browning agents. *Food Science and Applied Biotechnology*, *3*(1), 16–21.

Turker, A. U., & Usta, C. (2008). Biological screening of some Turkish medicinal plant extracts for antimicrobial and toxicity activities. *Natural Product Research*, *22*, 136–146.

Turkmen, N., Sari, F., & Velioglu, Y. S. (2006). Effects of extraction solvents on concentration and antioxidant activity of black and black mate tea polyphenols determined by ferrous tartrate and Folin–Ciocalteu methods. *Food Chemistry*, *99*, 835–841.

Upadhyay, R., Ramalakshmi, K., & Rao, L. J. M. (2012). Microwave-assisted extraction of chlorogenic acids from green coffee beans. *Food Chemistry*, *130*(1), 184–188.

Verma, V., Singh, R., Tiwari, R. K., Srivastava, N., & Verma, S. (2012). Antibacterial activity of extracts of Citrus, Allium and Punica against food borne spoilage. *Asian Journal of Plant Science and Research.*, *2* (4), 503–509.

Wang, Y. Z., Fu, S. G., Wang, S. Y., Yang, D. J., & Wu, Y. H. S. (2018). Effects of a natural antioxidant, polyphenol-rich rosemary (*Rosmarinus officinalis* L.) extract, on lipid stability of plant-derived omega-3 fatty-acid rich oil. *LWT—Food Science and Technology*, *89*, 210–216.

Wei, Q., Wolf-Hall, C., & Hall, C. A. (2009). Application of raisin extracts as preservatives in liquid bread and bread system. *Journal of Food Science*, *74*, M177–M184.

Yeh, H. Y., Chuang, C. H., Chen, H. C., Wan, C. J., Chen, T. L., & Lin, L. Y. (2014). Bioactive components analysis of two various gingers (*Zingiber officinale* Roscoe) and antioxidant effect of ginger extracts. *LWT—Food Science and Technology*, *55*, 329–334.

Zocca, F., Lomolino, G., & Lante, A. (2011). Dog rose and pomegranate extracts as agents to control enzymatic browning. *Food Research International*, *44*(4), 957–963.

CHAPTER

Plant extracts as nutrient enhancers

7

Nirmal Kumar Meena[1], Kanica Chauhan[2], Manohar Meghwal[3] and Anju Jayachandran[3]

[1]Department of Fruit Science, College of Horticulture and Forestry, Jhalawar, India [2]Department of Forest Products and Utilization, College of Horticulture and Forestry, Jhalawar, India [3]Department of Fruit Science, College of Agriculture, Thrissur, India

7.1 Introduction

The plant extracts represented by the medicinal, spice, and aromatic plants participate in the daily health and food actions through its numerous active principles and biocomplexes extracted from different methods and processes involving different solvents, allowing the study of the extracts in the condition of raw or purified drug, starting from nature or green plants or from dehydrated plants (Moreira et al., 2005; Passos et al., 2009). The plant extracts are used in the pharmaceutical and cosmetic industries, due to their medicinal and aromatic properties, and some species are used in cooking (Barbieri and Stumpf, 2005; Handa, 2008; Mir et al., 2019). These are primary raw material for folk medicine. The plant extracts contain several vitamins, minerals, antioxidants (AOX), polyphenols, alkaloids, and terpenes (Fig. 7.1).

Plant extracts are an important source of phytonutrients that are beneficial to human health as they have been used since ancient time in traditional drugs (Cunningham, 2015, Loizzo, 2016). These phytochemicals such as vitamins, minerals, polyphenols, AOX, alkaloids, and terpenes possess antioxidant activity (Wong and Chye., 2009), antibacterial (Nair et al., 2005), antifungal (Khan and Wassilew, 1987), antidiabetic (Kumar et al., 2008a; Singh and Gupta, 2007), antiinflammatory (Kumar et al., 2008b), antiarthritic (Kumar et al., 2008c), and radio-protective activities (Jagetia et al., 2005). Owing to these properties, they are largely used for medicinal purpose. Besides, they are important to regulate certain physiological and cellular processes.

The use of plant extracts are becoming more popular in the food industry due to the presence of certain bioactive compounds such as polyphenols, carotenoids, flavonoids, and several other secondary metabolites, which have AOX-like activities against low-density lipoproteins and DNA oxidative changes (Kiokias et al., 2018; Proestos and Varzakas, 2017; Proestos, 2020; Tonthubthimthong et al., 2001). Nevertheless, these bioactive compounds involve in many physiological functions like growth, cell enlargement, tissues formation, and other defense activities. These are extracted synthetically and widely used in food industry to develop functional foods. Plant extracts have been used in cosmetics, pharmaceuticals, and more recently in nutraceutical industries. However, characterization and proper identification of individual compounds are necessary to develop de novo products.

Plant Extracts: *Applications in the Food Industry*. DOI: https://doi.org/10.1016/B978-0-12-822475-5.00003-X

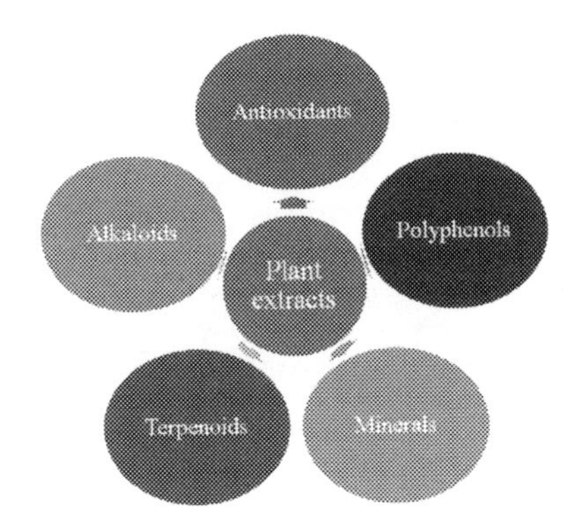

FIGURE 7.1

Plants extracts and inherent bioactive constituents.

The phytochemical investigation of a plant may involve the following steps: authentication and extraction of the plant material, separation and isolation of the constituents of interest, characterization of the isolated compounds, and quantitative evaluation (Evans, 2008). Extraction, isolation, purification, and characterization again become tedious job and vary from plant to plant and part. Certain extraction methods such as using solvents, microwave-assisted extraction, ultrasonic extraction have been identified (Altemimi et al., 2017). More recently developed instruments like high performance liquid chromatography (HPLC), Mass spectroscopy, nuclear magnetic resonance, UV−visible infrared technique, etc. can identify the isolated compound (Altemimi et al., 2017). However, more epidemiological studies are needed to validate the strong mechanism and functioning of these bioactive compounds in human health. This chapter provides an overview of inherent bioactive compounds in plant extracts and their role in human health.

7.2 Plant extracts as sources of vitamins

Plant-derived vitamins are of great interest because of their impact on human health. They are essential for metabolism because of their redox chemistry and role as enzymatic cofactors, not only in animals but also in plants. Several vitamins have strong antioxidant potential, including both water-soluble (vitamins B and C) and lipid-soluble (vitamins A, E and K) compounds. As one of the seven major nutrients, vitamins play important roles in the body. Vitamins are involved in the processes of normal metabolism and cell regulation, and they are necessary for growth and development, thus they are chemicals that we all need to stay healthy (Glavinic et al., 2017; Khayat et al., 2017). There are thirteen vitamins that are recognized as playing roles in human nutrition (Eggersdorfer et al., 2012). Based on their solubility, these vitamins can be divided into fat-soluble vitamins and water-soluble vitamins. The former contains vitamin A, D, E, and K, while the latter group includes the B-complex and C vitamins. A number of biological functions in the body have been associated with the fat-soluble vitamins (Lounder et al., 2017). Once the amount of vitamins cannot meet the body's needs, the vitamins must be supplied from the diet.

It has been reported that vitamin B1 (thiamin) is lost in food due to leaching and blenching on increased temperature (Serafini et al., 2002). According to a German study fruits and vegetables looses 31% of thiamin due to over cooking about 9% loose is by microwave treatment (Aurea & Samuel, 1998; Hertog et al., 1993). Plant extracts are the key material for traditional drug preparation and folk medicines. These extracts contain many vitamins and other phytochemicals which have therapeutic uses. Sunarić et al. (2020) isolated thiamine and riboflavin from different wild plant extracts with HPLC fluorescence detection. They reported that riboflavin concentration was ranged from 0.84–20.4 μg/g of dry extract whereas thiamine level was 0.06–0.67 μg/g. the highest riboflavin content was reported in wild garlic extract followed by rose-hip extract. Thiamine content was higher in elderberry extract. Datta et al. (2019) determined water-soluble vitamins in different edible plants and their extracts. They found that vitamin B group was the dominant in all species and B1, B2, B6 and B9 were present in all the extracts. They also reported that Vitamin B5 was not detected in any edible plant whereas ascorbic acid was reported to be available in *Asystasia ganjetica*, *Oldenlandia corymbosa* with maximum value in *Achyranthes aspera* (151.75 mg/100 g dry plant material). Similarly, *Ipomoea aquatica* had highest vitamin B1, B6 and B9. Ascorbic acid is well known for its AOX properties and its anti infection properties. It reported that *Dryopteris cochleata* leaf extract showed higher ascorbic acid acetone and ethylacetate extracts (157.37 and 156.04 μg/mL, respectively) (Kathirvel & Sujatha, 2016).

7.3 Plant extracts as sources of minerals

Mineral elements also are needed in minute quantities for the proper functioning of the human system, health growth, and development (Igwenyi et al., 2014). Naturally grown herbs and plants also have plenty of macro and trace elements, which are extremely valuable for our body and good health. They play an important role in cell metabolism and antioxidant system (Stern et al., 2007). It has been well documented that intake of minerals such as Ca, Mg, K, Fe, etc. reduce the risk of stroke, hypertension and also involved in oxygen transport system as a enzyme cofactor (Bongoni, et al., 2013; Janz et al., 2013; Larsson et al., 2008). Plant minerals such as selenium, copper, Zn, and Mn are well known for their antioxidant functions (Nikmaram et al., 2018). Calcium is one of the mineral believed to be an important factor governing fruit storage quality (Lechaudel et al., 2005). Ca is the main constituent of the skeleton and is important for regulating many vital cellular activities such as nerve and muscle function, hormonal actions, blood clotting, and cellular mortality (Yagi et al., 2013). Calcium is essential for healthy bones, teeth and blood (Charles, 1992). Phosphorous maintain blood sugar level, normal heart contraction dependent on phosphorous (Linder, 1991) also important for normal cell growth and repair. It helps in the process of ossification of bones by getting deposited in the form of calcium phosphate (Indrayan et al., 2005). Iron is the most well known in biological system. It performs a wide range of biological functions. Iron occupies a unique role in the metabolic process. The role of iron in the body is clearly associated with hemoglobin and the transfer of oxygen from lungs to the tissue cells (Janz et al., 2013; Prajna & Rama, 2015).

Xu et al. (2008) investigated that prolonged time of extraction and temperature could yield more minerals than the phenols and AOX in Pokan and Satsuma mandarin. They also reported that K was the chief mineral presented in the citrus peel extracts followed by Ca and Mg and Satsuma mandarin contained higher minerals element than the Ponkan peel extract. Monisha and Ragavan (2015) estimated mineral content in polyherbal extracts of various plants and revealed that Fe (4.6 mg/g), P (3.4 mg/g), Ca (2.0 mg/g), and K (1.15 mg/g) were present in the extract. Staszowska-Karkut and Materska (2020) reported mineral content in berry plants extracts and found that minerals such as potassium (K), calcium (Ca), magnesium (Mg), phosphorus (P), sodium (Na), and iron (Fe) copper (Cu), zinc (Zn), manganese (Mn), and boron (B) are present in the leaf extracts. Black currant leaf extract produced highest Ca content followed by Mg, P and Fe when compared with raspberry and Aronia extracts. Raspberry extracts was richest in K, B, and Na whereas Aronia in Mn. Nour et al. (2014) found that time of harvesting greatly influences the level of minerals and they found maximum mineral content (especially Ca) in the month of June in raspberry extract. It has been also documented that leaves contain more amount of minerals than the fruits (Janz et al., 2013). Guenane et al. (2020) studied the mineral content from the Algerian medicinal plants and remarkably reported the high amounts of K, Ca, Na, and Mg. The lowest content of Na that is 7168.46 ± 216.03 mg/kg was in *Marrubium vulgare* and maximum concentration was estimated at 24,395.16 ± 34.41 mg/kg in *Malva parviflora*. The calcium contents varied between 3362.38 ± 442.00–110,354.91 ± 3446.46 mg/kg, the highest calcium content was obtained for *M. parviflora* (110,354.91 ± 3446.46 mg/kg), followed by *Artemisia absinthium* (50,200.38 ± 2529.01 mg/kg) and *Scorzonena undulata* (49,648.51 ± 3311.26 mg/kg).

All the plant parts have nutritional qualities, which when used in the right proportions could be of tremendous benefit to the body. Mineral analysis conducted by Imelouane et al. (2011) showed that the *Thymus vulgaris* contained calcium (313,044 mg/kg), magnesium (53,873 mg/kg), potassium (279,491 mg/kg), iron (27,095 mg/kg), zinc (415.33 mg/kg), and manganese (731.41 mg/kg). Ponmari and Balasubiramanian (2017) studied five medicinally important plants such as *Glycyrrhiza glabra* (L.), *Gymnema sylvestre* (R.Br.), *Solanum trilobatum* (L.), *Alpinia calcarata* (Rox.), and *Centella asiatica* (L) and revealed that all the medicinal plants possess the highest Mg and lowest Zn content was observed in all the five studied plants except *Alpinia calcarata*. Mineral content in *Astragalus* spp. extracts hav been estimated to differentiate the composition at different growth stages. They have reported higher iron content at vegetative stage 65.29 and 30.19 mg Fe/100 g, in *Astragalus glycyphyllos* and *Astragalus cicer*, respectively, whereas Fe concentration in flowering plants amounted to 21.72 and 14.94 mg/100 g. The study suggests that at early phonological stages of the plants extract contain higher macro and micromineral contents (Butkutė et al., 2018). Sugar maple tree bark extracts showed higher mineral content (K, Ca, Mg, P, Na, Fe, and Cu) and lower Zn and Mn level than the red maple bark extract (Bhatta et al., 2018). Biel et al. (2020) evaluated mineral profile of globe artichoke leaf extracts and revealed that extract contained K (506.3 mg/100 g DM), P (414 mg/100 g DM), Ca (386.9 mg Ca/100 g DM), Mg in descending order, whereas Zn, Fe, Cr, and Mn were prominent among micronutrients. Due to these, artichoke extract is increasing in demand globally. The content of mineral elements in plants depends to a high degree on the soils abundance, cultivation techniques, soil fertility, nutritional status and extraction (Kruczek, 2005; Xu et al., 2008). Plant extract contains abundant minerals in sufficient quantity and can be used directly as infusion of as food additives.

7.4 Plant extracts as sources of antioxidants

The plants and their products are found throughout human history as herbal supplements as botanicals, nutraceuticals, and drugs (Ekor, 2014). In whole population of the world, about 60%–80% of the population still relies on conventional medicine for the healing of familiar diseases (Ravishankar & Shukla, 2007). Folklore medicine are as a source of primary health care and the chief reason for the use of folk medicine is the accessibility, affordability and cultural beliefs (Tag et al., 2012). Plants have been found of great importance due to their medicinal and nutritional properties with a primary source of bioactive compounds. The use of synthetic and natural food AOX regularly in medicine and foods particularly those having fats and oils to shield the food from oxidation. Butylated hydroxytoluene and butylated hydroxyanisole are the synthetic and natural food AOX which have been used extensively in cosmetic, food and therapeutic industries. But, owing to their instability at high temperatures, high volatility, synthetic antioxidant's carcinogenic behavior, users' inclinations led to shift in the consideration of producers or manufacturers from man-made to natural AOX (Papas, 1999). A variety of medicinal plants extracts have been reported to reveal antioxidant activity, including *Allium sativum, Zingiber officinale, Crocus sativus, Dodonaea viscose, Barleria noctiflora, Anacardium occidentale, Datura fastuosa, Caesalpinia bonducella*, and many more. Numerous AOX identified as active oxygen scavengers or free radicals, obtained naturally from the plant sources are used in food, cosmetic and remedial purposes proved to be brilliant alternatives for man-made AOX because of their inexpensiveness, and have no any harmful effect on human body.

To defy the detrimental effects of reactive oxygen species (ROS), plants have a powerfully built enzymatic and nonenzymatic scavenging pathway. Enzymes included are catalase, superoxide dismutase, ascorbate peroxidase, glutathione reductase, glutathione-S-transferase, dehydroascorbate reductase, monodehydroascorbate reductase, peroxidases, and glutathione peroxidase. Nonenzymatic compounds include glutathione, carotenoids, tocopherols, and ascorbate (AsA). There are unambiguous, well-synchronized ROS generating and scavenging systems present in different organelles of the plant cells. Lesser levels of ROS comparatively act as signaling essences that arouses abiotic stress tolerance by altering the expression of resistant genes. In plants, elevated levels of AOX have been accounted to demonstrate better resistance to different types of environmental stresses (Hasanuzzaman et al., 2012).

Plant produces many nonnutritive secondary metabolites, which have AOX-like effects. AOX are the compounds that protect human against several diseases by scavenging free radical activity. They also protect cell membrane and macromolecules by same mechanism (Wu et al., 2017a). Studies have been investigated that plant extracts like curcumin and ellagic acid are used as food supplements and have antimicrobial activities thus prevent the metabolism of aflatoxin B1 (AFB1) and increase the activity of glutathione-S-transferase involved in the detoxification of xenobiotics (Makhuvele et al., 2020). Plant extracts and flower extracts of *Astragalus* spp. contained AOX level ranges from 7.52 to 35.64 μmol/g (Butkutė et al., 2018). They also showed the highest antioxidant capacity in flower extracts followed by leaves and tem extracts. *A. cicer* extracts had higher 2,2-diphenyl-1-picrylhydrazyl (DPPH) scavenging activity than the *A. glycyphyllos* (Butkutė et al., 2018). Several studies showed that the extracts of different spice plants highly effective against food born diseases causing bacteria's (Prashanth et al., 2001; Mahfuzul Hoque et al., 2007; Negi et al., 2003). The study conducted by Bhatta et al. (2018) reported that red maple bark extracts showed higher antioxidant activity than that of sugar maple bark extracts. A study by Biel et al. (2020) was conducted to find out AOX activity of globe artichoke plant extracts. The study revealed that the ABTS++ assay showed higher antioxidant and very high radical scavenging capacity (79.74%) with AOX capacity at 1060.8 Trolox/1 g DM. Some researchers investigated AOX properties of *Bletilla striata* plant extracts and its polysaccharides in vitro and demonstrated that it has definite DPPH radical, hydroxyl radical, and superoxide anion systems ultrasonic microwave synergistic extraction had higher DPPH and hydroxyl radical activity (Cai et al., 2016; Qu et al., 2016). Extracts from *Kadsura* spp. had shown rich AOX value and key enzyme inhibition properties, thus fruits and other parts extracts could be utilized in food application (Sritalahareuthai et al., 2020). It has been reported that content of AOX positively correlated with the extraction temperature and elevated extraction temperature helps in forming browning compounds as a result of Maillard reaction, thus the level of AOX increases (Anese, Nicoli, et al., 1999; Anese, Manzocco, et al., 1999; Ju & Howard, 2005; Nicoli et al., 1997; O'Brien et al., 1989). Similarly, higher AOX were observed in grape skin extracts when value determined at higher temperature (Ju & Howard, 2005). Datta et al. (2019) performed a study on six medicinal herbs extracts to differentiate the AOX value from various solvent extraction methods and revealed that 70% hydroethanol was proved to better in terms of higher AOX yield followed by methanol. They also reported that Maximum DPPH radical scavenging ability was observed *A. ganjetica* (39.14%) and *A. aspera* (28.051%) whereas FRAP micromole Trolox equivalent (TE)/g was higher in *A. aspera* than the *A. ganjetica*.

7.5 Plant extracts as sources of polyphenols

Polyphenols are important secondary metabolites in plants, which have strong free radical scavenging capacity. Owing to their structural arrangement especially (−OH) group makes them antimicrobial in nature (Gyawali & Ibrahim, 2014). Fruits peel, leaves, bark, and root extracts are rich source of polyphenols and contain in a wide range. Phenolic compounds from the natural extracts gained significance importance due to their antioxidant potential and effective against different diseases. Medicinal herbs, tea, coffee, artichoke, leafy vegetables and many fruit plant extracts contain the strong polyphenolic content and its derivatives. Many of them are still in practice as folk medicines in different parts of the world. However, these are more common in rural areas due to their food supplement and primary medicine. Besides, these play crucial role in plant developmental processes and help to plant cope during adverse climatic conditions especially during stresses. Flavonoids are indispensable component found in variety of plants, fruits, vegetables, herbs and their extracts. Current trends and research outcomes on flavonoids suggest that they exert their important role in functioning of lowering down the coronary heart diseases, cardiovascular mortality rate and have great potential to modulate key cellular enzymatic functions (Panche et al., 2016). Tannins are polyphenolics compounds and are classified into hydrolysable and condensed type on the basis of their biological activity and functions. Tannins possess antioxidant value, free radical scavenging activity, antiulcerogenic activities, and antimicrobial and gastroprotective properties (Amarowicz et al., 2004; Ho et al., 2001; Koleckar et al., 2008). Condensed tannins are more common and abundant in plants and they represent value of common commercial tannin. Several tannins are found in vegetables and are being used as folk medicine. Rubus is another common source of ellagitannin. Some of the tannin like tannic acid and gallic acid is allowed to add in food products as a food additive. Plant phenols are classes of variable organized natural products that are well known to have important antimicrobial and antioxidant activities for their beneficial effects on health (Prashith et al., 2010; Sahu & Mahato, 1994). The primary metabolite is also a precursor to bioactive compounds used as medicinal drugs (Ebi & Ofoefule, 2000). In general, the bioavailability of polyphenols, or the quantity of polyphenols that are consumed unchanged, determines their biological function. After being absorbed, polyphenols could also pass through the gastrointestinal system, thereby affecting the intestinal micro biota. This can have two consequences: first, the active form of polyphenols is modified; Second, the composition of the intestinal micro biota is changed, which is likely to inhibit pathogenic bacteria and enrich beneficial bacteria. Therefore polyphenols have a significant influence on human health (Abbas et al., 2017).

The phloroglucinol and pyrogallol compounds isolated along with ferulic, vanillic, p-coumaric, and caffeic acids constitute the antioxidant activity of the plant (Mazumder et al., 2003). Shikimic acid, gallic acid, B-sitosterol, tannic acid, chebulic acid triethyl ester, gallic acid ethylester, and ellagicethaedioic acid has been confirmed to be in *Terminalia chebula* (Ates & Erdourul, 2003). There are several components in the Terminalia plant, such as tannins, flavonoids, sterols, amino acids, fructose, resin, and fixed oils. Compounds such as anthraquinones, 4, 2, 4 chebulyl-glucopyranose, terpinene, and terpineneol are also found (Srivastav et al., 2010). They can be divided into several groups. Phenols and phenolic acids vary from having several substitutions and hydroxylations to becoming a basic phenol ring with a single substitution such as cinnamic and caffeic acids. The site and degree of hydroxylation are shown to interact with the toxicity of the secondary metabolite. The further oxidized the structure is, the metabolite appears to be more inhibitory (Cowan, 1999). Phenolic inhibition pathways involve inhibiting enzymes. This inhibition is suggested to take place by reactions to the proteins with sulfhydryl groups (Coppo & Marchese, 2014; Cowan, 1999).

It was observed that ethanolic extracts of *Syrian propolis* extracts contain several active compounds including phenolic acids and phenolic aldehydes as well as flavonoids and quinones (Harfouch et al., 2016). The existence of five phenolic compounds, including phenolic acids, cinnamic acid and p-coumaric acid, was confirmed by phytochemical screening using HPLC and ferulic acid as well as catechin and sinapic acid in *Allium ampeloprasum* var. porrumethanolic leaf extract by (Alamri & Moustafa, 2012). The studies confirmed the antibacterial activities in almost all types of polyphenols which were extracted as solvent extraction.

Maple bark is a rich source of phenolics compounds and has glucidase inhibitory and anticancer activities (Yuan et al., 2011, 2012). Several plant parts and their extracts have been used o characterization the content of total phenolics. A diverse value of total phenolic content (TPC) in *Annona crassiflora* has been determined from different extraction methods and different parts (Arruda et al., 2017, 2018; Arruda & Pastore, 2019). Phenolic compounds have been investigated to different AOX activities, anticancer properties, antimicrobial and antiinflammatory properties (Caleja et al., 2017). Tusevski et al. (2014) reported a TPC (15.93 mg GAE/g) in liquorice milkvetch. Similarly, Butkutė et al. (2018) reported that *A. glycyphyllos* extracts from leaves and flowers contained higher amount of phenolic compounds (25.99 and 23.71 mg GAE/g, respectively). Bark extracts from red maple tree shows significant higher amount of TPC (40.12 ± 0.86 g GAE/100 g dry extract) than the white extracts. The study confirmed that the TPC in maple tree was higher than that of some tropical fruits by products in which TPC ranges from (0.37–0.46 g of GAE/100 g dry matter) (Selani et al., 2016). However, green tea leaf extract shows higher TPC (29.8 g GAE/100 g) than the white maple bark extract (Yin et al., 2012). Each and every plant has different amount of phenolic content and its composition and there are several multiple approaches used to characterization and isolation of composition. Many researchers determined the phenolic content in artichoke leaf extract and found 2795 mg CAE (chlorogenic acid equivalents)/100 g DM by Biel et al. (2020), in green globe leaf extract it varies from 8760 to 9561 mg CAE/100 g DM (Wang et al., 2003) and 3167 mg caffeic acid equivalent/100 g DM by Sałata and Gruszecki (2010). Several researchers compared the different solvent extraction process (including water extract) in numerous plants and herbs extracts and found promising results when used methanol as extract solvent (Ademoyegun et al., 2013; Gouveia & Castilho, 2012). TPC from twenty five leafy vegetables was determined by Ademoyegun et al. (2013) and the highest TPC content 164.52 mg GAE/100 g DM was found in methanol extract from *Sesamum radiatum*. It is a well-established fact that polyphenols have very strong antioxidant potential as compare to carotenoids and ascorbic acids (Biel et al., 2020) thus the plant extracts derived polyphenols well recognized with numerous therapeutic uses. As mentioned earlier, the extraction procedure makes differences in the value. A study conducted by Xu et al. (2008) in that hot water extraction of total phenol acids were carried out of citrus peel extracts of ponkan and Satsuma mandarin. They reported that ferulic acid was the dominant phenolic acid in both the citrus peel extracts, however, extract of Ponkan had higher content of caffeic and p-coumaric acid. Grape extracts and seed extracts is very rich in polyphenols and had been reported to be beneficial in reducing the risk of cardiovascular diseases and type-2 diabetes. It has been reported that cinnamon extract induces the glycogen synthase and insulin receptor kinase, triggers glucose uptake with inhibitory action on glycogen synthase kinase-3 beta anddephosphorylation of the insulin receptor thus enhance insulin sensitivity (Lin et al., 2016). Coffee extract rich in chlorogenic acid that has beneficial effects on type -2 diabetes, reason being chlorogenic acid inhibits Na + -dependent glucose transporters, SGLT1 and SGLT2 by interacting with the absorption of glucose from the intestine (Bassoli et al., 2008; Greenberg et al., 2006; Johnston et al., 2005; McCarty, 2005). The study also suggests that elevated extraction time may reduce the amount of TPC and at 100oC some of phenols get destroyed (Xu et al., 2008). Tea leaves extract contains an excellent amount of polyphenols including catechin and gallic acid. Salehi et al. (2019) reviewed the phytochemical composition of berberis plant extract and reported various tannins, polyphenols, essential oils and their application in food industry. Chemical structure of some important phenolic are depicted in Fig. 7.2.

7.6 Plant extracts as sources of alkaloids

Plants have always been a cornerstone for the traditional medicine systems, and they have brought continuous remedies to the mankind for thousands of years. It is observed that plants synthesize many secondary metabolites that support the plants to survive and reproduce (Jing et al., 2014). These secondary metabolites are alkaloids, phenols, steroids, glycosides, tannins, terpenoids, and phytoalexins. Among these, the most important group of secondary metabolites are the alkaloids

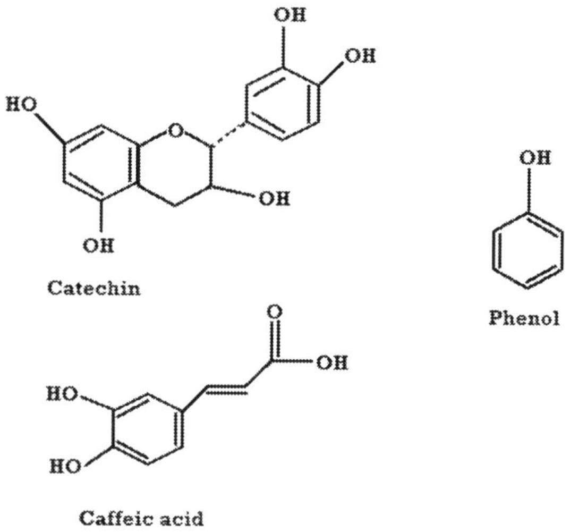

FIGURE 7.2

Chemical structure of catechin, caffeic acid, and basic phenols.

that are known to possess ample therapeutic properties (Roy, 2017). Alkaloids have strong biological effects on human as well as on animals in very small doses. They are found not only in food and drinks but also available as stimulant drugs which showed antiinflammatory, anticancer, analgesics, local anesthetic and pain relief, antimicrobial, antifungal and many other activities. Alkaloids are beneficial as ingredients for foods, supplements, pharmaceuticals, and in several other applications in human life (Kurek, 2019).

Alkaloids are a group of naturally occurring chemical compounds that contain mostly basic nitrogen atom. They are also some properties with neutral (McNaught, D. & Wilkinson, 1997) and even weak acid properties (Manske, 1965). These nitrogen atoms are usually situated in a ring/cyclic structure. The word "alkaloid" is derived from a Latin word "alkali" which is used to describe any nitrogen—containing base. These alkaloids are usually organic bases and can react with acids to form salts (Roy, 2017). Morphine was the first alkaloid to be isolated from opium poppy in crystalline form by a German chemist, Friedrich Serturner in 1804. Generally based on the structures (heterocyclic ring system), alkaloids can be classified into different classes such as indoles, quinolines, isoquinolines, pyrrolidines, pyrrolizidines, tropanes, terpenoids, and steroids. Type of alkaloids and their source are mentioned in Table 7.1.

No such classification exists, but they can be differentiated on the basis of structural pattern like indole alkaloids or precursor such as benzylisoquinoline, tropane, pyrrolizidine, or purine alkaloids (Kennedy & Wightman, 2011). Alkaloids contributed majority of neurotoxins, and traditional psychedelics such as atropine, scopolamine, and hyoscyamine, from the belladonna (*Atropa belladonna*) plant and traditional drugs such as nicotine, caffeine, methamphetamine (ephedrine), cocaine, and opiates from a group of plants for mankind purposes (Goldman, 2001; Zenk & Juenger, 2007). It has been confirmed by various studies that plant based alkaloids are toxic to mammals except caffeine and nicotine which are widely consuming by tea, coffee and tobacco

Table 7.1 Availability of different types of alkaloids in plant extracts.

Name plants	Alkaloids present	References
Annona crassiflora	Aporphine, atherospermidine, liriodenine, stephalagine, annonine	Arruda et al. (2017); Goncalves et al. (2006); Pereira et al. (2017)
Atropa belladonna	Atropine, scopolamine, and hyoscyamine	Goldman (2001)
Barberis vulgaris	Berberine, berberrubine	Rahimi-Madiseh et al. (2017); Yu et al. (2018)
Cammelia sinensis	Caffeine, thearuflavin, catechine	Chen et al. (2008); Guo et al. (2008); Kennedy and Wightman (2011)
Catharanthus roseus	Vincristine, vinblastine	Moudi et al. (2013)
Chondrodendron tomentosum	Tubocurarine	Moudi et al. (2013)
Cinchona officinalis	Quinine	Achan et al. (2011)
Cinchona pubescens	Quinine, quinidine, cinchonine, acronycine, melicopine, melicopidine, and acronycidine	Moore et al. (1995); Tillequin (1997)
Coffea spp.	Caffeine, theobromine, theophylline	Ashihara et al. (2008); Desgagné-Penix (2017); Koyama et al. (2003)
Cola acuminate	Kolatine, kolateine	Ashihara et al. (2008); Desgagné-Penix (2017)
Nelumbo nucifera	Neferine, nuciferine, nornuciferine, *N*-methylasimilobine	Morikawa et al. (2016); Zhang et al. (2015)
Nicotiana tabacum	Nicotine	Kennedy and Wightman (2011); Zulak et al. (2006)
Opium poppy	Morphine, codeine	Hussain et al. (2018)
Papaver somniferum	Morphine	Hussain et al. (2018)
Piper nigrum and Piper longum	Piperine, piperic acid, piperlonguminine, Pellitorine, piperolein	Takooree et al. (2019)
Rhizoma coptidis	Berberine, palmatine, berberrubine	Chen et al. (2008); Kupeli et al. (2002); Kuo et al. (2004); Li et al. (2014)
Solanum tuberosum	ά-Solanidine, ά-solanine	Jadhav et al. (1981)
Theobroma cacao	Caffeine, theobromine	Zulak et al. (2006)

plants (Kennedy & Wightman, 2011). Many types of classifications have been proposed by researchers. The most popular classification among them divides the entire class of compounds into three categories (Eagleson, 1994), which are as follows:

1. True—alkaloids are compounds that is derived from amino acid and possess a heterocyclic ring with nitrogen.

   For example, atropine, nicotine, etc.

2. Proto—alkaloids are compounds that are derived from amino acid and also contain a nitrogen atom which is not a part of heterocyclic ring structure.
   For example, adrenaline, ephedrine, etc.
3. Pesudo—alkaloid are compounds that is not derived from amino acid.

For example, caffeine, theobromine, etc.

Some studies have suggested that tea extracts are very rich in different alkaloids. Green and black tea leaf extract contains caffeine, catechin, theaflavin, and saponins (Chen et al., 2008; Guo et al., 2008; Vignoli et al., 2011). Caffeine concentration reported to be higher in green tea leaves (Ramdani et al., 2018). Quinine is an alkaloid obtained from the bark of *Cinchona officinalis* belonging to the family Rubiaceae. It is used as a powerful antimalarial drug (Achan et al., 2011). Colchicine is obtained from the plants of Liliaceae family, which is used for the treatment of gout (Kurek, 2019). Atropine belongs to tropane group of alkaloids that is obtained from *A. belladonna* of Solanaceae family (Kurek, 2019). Tubocurarine is an alkaloid, obtained from *Chondrodendron tomentosum* which acts as a muscle relaxant and is an ingredient of poison curare. Alkaloids vincristine and vinblastine are obtained from the pink periwinkle plant extracts, *Catharanthus roseus* belonging to family Apocynaceae (Moudi et al., 2013). Morphine is one of the most recognized alkaloids that have been used and is still intended for medical purposes. It is present in dried latex of unripe capsules of *Papaver somniferum* (Hussain et al., 2018). Codeine is a derivative of morphine from opium poppy that possesses excellent analgesic property. Morphine (10%) and codeine (0.5%) are present in opium.

Pereira et al. (2017) first time isolated and characterized the aporphine alkaloid namely stephalagine from the *A. crassiflora* peel extract. It has been reported that *Astragalus* spp. extract contains neurotoxin indolizidine alkaloid, swainsonine (Kristanc & Kreft, 2016). Similarly, Goncalves et al. (2006) characterized and isolated two alkaloids atherospermidine and liriodenine from the plant stem. Arruda et al. (2017) found alkaloid stephalagine content up to 30 mg/kg DW in *A. crassiflora*. However, annonine is another chief alkaloid present in Annona leaves due to geographical variations. Chen et al. (2008) analyzed the berberine (a isoquinoline alkaloid) from *Rhizoma coptidis* herb. Piperine, a well-known alkaloid from *Piper nigrum* and *Piper longum* has been using since many centuries. Berberine and berberrubine also extracted from *Barberis vulgaris* and used to cure jaundice, toothache, asthma, skin pigmentation, etc. (Rahimi-Madiseh et al., 2017; Yu et al., 2018). Zhang et al. (2015) reported that neferine alkaloid is found abundantly in *Nelumbo nucifera*. Takooree et al. (2019) reported that major alkaloids in *P. nigrum* are piperine, piperic acid, piperlonguminine, pellitorine, piperolein B, piperettine, etc. are present which show the biological activities. ά-Solanidine and ά-solanine are major glycoalkaloids distributed in potato (Jadhav et al., 1981).

Ferré (2008) found that caffeine is a competitive antagonist in mammals of inhibitory adenosine A_1 and A_2 receptors, which leads to activation via increased dopaminergic and glutamatergic activity. It is most commonly and widely used psychoactive compound due to its stimulatory effect. Caffeine is active constituents found in many herbs and plants such as tea (*Camellia sinensis*), guarana (*Paulinia cupana*), maté (*Ilex paraguariensis*), and cocoa (*Theobroma cacao*). Tobacco plant (*Nicotiana tabaccum*) produces nicotine (a pyridinealkaloid) which has insecticide and antiparasite properties (Zulak et al., 2006). Milugo et al. (2013) reported antagonistic correlation between alkaloids and saponins in *Rauvolfia caffra* plant extracts. They reported that leaf extracts containing saponins along with alkaloids, steroids, terpenoids glycosides showed poorest AOX activity (15%) whereas without saponins these all compounds in fraction showed strongest AOX a in activity (42.39%). Several studies revealed that alkaloids obtained from *R. coptidis* namely berberine and palmatine have antiinflammatory activity and widely used as traditional medicine in China (Kupeli et al., 2002; Kuo et al., 2004). Li et al. (2014) had done quantitative analysis of total alkaloids and their seasonal variation in *R. coptidis* plant extracts taken in various months. The result confirmed that bebrberine was the chief alkaloid throughout the growing period ranged from 35.22 to 79.45 mg/g DW followed by palmatine ranged (9.92 to 23.99 mg/g DW) coptisine (17.09–41.85 mg/gDW), jateorrhizine (2.98–6.88 mg/g DW), epiberberine (7.52–21.08 mg/g DW), and columbamine (5.02–10.84 mg/g DW). In addition to that, authors also reported that season has great impact on the quantity of alkaloids and saple collected during spring season had the higher amount of total alkaloids followed by July and October (Li et al., 2014). Physiological role of alkaloids is mainly due to defense against pathogen and herbivores and equally have several medicinal benefits if intake in permissible limit. Desgagné-Penix (2017) reviewed the distribution of alkaloids in woody plants. Isah (2016) reported that a tree *Taxus brevifolia* yields an important alkaloid taxol (Paclitaxel) which has anticancerous properties. Purine alkaloids such as caffeine, theobromine and theophylline were abundantly found in coffee bean extracts, cocoa beans, kolatine, and kolateine in *Cola acuminate* and other nonwoody species like tea, gurana, etc. (Koyama et al., 2003; Ashihara et al., 2008; Desgagné-Penix, 2017). Quinoline alkaloids (quinine, quinidine, cinchonine, acronycine, melicopine, melicopidine, acronycidine, etc.) are extracted from quinine tree (*Cinchona pubescens*), species of *Acronychia, Sarcomelicope* and *Acronychia acidula* (Epifano et al., 2013; Moore et al., 1995; Tillequin, 1997). Mulberry and coca bush also produce tropane alkaloids (Moore et al., 1995). Pomegranate tree and fruit extracts also rich in pyridine alkaloids. Zanthoxylum a member of Rutaceae family produces canthin-6-one alkaloid from bark and peel extracts (Cebrián-Torrejón et al., 2012). Plant extracts of *Acacia rigidula* possess indole category of alkaloids (including bcarbolines, harmine, harmaline, elaeagnine, tryptoline, etc.) (Clement et al., 1998). A quantitative analysis with mass liquid spectroscopy has been carried out to isolate alkaloids in flower extract of *Nelumbo nucifera* (Morikawa et al., 2016). They have isolated nuciferine, nornuciferine, *N*-methylasimilobine, asimilobine, pronuciferine, and armepavine alkaloids. Due to the structural diversity and alkalinity, their fractionation and characterization faced practical difficulties. Recent chromatographic techniques could be used for efficient quantification and characterization. Alkaloids due to their wide array and adaptability, represents a path for pharmaceutical and nutraceutical industries. Many vegetables, herbs and bitter fruits contain high amount of glycoalkaloids but need care when consuming such foods.

7.7 Plant extracts as sources of terpenes

Terpene is an isoprene-based naturally occurring compound having medicinal potential and found in plants and animals. Terpenes belong to hydrocarbon groups that consist of 5-carbon isoprene (C5H8) units as their basic building block. However, only few terpenes have been investigated for their medicinal uses (Franklin et al., 2000). Among the all, 67% potentiates belong to monoterpenes and sisquiterpenes. Terpene has activities such as antiinflammatory and also prevent inflammatory diseases (Franklin et al., 2000). Most common plants like tea, cannabis, citrus fruits, salvia, etc. contain abundant amount of medicinal terpene. Some important terpenes like p-menthane monoterpenoids, cannabinoids, etc. synthesized through 2-C-methyl-d-erythritol-4-phosphate pathway whereas thapsigargin and artemisinin produced through the mevalonate pathway. There are some common plant terpenes which are integral part of human diet. Terpenes are diverse group of lipid-soluble compounds ubiquitously synthesized via mevalonate and deoxy-d-xylulose pathways (Rohmer, 1999). These comprise one or 5-carbon isoprene units. Likewise, terpenoids don't have separate classification system but they are classified on the basis of number of isoprene units such as hemiterpene (1 unit), monoterpene (2 units), susquiterpene (3 units), and so on. Terpenes show the toxicity to the insects but they are less toxic to mammals (Rattan, 2010). Nowadays, recent studies showed that terpenoids are also present in our food system and contribute flavors and essential components to our diet. *Ginkgo biloba* leaf extracts have been used since long back and it reported that extracts of *G. biloba* yield species specific terpenes such as bilobalide and ginkgolides A, B, C, and J which show insecticidal, antifeedant, and antimicrobial activities (Ahn et al., 1997; Kleijnen & Knipschild, 1992; Matsumoto & Sei, 1987; Mazzanti et al., 2000). It is also reported that genus *Carissa* contains good amount of sesquiterpenes (Kirira et al., 2006; Wangteeraprasert & Likhitwitayawuid, 2009). Kumar et al. (2018) reported that bark and leaf extracts of Arjun tree contains terpenoids in ethanolic extracts. Some forest tree groups such as conifer oleoresin contains monoterpenes such as pinene and camphor; diterpenes such as taxadiene and phytane, sesquiterpenes like nerolidol (Martin et al., 2003). Several studies reported that terpenes from some forest species show antiinflammatory activities. A terpenes namely α-Pinene found in coniferous and rosemary showed such effects (Cho et al., 2017). The Mentha plant contained several p-menthane monoterpenoids. There are two types cannabinoids from *Cannabis sativa*, that is, D9-tetrahydrocannabinol and cannabidiol reported for their psychoactive and pain relieving properties, respectively (Bergman et al., 2019). It has been reported that wormwood (Art*emisia annua* L.) a plant of Asteraceae family produced sesquiterpene endoperoxide artemisinin which is highly effecting against malarial diseases (Meshnick et al., 1996; Nosten & White, 2007). Similarly, Thapsigargin from *Thapsia garganica*, Ingenol 3-mebutate from *Euphorbia peplus and Euphorbia lathyrus* have been isolated and used in various treatments (Siller et al., 2009). Antibacterial mode of terpenes is still unknown and unclear however, some researchers tried to elucidate the possible mechanism, that is, oxygen uptake and oxidative phosphorylation, which are crucial for microbes survival (Griffin et al., 1999). Diterpenes show its efficacy in combination therapy with antibiotics (Gupta et al., 2016). Among them, beta-carotene, phylloquinone, and tocopherols constitute pro vitamin activity while some phytosterols and essential oils provide health benefits by acting as AOX.

7.8 Conclusion

Plant extracts of various medicinal, fruits, and vegetable herbs are rich in phytonutrients, thus these have a great potential to cure many diseases. Vitamins and minerals are the primary nutrients present abundantly in various plant extracts and furnish various functions alone or in combination with other phytochemicals. Phenolic compounds, flavonoids, terpenes, etc. act as strong AOX and capable of scavenging free radicals thus urgent need to explore such fractions for future medicinal purposes. Alkaloids are also becoming integral part of diet due high drug potential and therapeutic uses. Identification and characterization of these bioactive compounds may be difficult through traditional extraction processes, therefore the use of advanced methods such as ultra heat extraction, liquid chromatography, HPLC, and mass spectroscopic determination could enhance the efficacy and efficiency. The use of plant extracts for their bioactive compounds is a new era for commercial pharmaceutical and nutraceutical industries and need to be enhanced for mankind.

References

Abbas, M., Saeed, F., Anjum, F. M., Afzaal, M., Tufail, T., Bashir, M. S., ... Suleria, H. A. R. (2017). Natural polyphenols: An overview. *International Journal of Food Properties*, *20*(8), 1689−1699.

Achan, J., Talisuna, A. O., Erhart, A., Yeka, A., Tibenderana, J. K., Baliraine, F. N., ... D'Alessandro, U. (2011). Quinine, an old anti-malarial drug in a modern world: Role in the treatment of malaria. *Malaria Journal*, *10*(1), 1−12.

Ademoyegun, O. T., Akin-Idowu, P. E., Ibitoye, D. O., & Adewuyi, G. O. (2013). Phenolic contents and free radical scavenging activity in some leafy vegetables. *International Journal of Vegetable Science*, *19*(2), 126−137.

Ahn, Y. J., Kwon, M., Park, H. M., & Han, C. K. (1997). Potent insecticidal activity of *Ginkgo biloba* derived trilactone terpenes against *Nilaparvata lugens*, In Edn. Young J. Ahn, M. Kwon, Hyung M. Park, and Chang K. Han *Phytochemicals for Pest Control*, *7*, 90−105.

Alamri, S. A., & Moustafa, M. F. (2012). Antimicrobial properties of 3 medicinal plants from Saudi Arabia against some clinical isolates of bacteria. *Saudi Medical Journal*, *33*(3), 272−277.

Altemimi, A., Lakhssassi, N., Baharlouei, A., Watson, D. G., & Lightfoot, D. A. (2017). Phytochemicals: Extraction, isolation, and identification of bioactive compounds from plant extracts. *Plants*, *6*(4), 42.

Amarowicz, R., Troszyńska, A., Baryłko-Pikielna, N., & Shahiid, F. (2004). Polyphenolics extracts from legume seeds: Correlations between total antioxidant activity, total phenolics content, tannins content and astringency. *Journal of Food Lipids*, *11*(4), 278−286.

Anese, M., Manzocco, L., Nicoli, M. C., & Lerici, C. R. (1999). Antioxidant properties of tomato juice as affected by heating. *Journal of the Science of Food and Agriculture*, *79*(5), 750−754.

Anese, M., Nicoli, M. C., Massini, R., & Lerici, C. R. (1999). Effects of drying processing on the Maillard reaction in pasta. *Food Research International*, *32*(3), 193−199.

Arruda, H. S., & Pastore, G. M. (2019). Araticum (*Annona crassiflora* Mart.) as a source of nutrients and bioactive compounds for food and non-food purposes: A comprehensive review. *Food Research International*, *123*, 450−480.

Arruda, H. S., Pereira, G. A., & Pastore, G. M. (2017b). Optimization of extraction parameters of total phenolics from *Annona crassiflora* Mart. (araticum) fruits using response surface methodology. *Food Analytical Methods*, *10*(1), 100−110. Available from https://doi.org/10.1007/s12161-016-0554-y.

Arruda, H. S., Pereira, G. A., de Morais, D. R., Eberlin, M. N., & Pastore, G. M. (2018). Determination of free, esterified, glycosylated and insoluble-bound phenolics composition in the edible part of araticum fruit (*Annona crassiflora* Mart.) and its byproducts by HPLC-ESI-MS/MS. *Food Chemistry*, *245*, 738. Available from https://doi.org/10.1016/j.foodchem.2017.11.120, 749.

Ashihara, H., Sano, H., & Crozier, A. (2008). Caffeine and related purine alkaloids: Biosynthesis, catabolism, function and genetic engineering. *Phytochemistry*, *69*(4), 841−856.

Ates, D. A., & Erdourul, O. T. (2003). Antimicrobial activities of various medicinal and commercial plant extracts. *Turkish Journal of Biology*, *27*, 157−162.

Aurea, M. A., & Samuel, O. A. (1998). Fat and fatty acid concentrations in some green vegetables. *Journal of Food Composition and Analysis*, *11*, 375–380.

Barbieri, R. L., & Stumpf, E. R. T. (2005). Origem, evolução e história das rosas cultivadas. *Current Agricultural Science and Technology*, *11*(3), 267–271.

Bassoli, B. K., Cassolla, P., Borba-Murad, G. R., Constantin, J., Salgueiro-Pagadigorria, C. L., Bazotte, R. B., ... de Souza, H. M. (2008). Chlorogenic acid reduces the plasma glucose peak in the oral glucose tolerance test: Effects on hepatic glucose release and glycaemia. *Cell Biochemistry and Function: Cellular Biochemistry and its Modulation by Active Agents or Disease*, *26*(3), 320–328.

Bergman, M. E., Davis, B., & Phillips, M. A. (2019). Medically useful plant terpenoids: Biosynthesis, occurrence, and mechanism of action. *Molecules (Basel, Switzerland)*, *24*(21), 3961.

Bhatta, S., Ratti, C., Poubelle, P. E., & Stevanovic, T. (2018). Nutrients, antioxidant capacity and safety of hot water extract from sugar maple (*Acer saccharum* M.) and red maple (*Acer rubrum* L.) bark. *Plant Foods for Human Nutrition*, *73*(1), 25–33.

Biel, W., Witkowicz, R., Piątkowska, E., & Podsiadło, C. (2020). Proximate composition, minerals and antioxidant activity of artichoke leaf extracts. *Biological Trace Element Research*, *194*(2), 589–595.

Bongoni, R., Steenbekkers, L. P. A., Verkerk, R., van Boekel, M. A. J. S., & Dekker, M. (2013). Studying consumer behavior related to the quality of food: A case on vegetable reparation affecting sensory and health attributes. *Trends in Food Science and Technology*, *33*(2), 145-139.

Butkutė, B., Dagilytė, A., Benetis, R., Padarauskas, A., Cesevičienė, J., Olšauskaitė, V., & Lemežienė, N. (2018). Mineral and phytochemical profiles and antioxidant activity of herbal material from two temperate *Astragalus* species. *Biomed Research International*, 2018.

Cai, J. Y., Xiong, J. W., Huang, Y. F., Zhao, Y. Y., Zhang, C. Y., Liu, W. W., & Wei, K. H. (2016). Study on ultrasonic-microwave synergistic extraction of polysaccharose from *Bletilla striata* and its antioxidant activity. *Sci. Technol. Food Ind*, *22*, 274–284.

Caleja, C., Ribeiro, A., Barreiro, M. F., & Ferreira, C. F. R. (2017). Phenolic compounds as nutraceuticals or functional food ingredients. *Current Pharmaceutical Design*, *23*(19), 2787–2806. Available from https://doi.org/10.2174/1381612822666161227153906.

Cebrián-Torrejón, G., Kahn, S. A., Ferreira, M. E., Thirant, C., de Arias, A. R., Figadère, B., ... Poupon, E. (2012). Alkaloids from Rutaceae: Activities of canthin-6-one alkaloids and synthetic analogues on glioblastoma stems cells. *Medicinal Chemistry Communications.*, *3*(7), 771–774.

Charles, P. (1992). Calcium absorption and calcium bioavailability. *Journal of Internal Medicine*, *231*(2), 161–168.

Chen, Q., Zhao, J., Liu, M., Cai, J., & Liu, J. (2008). Determination of total polyphenols content in green tea using FT-NIR spectroscopy and different PLS algorithms. *Journal of Pharmaceutical and Biomedical Analysis*, *46*(3), 568–573.

Cho, K. S., Lim, Y. R., Lee, K., Lee, J., Lee, J. H., & Lee, I. S. (2017). Terpenes from forests and human health. *Toxicological research*, *33*(2), 97–106.

Clement, B. A., Goff, C. M., & Forbes, T. D. A. (1998). Toxic amines and alkaloids from Acacia rigidula. *Phytochemistry*, *49*(5), 1377–1380.

Coppo, E., & Marchese, A. (2014). Antibacterial activity of polyphenols. *Current Pharmaceutical Biotechnology*, *15*(4), 380–390.

Cowan, M. M. (1999). Plant products as antimicrobial agents. *Clinical Microbiology Reviews*, *12*(4), 564–582.

Cunningham, E. (2015). What nutritional contribution do edible flowers make? *Journal of the Academy of Nutrition and Dietetics*, *15*(5), 856.

Datta, S., Sinha, B. K., Bhattacharjee, S., & Seal, T. (2019). Nutritional composition, mineral content, antioxidant activity and quantitative estimation of water soluble vitamins and phenolics by RP-HPLC in some lesser used wild edible plants. *Heliyon*, *5*(3), e01431.

Desgagné-Penix, I. (2017). Distribution of alkaloids in woody plants. *Plant Science Today*, *4*(3), 137–142.

Eagleson, M. (1994). *Concise encyclopedia chemistry* (p. 254) Newyork, USA: Sagwan Press.

Ebi, G. C., & Ofoefule, S. I. (2000). Antimicrobial activity of Pterocarpusosun stems. *Fitoterapia*, *71*(4), 433–435.

Eggersdorfer, M., Laudert, D., Letinois, U., McClymont, T., Medlock, J., Netscher, T., & Bonrath, W. (2012). One hundred years of vitamins—A success story of the natural sciences. *Angewandte Chemie International Edition*, *51*(52), 12960–12990.

Ekor, M. (2014). The growing use of herbal medicines: Issues relating to adverse reactions and challenges in monitoring safety. *Frontiers in Pharmacology*, *4*, 177.

Epifano, F., Fiorito, S., & Genovese, S. (2013). Phytochemistry and pharmacognosy of the genus Acronychia. *Phytochemistry*, *95*, 12–18.

Evans, S. (2008). Changing the knowledge base in Western herbal medicine. *Social Science and Medicine*, *67* (12), 2098–2106.

Ferré, S. (2008). An update on the mechanisms of the psychostimulant effects of caffeine. *Journal of Neurochemistry*, *105*(4), 1067–1079.

Franklin, L.U., Cunnington, G. D., & Young, D. E. (2000). *U.S. Patent No. 6,130,253*. Washington, DC: U.S. Patent and Trademark Office.

Glavinic, U., Stankovic, B., Draskovic, V., Stevanovic, J., Petrovic, T., Lakic, N., & Stanimirovic, Z. (2017). Dietary amino acid and vitamin complex protects honey bee from immunosuppression caused by *Nosema ceranae*. *PLoS One*, *12*(11), e0187726.

Goldman, P. (2001). Herbal medicines today and the roots of modern pharmacology. *Annals of Internal Medicine*, *135*(8_Part_1), 594–600.

Goncalves, M. A., Lara, T. A., & Pimenta, L. P. S. (2006). *Oxaporphynic alkaloids of Annona wood crassiflora Mart; Alcaloides oxaporfinicos da madeira de Annona crassiflora Mart.*

Gouveia, S. C., & Castilho, P. C. (2012). Phenolic composition and antioxidant capacity of cultivated artichoke, *Madeira cardoon* and artichoke-based dietary supplements. *Food Research International*, *48*(2), 712–724.

Greenberg, J. A., Boozer, C. N., & Geliebter, A. (2006). Coffee, diabetes, and weight control. *The American Journal of Clinical Nutrition*, *84*(4), 682–693.

Griffin, S. G., Wyllie, S. G., Markham, J. L., & Leach, D. N. (1999). The role of structure and molecular properties of terpenoids in determining their antimicrobial activity. *Flavour and Fragrance Journal*, *14*(5), 322–332.

Guenane, H., Mechraoui, O., Bakchiche, B., Djedid, M., Gherib, A., & Benalia, M. (2020). Antibacterial, antioxidant activities and mineral content from the Algerian medicinal plants. Scientific study and research. *Chemistry and Chemical Engineering, Biotechnology, Food Industry*, *21*(2), 175–194.

Guo, Y. Q., Liu, J. X., Lu, Y., Zhu, W. Y., Denman, S. E., & McSweeney, C. S. (2008). Effect of tea saponin on methanogenesis, microbial community structure and expression of mcrA gene, in cultures of rumen micro-organisms. *Letters in Applied Microbiology*, *47*(5), 421–426.

Gupta, V. K., Tiwari, N., Gupta, P., Verma, S., Pal, A., Srivastava, S. K., & Darokar, M. P. (2016). A clerodane diterpene from *Polyalthia longifolia* as a modifying agent of the resistance of methicillin resistant *Staphylococcus aureus*. *Phytomedicine: International Journal of Phytotherapy and Phytopharmacology*, *23* (6), 654–661.

Gyawali, R., & Ibrahim, S. A. (2014). Natural products as antimicrobial agents. *Food Control*, *46*, 412–429.

Handa, S. S. (2008). An overview of extraction techniques for medicinal and aromatic plants. In S. S. Handa, S. P. S. Khanuja, G. Longo, & D. D. Rakesh (Eds.), *Extraction technologies for medicinal and aromatic plants* (1, p. 260). Trieste: United Nations Industrial Development Organization and the International Centre for Science and High Technology (ICS - UNIDO).

Harfouch, R. M., Mohammad, R., & Suliman, H. (2016). Antibacterial activity of Syrian propolis extract against several strains of bacteria in vitro. *World Journal of Pharmaceutical Sciences*, *6*, 42–46.

Hasanuzzaman, M., Hossain, M. A., da Silva, J. A. T., & Fujita, M. (2012). *Plant response and tolerance to abiotic oxidative stress: Antioxidant defense is a key factor. Crop stress and its management: Perspectives and strategies* (pp. 261–315). Netherlands: Springer.

Hertog, M. G. L., Feskens, E. J. M., Hollman, P. C. H., Katan, M. B., & Kromhout, D. (1993). Dietary antioxidant flavonoids and risk of coronary heart disease: The Zutphen Elderly Study. *Lancet*, *342*, 1007–1011.

Ho, K. Y., Tsai, C. C., Huang, J. S., Chen, C. P., Lin, T. C., & Lin, C. C. (2001). Antimicrobial activity of tannin components from *Vaccinium vitis-idaea* L. *Journal of Pharmacy and Pharmacology*, *53*(2), 187–191.

Hussain, G., Rasul, A., Anwar, H., Aziz, N., Razzaq, A., Weil, W., ... Lil, X. (2018). Role of plant derived alkaloids and their mechanism in neurodegenerative disorders. *International Journal of Biological Science*, *3*, 341–357.

Igwenyi, I. O., Agwor, A. S., Nwigboji, I. U., Agbafor, K. N., & Offor, C. E. (2014). Proximate analysis, mineral and phytochemical composition of *Euphorbia hyssopifolia*. *Steroids*, *18*, 6–27.

Imelouane, B., Tahri, M., Elbastrioui, M., Aouinti, F., & Elbachiri, A. (2011). Mineral contents of some medicinal and aromatic plants growing in eastern Morocco. *Journal of Materials and Environmental Science*, *2*(2), 104–111.

Indrayan, A. K., Sharma, S., Durgapal, D., Kumar, N., & Kumar, M. (2005). Determination of nutritive value and analysis of mineral elements for some medicinally valued plants from Uttaranchal. *Current Science*, 1252–1255.

Isah, T. (2016). Anticancer Alkaloids from trees: Development into drugs. *Pharmacogn Review*, *10*(20), 90–99. Available from https://doi.org/10.4103/0973-7847.194047.

Jadhav, S. J., Sharma, R. P., & Salunkhe, D. K. (1981). Naturally occurring toxic alkaloids in foods. *CRC Critical Reviews in Toxicology*, *9*(1), 21–104.

Jagetia, G. C., Baliga, M. S., & Venkatesh, P. (2005). Influence of seed extract of *Syzygium cumini* (jamun) on mice exposed to different doses of γ-radiation. *Journal of Radiation Research*, *46*(1), 59–65.

Janz, T. G., Johnson, R. L., & Rubenstein, S. D. (2013). Anemia in the emergency department: Evaluation and treatment. *Emerg. Med. Pract.*, *15*, 1–15.

Jing, H., Liu, J., Liu, H., & Xin, H. (2014). Histochemical investigation and kinds of alkaloids in leaves of different developmental stages in *Thymus quinquecostatus*. *The Scientific World Journal*, *2014*.

Johnston, K., Sharp, P., Clifford, M., & Morgan, L. (2005). Dietary polyphenols decrease glucose uptake by human intestinal Caco-2 cells. *FEBS Letters*, *579*(7), 1653–1657.

Ju, Z., & Howard, L. R. (2005). Subcritical water and sulfured water extraction of anthocyanins and other phenolics from dried red grape skin. *Journal of Food Science*, *70*(4), S270–S276.

Kathirvel, A., & Sujatha, V. (2016). Phytochemical studies, antioxidant activities and identification of active compounds using GC–MS of *Dryopteris cochleata* leaves. *Arabian Journal of Chemistry*, *9*, S1435–S1442.

Kennedy, D. O., & Wightman, E. L. (2011). Herbal extracts and phytochemicals: Plant secondary metabolites and the enhancement of human brain function. *Advances in Nutrition*, *2*(1), 32–50.

Khan, M., & Wassilew, S. W. (1987). In H. Schmutterer, & K. R. S. Asher (Eds.), *Natural pesticides from the neem tree and other tropical plants* (pp. 645–650). Germany: Digitalverlag GmbH.

Khayat, S., Fanaei, H., & Ghanbarzehi, A. (2017). Minerals in pregnancy and lactation: A review article. *Journal of Clinical and Diagnostic Research: JCDR*, *11*(9), QE01.

Kiokias, S., Proestos, C., & Oreopoulou, V. (2018). Effect of natural food antioxidants against LDL and DNA oxidative changes. *Antioxidants*, *7*(10), 133.

Kirira, P. G., Rukunga, G. M., Wanyonyi, A. W., Muregi, F. M., Gathirwa, J. W., Muthaura, C. N., ... Ndiege, I. O. (2006). Anti-plasmodial activity and toxicity of extracts of plants used in traditional malaria therapy in Meru and Kilifi Districts of Kenya. *Journal of Ethnopharmacology*, *106*(3), 403–407.

Kleijnen, J., & Knipschild, P. (1992). Ginkgo-biloba. *Lancet, 340*(8828), 1136–1139.

Koleckar, V., Kubikova, K., Rehakova, Z., Kuca, K., Jun, D., Jahodar, L., & Opletal, L. (2008). Condensed and hydrolysable tannins as antioxidants influencing the health. *Mini Reviews in Medicinal Chemistry, 8* (5), 436–447.

Koyama, Y., Tomoda, Y., Kato, M., & Ashihara, H. (2003). Metabolism of purine bases, nucleosides and alkaloids in theobromine-forming *Theobroma cacao* leaves. *Plant Physiology and Biochemistry, 41*(11–12), 977–984.

Kristanc, L., & Kreft, S. (2016). European medicinal and edible plants associated with subacute and chronic toxicity part II: Plants with hepato-, neuro-, nephro- and immunotoxic effects. *Food and Chemical Toxicology, 92*, 38–49.

Kruczek, A. (2005). Effect of row fertilization with different kinds of fertilizers on the maize yield. *Acta Scientiarum Polonorum. Agricultura (Poland), 4*(2), 37–46.

Kumar, A., Krishan, M. R. V., Aravindan, P., Jayachandran, T., Deecaraman, M., Ilavarasan, R., ... Padmanabhn, A. (2008a). Anti inflammatory activity of *Syzygium cumini* seed. *African Journal of Biotechnology, 7*, 941–943.

Kumar, A., Krishan, M. R. V., Aravindan, P., Jayachandran, T., Deecaraman, M., Ilavarasan, R., & Padmanabhan, A. (2008b). Anti diabetic activity of *Syzygium cumini* seed and its isolate compounds against streptozotocin induced diabetic rats. *J. Med. Plants. Res., 2*, 246–249.

Kumar, K. E., Mastan, S. K., Reddy, K. R., Reddy, G. A., Raghunandan, N., & Chaitanya, G. (2008c). Anti arthritic property of methanolic extract of *Syzygium cumini* seed. *International Journal of Integrated. Biology, 4*, 55–60.

Kumar, V., Sharma, N., Sourirajan, A., Khosla, P. K., & Dev, K. (2018). Comparative evaluation of antimicrobial and antioxidant potential of ethanolic extract and its fractions of bark and leaves of *Terminalia arjuna* from north-western Himalayas, India. *Journal of traditional and complementary medicine, 8*(1), 100–106.

Kuo, C. L., Chi, C. W., & Liu, T. Y. (2004). The anti-inflammatory potential of berberine in vitro and in vivo. *Cancer Letters, 203*(2), 127–137.

Kupeli, E., Koşar, M., Yeşilada, E., Baser, K. H. C., & Baser, C. (2002). A comparative study on the anti-inflammatory, antinociceptive and antipyreticeffects of isoquinoline alkaloids from the roots of Turkish Berberis species. *Life Sciences, 72*, 645–657.

Kurek, J. (2019). *Alkaloids: Their importance in nature and for human life* (pp. 1–8). Poznan, Poland: Intech open.

Larsson, S. C., Virtanen, M. J., Mars, M., Männistö, S., Pietinen, P., Albanes, D., & Virtamo, J. (2008). Magnesium, calcium, potassium, and sodium intakes and risk of stroke in male smokers. *Archives of Internal Medicine, 168*(5), 459–465.

Lechaudel, M., Joas, J., Caro, Y., Genard, M., & Jannoyer, M. (2005). Leaf: Fruit ratio and irrigation supply affect seasonal changes in minerals, organic acids and sugars of mango fruit. *Journal of the Science of Food and Agriculture, 85*(2), 251–260.

Li, X., Qu, L., Dong, Y., Han, L., Liu, E., Fang, S., & Wang, T. (2014). A review of recent research progress on the Astragalus genus. *Molecules (Basel, Switzerland), 19*(11), 18850–18880.

Lin, G. M., Chen, Y. H., Yen, P. L., & Chang, S. T. (2016). Antihyperglycemic and antioxidant activities of twig extract from *Cinnamomum osmophloeum. Journal of Traditional and Complementary Medicine, 6*(3), 281–288.

Linder, M. C. (1991). *Nutritional, biochemistry and metabolism with clinical applications* (pp. 191–212). Norwalk, CT: Appleton and Lange, Vol. 2.

Loizzo, M. R. (2016). Edible flowers: A rich source of phytochemicals with antioxidant and hypoglycemic properties. *Journal of Agricultural and Food Chemistry, 12*, 2467–2474.

Lounder, D. T., Khandelwal, P., Dandoy, C. E., Jodele, S., Grimley, M. S., Wallace, G., ... Davies, S. M. (2017). Lower levels of vitamin A are associated with increased gastrointestinal graft-vs-host disease in children. *Blood, The Journal of the American Society of Hematology*, *129*(20), 2801–2807.

Mahfuzul Hoque, M. D., Bari, M. L., Inatsu, Y., Juneja, V. K., & Kawamoto, S. (2007). Antibacterial activity of guava (*Psidium guajava* L.) and neem (*Azadirachta indica* A. Juss.) extracts against foodborne pathogens and spoilage bacteria. *Foodborne Pathogens and Disease*, *4*(4), 481–488.

Makhuvele, R., Naidu, K., Gbashi, S., Thipe, V. C., Adebo, O. A., & Njobeh, P. B. (2020). The use of plant extracts and their phytochemicals for control of toxigenic fungi and mycotoxins. *Heliyon*, *6*(10), e05291.

Manske, R. H. F. (1965). *The alkaloids. Chemistry and physiology*. New York, NY: Academic Press, USA, 356p.

Martin, D. M., Gershenzon, J., & Bohlmann, J. (2003). Induction of volatile terpene biosynthesis and diurnal emission by methyl jasmonate in foliage of Norway spruce. *Plant Physiology*, *132*, 1586–1599.

Matsumoto, T., & Sei, T. (1987). Antifeedant activities of *Ginkgo biloba* L. components against the larva of *Pieris rapae* crucivora. *Agricultural and Biological Chemistry*, *51*(1), 249–250.

Mazumder, U. K., Gupta, M., Manikandan, L., Bhattacharya, S., Haldar, P. K., & Roy, S. (2003). Evaluation of anti-inflammatory activity of *Vernonia cinerea* Less. extract in rats. *Phytomedicine: International Journal of Phytotherapy and Phytopharmacology*, *10*, 185–188.

Mazzanti, G., Mascellino, M., Battinelli, L., Coluccia, D., Manganaro, M., & Saso, L. (2000). Antimicrobial investigation of semipurified fractions of *Ginkgo biloba* leaves. *Journal of Ethnopharmacology*, *71*, 83–88.

McCarty, M. F. (2005). Nutraceutical resources for diabetes prevention—An update. *Medical Hypotheses*, *64* (1), 151–158.

McNaught, D., & Wilkinson, A. (1997). *IUPAC: Compendium of chemical terminology*. New Jersey, USA: Blackwell Scientific Publications, Oxford.

Meshnick, S. R., Taylor, T. E., & Kamchonwongpaisan, S. (1996). Artemisinin and the antimalarial endoperoxides: From herbal remedy to targeted chemotherapy. *Microbiological Reviews*, *60*(2), 301–315.

Milugo, T. K., Omosa, L. K., Ochanda, J. O., Owuor, B. O., Wamunyokoli, F. A., Oyugi, J. O., & Ochieng, J. W. (2013). Antagonistic effect of alkaloids and saponins on bioactivity in the quinine tree (*Rauvolfia caffra* sond.): Further evidence to support biotechnology in traditional medicinal plants. *BMC Complementary and Alternative Medicine*, *13*(1), 1–6.

Mir, S. A., Shah, M. A., Ganai, S. A., Ahmad, T., & Gani, M. (2019). Understanding the role of active components from plant sources in obesity management. *Journal of the Saudi Society of Agricultural Sciences*, *18* (2), 168–176.

Monisha, S., & Ragavan, B. (2015). Investigation of phytochemical, mineral content, and physiochemical property of a polyherbal extract. *Asian Journal of Pharmaceutical and Clinical Research*, *8*(3), 238–242.

Moore, J. M., Moore, J. F., Fodor, G., & Jones, A. B. (1995). Detection and characterization of cocaine and related tropane alkaloids in coca leaf, cocaine, and biological specimens. *Forensic Science Review*, *7*(2), 77–101.

Moreira, M. R., Ponce, A. G., Del Valle, C. E., & Roura, S. I. (2005). Inhibitory parameters of essential oils to reduce a foodborne pathogen. *LWT-Food Science and Technology*, *38*(5), 565–570.

Morikawa, T., Kitagawa, N., Tanabe, G., Ninomiya, K., Okugawa, S., Motai, C., ... Muraoka, O. (2016). Quantitative determination of alkaloids in lotus flower (flower buds of *Nelumbo nucifera*) and their melanogenesis inhibitory activity. *Molecules (Basel, Switzerland)*, *21*(7), 930.

Moudi, M., Go, R., Yien, C. Y., & Nazre, M. (2013). Vinca alkaloids. *International Journal of Preventive Medicine*, *4*, 1231–1234.

Nair, R., Kalariya, T., & Chanda, S. (2005). Antibacterial activity of some selected Indian medicinal flora. *Turkish Journal of Biology*, *29*, 41–47.

Negi, P. S., Jayaprakasha, G. K., & Jena, B. S. (2003). Antioxidant and antimutagenic activities of pomegranate peel extracts. *Food Chemistry*, *80*(3), 393–397.

Nicoli, M. C., Anese, M., Parpinel, M. T., Franceschi, S., & Lerici, C. R. (1997). Loss and/or formation of antioxidants during food processing and storage. *Cancer Letters*, *114*(1–2), 71–74.

Nikmaram, N., Budaraju, S., Barba, F. J., Lorenzo, J. M., Cox, R. B., Mallikarjunan, K., & Roohinejad, S. (2018). Application of plant extracts to improve the shelf-life, nutritional and health-related properties of ready-to-eat meat products. *Meat Science*, *145*, 245–255.

Nosten, F., & White, N. J. (2007). Artemisinin-based combination treatment of falciparum malaria. *The American Journal of Tropical Medicine and Hygiene*, *77*(6_Suppl), 181–192.

Nour, V., Trandafir, I., & Cosmulescu, S. (2014). Antioxidant capacity, phenolic compounds and minerals content of blackcurrant (*Ribes nigrum* L.) leaves as influenced by harvesting date and extraction method. *Industrial Crops and Products*, *53*, 133–139.

O'Brien, J., Morrissey, P. A., & Ames, J. M. (1989). Nutritional and toxicological aspects of the Maillard browning reaction in foods. *Critical Reviews in Food Science and Nutrition*, *28*(3), 211–248.

Panche, A. N., Diwan, A. D., & Chandra, S. R. (2016). Flavonoids: An overview. *Journal of Nutritional Science*, *5*.

Papas, A. M. (1999). Diet and antioxidant status. *Food and Chemical Toxicology*, *37*(9), 999–1007.

Passos, M. G., Carvalho, H., & Wiest, J. M. (2009). In vitro inhibition and inactivation of different extraction methods in *Ocimum gratissimum* L.("alfavacão," "alfavaca," "alfavaca-cravo")-Labiatae (Lamiaceae) against foodborne bacteria of interest. *Revista Brasileira de Plantas Medicinais*, *11*(1), 71–78.

Pereira, M. N., Justino, A. B., Martins, M. M., Peixoto, L. G., Vilela, D. D., Santos, P. S., ... Espindola, F. S. (2017). Stephalagine, an alkaloid with pancreatic lipase inhibitory activity isolated from the fruit peel of *Annona crassiflora* Mart. *Industrial Crops and Products*, *97*, 324–329.

Ponmari, M., & Balasubiramanian, K. K. (2017). Evaluation of mineral contents in some medicinal plants used by traditional healers. *International Journal of Research in Pharmacy and Pharmaceutical Sciences*, *2*, 30–34.

Prajna, P. S., & Rama, B. P. (2015). Phytochemical and mineral analysis of root of *Loeseneriella arnottiana* wight. *International Journal of Current Research in Biosciences and Plant Biology*, *2*(3), 67–72.

Prashanth, D., Amit, A., Samiulla, D. S., Asha, M. K., & Padmaja, R. (2001). α-Glucosidase inhibitory activity of *Mangifera indica* bark. *Fitoterapia*, *72*(6), 686–688.

Prashith, K. T. R., Vinayaka, K. S., Soumya, K. V., Ashwini, S. K., & Kiran, R. (2010). Antibacterial and antifungal activity of methanolic extract of *Abruspulchellus wall* and *Abrusprecatorius precatorius* Linn- A comparative study. *Int. J. Toxicology and Pharmacolgy. Resesearch*, *2*(1), 26–29.

Proestos, C. (2020). The benefits of plant extracts for human health. *Foods*, *9*(11), 1653.

Proestos, C., & Varzakas, T. (2017). Aromatic plants: Antioxidant capacity and polyphenol characterisation. *Foods*, *6*, 28.

Qu, Y., Li, C., Zhang, C., Zeng, R., & Fu, C. (2016). Optimization of infrared-assisted extraction of *Bletilla striata* polysaccharides based on response surface methodology and their antioxidant activities. *Carbohydrate Polymers*, *148*, 345–353.

Rahimi-Madiseh, M., Lorigoini, Z., Zamani-Gharaghoshi, H., & Rafieian-Kopaei, M. (2017). *Berberis vulgaris*: Specifications and traditional uses. *Iranian Journal of Basic Medical Sciences*, *20*(5), 569.

Ramdani, D., Chaudhry, A. S., & Seal, C.J. (2018, February). Alkaloid and polyphenol analysis by HPLC in green and black tea powders and their potential use as additives in ruminant diets. In *AIP conference proceedings* (Vol. 1927, No. 1, p. 030008). AIP Publishing LLC.

Rattan, R. S. (2010). Mechanism of action of insecticidal secondary metabolites of plant origin. *Crop protection*, *29*(9), 913–920.

Ravishankar, B., & Shukla, V. J. (2007). Indian systems of medicine: A brief profile. *African Journal of Traditional, Complementary and Alternative Medicines*, *4*(3), 319–337.

Rohmer, M. (1999). The discovery of a mevalonate-independent pathway for isoprenoid biosynthesis in bacteria, algae and higher plants. *Natural Product Reports*, *16*(5), 565–574.

Roy, A. (2017). A review on the alkaloids: An important therapeutic compound from plants. *Intl. J. Plant Biotech.*, *3*(2), 1–9.

Sałata, A., & Gruszecki, R. (2010). The quantitative analysis of polyphenolic compounds in different parts of the artichoke (*Cynara scolymus* L.) depending on growth stage of plants. *Acta Scientiarum Polonorum*, *9* (3), 175–181.

Sahu, N. P., & Mahato, S. B. (1994). Anti-inflammatory triterpenesaponins of *Pithecellobium dulce*: Characterization of an echinocystic acid bisdesmoside. *Phytochemistry*, *37*(5), 1425–1427.

Salehi, B., Selamoglu, Z., Sener, B., Kilic, M., Kumar Jugran, A., de Tommasi, N., Sinisgalli, C., Milella, L., Rajkovic, J., Morais-Braga, F. B., & Bezerra, F. C. (2019). Berberis plants—drifting from farm to food applications, Phytotherapy, and Phytopharmacology. *Foods*, *8*(10), 522.

Selani, M. M., Bianchini, A., Ratnayake, W. S., Flores, R. A., Massarioli, A. P., de Alencar, S. M., & Brazaca, S. G. C. (2016). Physicochemical, functional and antioxidant properties of tropical fruits co-products. *Plant Foods for Human Nutrition*, *71*(2), 137–144.

Serafini, M., Bugianesi, R., Salucci, M., Azzini, E., Raguzzini, A., & Maiani, G. (2002). Effect of acute ingestion of fresh and stored lettuce (*Lactuca sativa*) on plasma total antioxidant capacity and antioxidant levels in human subjects. *British Journal of Nutrition*, *88*(6), 615–623.

Siller, G., Gebauer, K., Welburn, P., Katsamas, J., & Ogbourne, S. M. (2009). PEP005 (Ingenol mebutate) gel, a novel agent for the treatment of actinic keratosis: Results of a randomized, double-blind, vehicle-controlled, multicentre, phase IIa study. *Australasian Journal of Dermatology*, *50*(1), 16–22.

Singh, N., & Gupta, M. (2007). Effect of ethanolic extract of *Syzygium cumini* seed powder on pancreatic islets of alloxen diabetic rats. *Indian J. Experimental Biol.*, *45*, 861–867.

Sritalahareuthai, V., Temviriyanukul, P., On-Nom, N., Charoenkiatkul, S., & Suttisansanee, U. (2020). Phenolic profiles, antioxidant, and inhibitory activities of *Kadsura heteroclita* (Roxb.) Craib and *Kadsura coccinea* (Lem.) AC Sm. *Foods*, *9*(9), 1222.

Srivastav, A., Chandra, A., Singh, M., Jamal, F., Rastogi, P., Rajendran, S. M., ... Lakshmi, V. (2010). Inhibition of hyaluronidase activity of human and rat spermatozoa in vitro and antispermatogenic activity in rats in vivo by *Terminalia chebula*, a flavonoid rich plant. *Reproductive toxicology*, *29*(2), 214–224.

Staszowska-Karkut, M., & Materska, M. (2020). Phenolic composition, mineral content, and beneficial bioactivities of leaf extracts from black currant (*Ribes nigrum* L.), raspberry (*Rubus idaeus*), and aronia (*Aronia melanocarpa*). *Nutrients*, *12*(2), 463.

Stern, B. R., Solioz, M., Krewski, D., Aggett, P., Aw, T. C., Baker, S., ... Keen, C. (2007). Copper and human health: Biochemistry, genetics, and strategies for modeling dose-response relationships. *Journal of Toxicology and Environmental Health, Part B*, *10*(3), 157–222.

Sunarić, S., Pavlović, D., Stanković, M., Živković, J., & Arsić, I. (2020). Riboflavin and thiamine content in extracts of wild-grown plants for medicinal and cosmetic use. *Chemical Papers*, *74*(6), 1729–1738.

Tag, H., Kalita, P., Dwivedi, P., Das, A. K., & Namsa, N. D. (2012). Herbal medicines used in the treatment of diabetes mellitus in Arunachal Himalaya, northeast, India. *Journal of Ethnopharmacology*, *141*(3), 786–795.

Takooree, H., Aumeeruddy, M. Z., Rengasamy, K. R., Venugopala, K. N., Jeewon, R., Zengin, G., & Mahomoodally, M. F. (2019). A systematic review on black pepper (*Piper nigrum* L.): From folk uses to pharmacological applications. *Critical Reviews in Food Science and Nutrition, 59(sup1)*, S210–S243.

Tillequin, F. (1997). Alkaloids in the genus Sarcomelicope. Recent research develop. *Phytochemistry*, *1*, 675–687.

Tonthubthimthong, P., Chuaprasert, S., Douglas, P., & Luewisutthichat, W. (2001). Supercritical CO_2 extraction of nimbin from neem seeds–an experimental study. *Journal of Food Engineering*, *47*(4), 289–293.

Tusevski, O., Kostovska, A., Iloska, A., Trajkovska, L., & Simic, S. G. (2014). Phenolic production and antioxidant properties of some Macedonian medicinal plants. *Central European Journal of Biology*, *9*(9), 888–900.

Vignoli, J. A., Bassoli, D. G., & Benassi, M. D. T. (2011). Antioxidant activity, polyphenols, caffeine and melanoidins in soluble coffee: The influence of processing conditions and raw material. *Food Chemistry*, *124* (3), 863–868.

Wang, M., Simon, J. E., Aviles, I. F., He, K., Zheng, Q. Y., & Tadmor, Y. (2003).). Analysis of antioxidative phenolic compounds in artichoke (*Cynara scolymus* L.). *Journal of Agricultural and Food Chemistry*, *51* (3), 601–608.

Wangteeraprasert, R., & Likhitwitayawuid, K. (2009). Lignans and a sesquiterpene glucoside from *Carissa carandas* stem. *Helvetica Chimica Acta*, *92*(6), 1217–1223.

Wong, J. Y., & Chye, F. Y. (2009). Antioxidant properties of selected tropical wild edible mushrooms. *Journal of Food Composition and Analysis*, *22*(4), 269–277.

Wu, Q., Wang, X., Nepovimova, E., Wang, Y., Yang, H., Li, L., ... Kuca, K. (2017a). Antioxidant agents against trichothecenes: New hints for oxidative stress treatment. *Oncotarget.*, *8*, 110708–110726.

Xu, G. H., Chen, J. C., Liu, D. H., Zhang, Y. H., Jiang, P., & Ye, X. Q. (2008). Minerals, phenolic compounds, and antioxidant capacity of citrus peel extract by hot water. *Journal of Food Science*, *73*(1), C11–C18.

Yagi, S., Rahman, A. E. A., ELhassan, G. O., & Mohammed, A. M. (2013). Elemental analysis of ten sudanese medicinal plants using X-ray fluorescence. *Journal of Applied and Industrial Sciences*, *1*(1), 49–53.

Yin, J. I. E., Becker, E. M., Andersen, M. L., & Skibsted, L. H. (2012). Green tea extract as food antioxidant. Synergism and antagonism with α-tocopherol in vegetable oils and their colloidal systems. *Food Chemistry*, *135*(4), 2195–2202.

Yu, X. T., Xu, Y. F., Huang, Y. F., Qu, C., Xu, L. Q., Su, Z. R., ... Chen, J. P. (2018). Berberrubine attenuates mucosal lesions and inflammation in dextran sodium sulfate-induced colitis in mice. *PLoS One*, *13*(3), e0194069.

Yuan, T., Wan, C., González-Sarrías, A., Kandhi, V., Cech, N. B., & Seeram, N. P. (2011). Phenolic glycosides from sugar maple (*Acer saccharum*) bark. *Journal of Natural Products*, *74*(11), 2472–2476.

Yuan, T., Wan, C., Liu, K., & Seeram, N. P. (2012). New maplexins F-I and phenolic glycosides from red maple (*Acer rubrum*) bark. *Tetrahedron*, *68*, 959–964. Available from https://doi.org/10.1016/j.tet.2011.11.062.

Zenk, M. H., & Juenger, M. (2007). Evolution and current status of the phytochemistry of nitrogenous compounds. *Phytochemistry*, *68*, 2757–2772.

Zhang, Y., Lu, X., Zeng, S., Huang, X., Guo, Z., Zheng, Y., ... Zheng, B. (2015). Nutritional composition, physiological functions and processing of lotus (*Nelumbo nucifera* Gaertn.) seeds: A review. *Phytochemistry Reviews*, *14*(3), 321–334.

Zulak, K. G., Liscombe, D. K., Ashihara, H., & Facchini, P. J. (2006). Alkaloids. Plant secondary metabolites: Occurrence. *Structure and Role in the Human Diet*, 102–136.

CHAPTER

Plant extracts as flavoring agents

8

Nikitha Modupalli, Lavanya Devraj and Venkatachalapathy Natarajan

Department of Food Engineering, National Institute of Food Technology Entrepreneurship and Management (formerly Indian Institute of Food Processing Technology), Thanjavur, India

8.1 Introduction

Flavor is a very important sensory aspect that acts as one of the driving factors for consumer preferences. The flavor of the food is due to the presence of volatile aroma compounds in the foods. These compounds create a chemesthesic sensation in the oral cavity, which can be perceived by the sensitive nerve endings in the eyes, nose, and buccal cavity (Menis-Henrique, 2020). Also, volatile compounds occur differently in different foods. The specific sensation of flavor from each food is due to the characteristic chemical compounds in that particular food product. Though the flavor is a combination of aroma and taste sensory perceptions, it has been widely established that any substance's aroma and flavor are directly related, independent of the taste. Owing to such importance of the flavor compounds in the food sector, the production and separation of the volatile aromatic compounds have gained widespread importance worldwide. The flavoring compounds can be used to impart or intensify the flavor of food commodities. There has been a rapid growth in the need for processing and preservation of foods to increase shelf-stability. According to the United States- Food and Drug Administration, flavoring agents can be defined as the substances that can be added to food and food substances to impart flavor, while flavor enhancers are the substances that are added to substitute, intensify, or modify an existing flavor or aroma of the food product (Burdock, 2019). Being the most volatile, the flavor compounds tend to be lost to a greater extent during thermal processing and preservation. This can improve the shelf-life of flavor-deficit products that cannot meet consumer standards. One solution for this problem is to add the flavoring compounds separately after the thermal processing of the product in specific quantities based on the required intensity. In a few products like baked goods, flavoring compounds form the basic ingredients required to produce goods (Taylor & Hort, 2007).

The flavoring compounds can be roughly classified into plant-based, artificial, and biotechnologically formed flavors. As suggested by the name, the plant-based flavors are separated from the plant-based sources rich in aromatic compounds, like vanilla, spices, and herbs. The artificial flavors are formulated by treating the chemical compounds to form aromatic substances by certain processes and techniques. The biotechnology flavors are obtained by the bioconversion of any biological material by micro-organisms such as fermentation. Certain foods like coffee, cocoa and fermented meat develop their respective characteristic flavors due

Plant Extracts: *Applications in the Food Industry*. DOI: https://doi.org/10.1016/B978-0-12-822475-5.00006-5

to a complex process in a combination involving plant material, fermentation and roasting processes (Klinjapo & Krasaekoopt, 2018). The flavor compounds make one of the base ingredients form the final product with the desired functionality and reach in such goods. All these reasons have aided the rapid growth of the flavoring industry and the need for more developments in the production of required flavor compounds. The production of flavor compounds has been traditionally done by separating the volatile aromatic chemicals from direct sources using different solid-liquid extraction (SLE) or distillation techniques. However, such naturally occurring flavor compounds tend to lose the flavor over storage and are less intense. To overcome this problem, synthetic flavoring compounds have been developed artificially in laboratory settings. These synthetic flavor compounds are highly intense, are required in low concentrations and can be stored for a long time without disintegration (Berger, 2015; Grumezescu & Holban, 2017). The term "artificial flavor" or "artificial flavoring" means any substance with flavoring properties, which is not derived from a spice, fruit or fruit juice, vegetable or vegetable juice, edible yeast, herb, bark, bud, root, leaf or similar plant material, meat, fish, poultry, eggs, dairy products, or fermentation products. Some of the most commonly used synthetic flavorings are vanilla, chocolate, citrus, butter, coffee etc. Also, the flavors of certain flowers like lavender and rose have been widely used in the food industries for several applications. Table 8.1 summarizes different synthetic flavoring compounds and their commonly used substrates.

Several of these synthetic counterparts of the flavoring exert certain side effects. Allergic reactions, attention deficit hyperactivity disease, and carcinogenicities are the known and suspected health risks of artificial flavors and have been banned by the regulatory agencies against their use in food production and processing (Ramesh & Muthuraman, 2018). The European Union has restricted most of the synthetic flavors for these reasons (Martins, Roriz, Morales, Barros, & Ferreira, 2016; Ramesh & Muthuraman, 2018). The organizations like Flavor and Extract Manufacturers Association (FEMA) and International Organization of Flavor Industry have laid down certain regulations for use of artificfial and synthetic flavors in food produts by exerting permissable limits on each compound based on their biological activity. The artificial compounds such as coumarin, estragole, safrole and isosafrole, p-ppropyl anisole, cinnamyl anthracite etc., have been permanently banned for usage in food products due to their adverse effects. Also, carryover flavors, those present in raw materials and carried over to the final product, are permitted to be used in the limited quantity. The European Food Safety Authority and World Health Organization Joint Expert Committee on Food Additives (WHO-JEFCA) have establisheed a protocol for evaluationg the safety of the additives. All the additives were subsequently categorized into 34 groups based on the consistent chemical, metabolic and biological behavior. In case of the compounds that degrade into fragment, the fragment with highest toxic score is considered and the compound would be placed in the respective chemical category. The chemicals of each category would be assigned with certain permissable limit depending on the toxicity of the group.

The natural flavorings are always safe for consumption and generally do not cause any health complications. The increase in consumer requirement for natural and organic foods has boosted the concept of natural flavorings and additives usage. The sharp rise in market trend has also prompted the increase in natural extracts for flavor purposes in bulk production of food.

Table 8.1 Common synthetic flavors and their use.

S. No	Synthetic flavoring agents	Flavor profile	Application
1.	Diacetyl	Buttery and fermented flavor	Alcoholic beverages, caramel, and pastries
2.	Ethyl decadienate	Fruity flavor	Fruit powders, juices, and beer
3.	Ethyl maltol	Caramelized and smoky flavor	Candy and confectionary
4.	Ethyl propionate	Pineapple and fruity flavor	Juices, candy, fruit powder, beer, and confectionary
5.	Isoamyl acetate	Pear and banana flavors	Juices, candy, fruit powder, beer, and confectionary
6.	Lutidene	Coffee and nutty flavors	Bread, coffee, tea, meat products, and cheese
7.	Methyl anthranilate	Grape flavor	Perfumes, candy, confectionary, fruit powders, juices, wines, non-alcoholic beverages and ready-to-drink beverages
8.	Acetanisole	Tea flavor, tomato fruity flavor	Tea, flavored beverages, and ready-to-drink beverages
9.	Methyl salicylate	Wine and apple flavor	Alcoholic beverages, and non-alcoholic beer like beverages
10.	Furfural mercaptan	Meat and strong flavor	Animal products, meat analogs
11.	2,3,5-Trimethyl thiazole	Nutty flavor	Chocolate and cocoa products, coffee, ready-to-drink beverages, nut based foods, and products with added nuts
12.	Ethyl butyrate	Strawberry, apple, and fruity flavors	Fruit-based foods and beverages
13.	Methyl sulfide	Meat, dairy, and smoky flavor	Processed dairy products and convenience foods, meat products, and meat analogs
14.	Octanal	Grape fruit, citus, and orange flavor	Citrus-based and fruity beverages, fruit powders and ready-to-drink products

8.2 Plant extracts used for flavoring

Humans practised the concept of synthesis of flavor and flavor-based components from the plants from ancient times to improve the aesthetic and organoleptic properties of food and beverages. Most of the plant origin flavors were extracted from spices due to aromatic and volatile/pungent components. Flavoring matters from the plant source were recognized as safe, which has been used as seasoning agents in Indian cuisines. The industrial production and commercialization of fragrances and flavors were started during the 19th century. Europe occupies 36% of the market share in producing essential oils and natural extracts, followed by America (32%) and the Asian Pacific region (26%). The wide array of flavor ingredients was isolated from the spices, for example, cinnamaldehyde from cinnamon oil and benzaldehyde from the bitter almond oil.

The plant-based food flavoring agents are naturally occurring polyphenolic compounds, organic esters, acids, alkaloids, and carotenoids. Many surveys have summarized the consumers' lesser keenness in purchasing products containing artificial additives, in contradiction to natural food additions (Drew, 1994). Also, many research types have proved that a good number of flavoring agents and their precursors exert health-benefiting properties (Ayseli & Ipek Ayseli, 2016; Schwab, Davidovich-Rikanati, & Lewinsohn, 2008). Many flavors have been detected and developed since 1960. Several chemical compounds are ubiquitous and occur as aromatic chemicals in many foods. Not all volatile compounds are aromatic. It is a fascinating challenge to identify the volatile chemical compound that contributes to a specific flavor in a complex food system. Different flavors from various plant sources are summarized in Table 8.2.

8.3 Production of plant-based flavors

Food flavorings have important influence on food evaluation and consumer preferences. As a result, flavor production has been sought after by the food industry for the past decades. Natural flavors like vanilla, banana (ethyl acetate), cocoa, coffee, pineapple, berries, orange, mango, mint, cardamom, etc., have been excessively used in traditional and contemporary foods. However, the natural flavors are mild and are required in large quantities to obtain the required concentration and intensity. This problem has paved the way for the rapid development of synthetic flavors that can be artificially produced using chemical compounds. These synthetic flavors are highly intense and can give the desired flavor in food products even at low concentrations. The production of synthetic flavors can be costing lesser than that of the natural flavors, and it can be done in bulk production, which is not the case for natural aroma chemicals. Due to these above-stated reasons, synthetic flavors have occupied a large market share in the last decade. But the growing inclination of the consumers towards natural foods, free of chemical additives, has been acting as a driving force to the growth of the natural food market. All these reasons added to the rise in consumer acceptance of natural flavors and extracts. This market trend pressurized the flavor production industry to increase the production scale, which was not successful using traditional production practices (Cravotto & Cintas, 2007). This led to the need for further innovations and research in developing the production methods of natural plant-based flavors on a larger scale. The last decade has been a golden era for the research and development in the flavor industry, from production methods to identifying plant flavor sources. Different extraction techniques for plant extracts have been elaborated on in the previous chapters.

The production of flavor compounds involves the separation of aromatic components from the matrix of the source. The aromatic compounds are generally multicomponent blends of different aromatic and non-aromatic compounds containing traces of raw materials, which can be effectively used as a flavoring agent with the desired final compound outcomes. The flavoring compounds can occur in both solid and liquid physical forms. Different processes can do the manipulation of extracted flavors to modify the state and properties of the compounds. The manipulation method needs to be chosen, considering that the aromatic property of the substances needs to be intact. Most of the extraction processes give the final product in liquid form. The final product concentration is one more important characteristic of the volatile compounds as it determines the amount to be used for achieving a desirable effect in food applications.

Table 8.2 Various plant-based flavoring compounds.

Flavor compound	Plant source (extract)	Flavor perception	References
Vanillin	Vanilla bean	Sweet, floral	Ayala-Zavala, González-Aguilar, and Del-Toro-Sánchez (2009)
Trigonelline	Coffee, fenugreek	Astringent, bitter	(Murlidhar & Goswami, 2012)
Chlorogenic acids	Coffee, papaya, wine, blueberry	Astringent, bitter	Ayseli and Ipek Ayseli (2016)
Pico-crocin	Saffron	Bitter	Akowuah and Htar (2014)
d-Limonene	Citrus oils	Citrus	Rowe (2012)
Linalool	Hops, orange, beer, citrus essential oil, and coffee oil	Citrus, floral	Ayala-Zavala et al. (2009), Boulogne, Petit, Ozier-Lafontaine, Desfontaines, and Loranger-Merciris (2012), Park (2015), Schwab et al. (2008)
Resveratrol	Grapes	Astringent	Singh, Liu, and Ahmad (2015)
Naringin	Oranges, citrus extracts	Bitter	Fan, Li, and Fan (2015)
Gingerol	Ginger roots	Pungent, earthy, and musky	Tzeng, Chang, and Liu (2014)
Cinnamaldehyde	Cinnamon bark	Earthy, bitter	(Unlu, Ergene, Unlu, Zeytinoglu, & Vural, 2010)
Curcumene, α-tumorene	Turmeric roots (*Curcuma longa*)	Earthy, spicy, and musky	(Angel, Menon, Vimala, & Nambisan, 2014)
α-Pinene	Rosemary leaves	Floral, fragrant	(Prakash, Kedia, Mishra, & Dubey, 2015)
Carvone D	Mentha leaves	Cool, minty, fresh, and floral	(de Sousa Barros, De Morais, Ferreira, & Vieira, 2015)
Eugenol	Clove, cinnamon	Spicy, woody, irritant	(Chatterjee & Bhattacharjee, 2015)
Geraniol	Geranium grass	Sweet, floral	(Ahmad & Viljoen, 2015)
2E-Deacanol	Coriander leaves	Grassy, earthy, fresh	(Matasyoh, Maiyo, Ngure, & Chepkorir, 2009)
Isoamyl alcohol	Banana	Banana	Wick et al. (1966)
Eugenol, methyl eugenol	Allspice	Earthy, spicy, woody	(Zabka, Pavela, & Slezakova, 2009)

8.3.1 Flavor extracts in liquid form

The liquid flavors can occur as low viscosity liquids, medium viscous liquids, suspensions, emulsions, or pastes. The viscosity of products can be decreased by adding any carrier substances and increased by slow and mild heating for long timings. The flavor components are generally very intense and need to be added with any carrier material for increasing the stability and scaling up the quantity. The aqueous flavors are commonly added with ethanol or glycerol, whereas lipophilic flavors, vegetable oils or triacetin are prominently used carriers. The process of increasing viscosity by applying mild heat can cause deterioration of the complex's flavor profile and needs to be carefully selected and monitored for suitable flavor complexes. The production of the flavor complexes is a very pristine process in which the aromatic chemical compounds are added slowly in a step-by-step process. Prominently, the low volatile flavors can be added in the initial steps, and high volatiles is added in the final steps to avoid excessive loss. Also, for liquid flavors, food additives like antioxidants and preservatives are added to improve the stability of the complexes.

8.3.2 Flavor extracts in solid form

The dry form of food flavors is an important method of flavor incorporation in certain food products like tea, coffee, spice blends, etc. In such cases, the food flavors can only be used in powder or solid form. The powdered flavor matrices also are more stable and can be used for delayed flavor delivery. The dry flavors can also be an effective way to avoid deterioration of flavor compounds in foods during further processing. The production of dry flavors can be done using spray drying or spray chilling, encapsulation, compaction, coacervation and melt extrusion.

8.3.2.1 Spray drying

Spray drying is one of the most common methods used to convert liquid foods into a solid form. The spray drying of flavoring matrices involves converting liquid flavor and carrier material into the slurry, atomization of slurry, and drying and encapsulation. The carrier materials used for flavoring compounds are maltodextrins and plant starches. The carrier substances are made into a slurry using any continuous phase like water, alcohol, etc., into which the aromatic compounds are added. The flavor-slurry is atomized into the drying tower along with hot air to evaporate the continuous phase. During this process, the carrier molecules form a membrane around the flavoring substances to form a solid, powder-like form. In the spray chilling process used for heat-sensitive materials, cold air is used to atomize the slurry instead of the heated gas. The chilled air will freeze the carrier material forming granule like particles with flavoring compounds (Okuro, Eustáquio de Matos, & Favaro-Trindade, 2013). In both spray drying and freezing methods, the particles are separated from the air stream suing cyclone separator, after which they are carefully packed so that they do not get into contact with air. The optimal flavor particle is generally hollow on the inside and is small and round. The spray-dried flavor particles are generally used to produce the tea and coffee blends as they tend to hold the flavor for longer shelf periods. The details about spray drying for the encapsulation of flavoring substances have been elaborated on in further sections of this chapter.

8.3.2.2 Plated or extended flavors

The plating or extension of flavors is one of the traditional and old methods to produce dry flavors. The method of plating involves the flavors to be adsorbed on to any dry or powdery carrier material. The most commonly used carrier materials are salt, maltodextrins, starches, or lactose. The flavor substances in liquid form are fed onto a bed of carrier material under continuous stirring, using nozzles or sprayers. The continuous movement in the bed is achieved by paddles or stirrers, which helps distribute the flavor substances throughout the carrier material. It is a very economical process with minimal equipment usage. However, this procedure does not yield a product of high shelf-stability.

8.3.2.3 Compaction

The compacted flavors are granulated particles of dry components of a diameter ranging from 0.5 to 5 mm. The most important food application of the granulated flavors is in the beverage industry (tea, coffee and instant cocoa drink mixes). This process uses the spray-dried powder, compacted firmly using rollers, after which it is broken into small granules of flavoring agents. It is advantageous to be used in granulated food matrices and instant mixes. It is also much easier to fortify the material with certain high-value substances like vitamins in granules rather than powder. It is widely used in the tea industry as powdered tea tends to lose flavor soon in tea bags and is not suitable for de-mixing and blending steps in tea processing.

8.4 Advanced technologies to assess the quality of plant-based flavorings

8.4.1 Isotopic ratio mass spectrometry

Every chemical reaction chain differs in one form or the other by the isotopes of the same element. The fact of zero-point energy in any chemical reaction directly reflects on the mass fractions of the atoms. These mass variations, especially in lighter elements like hydrogen or carbon, are useful in detecting specific material origin. In the case of the aroma chemicals, these variations can be successfully used to differentiate between plants, animal, biotechnological and synthetic origins of the volatiles. One of the easiest routes is identifying the type of plant-based photosynthesis method adapted, that is, C3, C4, or CAM pathways. These three pathways do not let carbon accumulation as ^{13}C, but at different levels. This can be used to identify the type of plant from which the volatile compound has originated. The difference in the levels of accumulation can be denoted by delta ^{13}C ($d^{13}C$), which is the ratio of ^{13}C: ^{12}C stable isotopic forms compared with the ratio of PeeDee belemnite mineral. The typical $d^{13}C$ ratio for C4 plants is -10 to -16, -23 to -32 for C3 plants and -12 to -30 for CAM plants. The $d^{13}C$ ratio for chemically synthesized flavors from fossil fuels ranges from -15 to -33 (Asche, Michaud, & Brenna, 2003). Vanilla is one of the most successfully distinguished flavor using this method. The original vanillin is produced from a tropical plant called the CAM pathway, containing an isotope ratio of -12. The vanillin from the petrochemical origin (synthetic flavor) shows an isotope ratio of -24, and the one derived from ferulic acid from the C3 rice plant shows a -31. Similar the carbon elemental isotopes, other isotopes like hydrogen and deuterium ratios can also be used similarly. However, these ratios can be affected by non-biological reactions like proton transfer at active sites (Rowe, 2012).

8.4.2 Radiocarbon dating

Radiocarbon dating is a method of estimating the age of any biological matter by measuring the ^{14}C isotope present in it. The measured isotope is compared against a standard, which is ^{14}N isotope generally. The ^{14}N isotope has an estimated half-life of 5700 years, which can be used to estimate the half-life period of ^{14}C isotope. The measurement is usually done using a scintillation counting protocol, which indicated the carbon isotope age. It indicates if the material is of recent biological origin or if it is a fossil fuel product. It can be a good test for aroma chemicals that are generally isolated from essential oils, like cinnamaldehyde (from cassia oil). Naturally occurring cinnamaldehyde can be differentiated from the chemically modified compound from benzaldehyde.

8.5 Encapsulation of plant extract flavorings

Generally, the flavoring agents are highly volatile substances and tend to degrade due to heat, light, oxygen, chemical reacting ability, and microbial infestation. The natural flavoring agents are more volatile and less stable than their synthetic counterparts. These are mostly added at the end of the processing line to avoid denaturation and other losses. Encapsulation can be defined as a process to entrap on substance (generally target components) into another substance (Nedovic, Kalusevic, Manojlovic, Levic, & Bugarski, 2011). The process of encapsulation is prominently carried out in biological processes to improve the bioavailability of high-value compounds. It helps in delivering the unstable components without undergoing any degradation or denaturation during the processing. Encapsulation of plant-based flavors is a beneficial method to avoid degradation and losses during usage. This process helps in several ways, like (1) improved handling convenience, (2) reduced environmental reactivity and (3) improved concentration and dispersion characteristics of active compound (Lesmes & McClements, 2009). Microencapsulation and nanoencapsulation are the most efficient encapsulation techniques that encompass and protect the flavors during food processing, and also promote the controlled release, easier handling of flavors, reduce flavor interaction with external environmental factors and transfer rate from the core to impart even flavor to the components (Gharsallaoui, Roudaut, Chambin, Voilley, & Saurel, 2007). The encapsulation technique, both micro or nanoscale, can protect the enclosed material against environmental degradation and also help in the controlled release of the material. The encapsulation technique converts the liquid flavor extracts into powder form, easy to store and handle. The encapsulated powder flavors can be used in several industries like chewing gum, powdered drinks, instant beverages, and bakery and confectionery products (Klinjapo & Krasaekoopt, 2018). The encapsulated material can be called core material, active substance, or internal phase, whereas the encapsulating material can be a shell, coat, membrane, or carrier phase (Gharsallaoui et al., 2007). The core material selection is based on the properties like solubility, glass transition, molecular weight, permeability, stability, film-forming, emulsifying and crystaization properties, and desired functionality like strength, flexibility, type of release and concentration of the encapsulates (Nedovic et al., 2011). The coating materials are the protective barriers between the core and environment, need to be food-grade, biodegradable, and non-reactant and had to be recognized as safe (GRAS). The major coating materials used in the food industry are:

- Carbohydrates: Starch, maltodextrins, amylose, cyclodextrins, and modified starches
- Plant extracts: Gum arabica, gum tragacanth, pectin, and galactomannans
- Proteins: Albumin, zein, gluten, gelatin, casein, whey protein isolate, soy protein isolate, and pea protein
- Lipids: Phospholipids, fatty acids, and glycerides

- Animal and microbial origin: Chitosan, xanthan, and dextran

Microencapsulation is a method of enclosing small quantities of any compound in a shell-like casing to protect from the external environment (Saifullah, Shishir, Ferdowsi, Tanver Rahman, & Van Vuong, 2019; Yousuf, Gul, Wani, & Singh, 2016). The encapsulation technique involves converting the core material into tiny droplets initially, which the wall material can encompass through different processes. The most important encapsulating processes are spray drying, freeze-drying, fluidized bed drying and solvent evaporation. Coacervation and extrusion are also recently sought techniques for microencapsulation of flavor compounds (Shishir, Xie, Sun, Zheng, & Chen, 2018). Nanoencapsulation is a technique similar to microencapsulation, except that it delivers the encapsulated powder of nanoscale (typically >μm). The nanoencapsulation is beneficial than the microlevel for reasons like enhanced bioavailability, targeted drug transport and release (Saifullah et al., 2019; (Shishir, Xie, Sun, Zheng, & Chen, 2018). Tables 8.3 and 8.4 elaborate on different types of micro and nanoencapsulation trends of various aromatic compounds.

The vanillin, the most commonly used flavor obtained from vanilla, is highly sensitive to heat and light. Encapsulating naturally extracted vanillin in β-cyclodextrin using the freeze-drying technique has shown that the encapsulated vanilla was more soluble in water than the free molecules. Also, it has shown greater stability against oxidation (Karathanos, Mourtzinos, Yannakopoulou, & Andrikopoulos, 2007). Noshad, Mohebbi, Koocheki, and Shahidi (2015) studied the effect of microencapsulation of vanillin using a spray drying technique. Response surface methodology was adopted to analyze the optimum process conditions of microencapsulation using soy protein and maltodextrin. The research findings state that spray drying of 8.5% maltodextrin with 0.36% vanillin at 184°C is the optimum parameters for moisture content, encapsulation efficiency and particle size. Cardamom is one of the most widely used flavors in confectionery, bakery, beverage, (coffee and tea) and the pharmaceutical industries. But the cardamom flavoring substances are highly

Table 8.3 Microencapsulation of flavors and aromatic compounds.

S. No	Plant extract	Carrier material	Encapsulation technique	References
1.	Turmeric extract	β-Cyclodextrin, Brown rice flour	Homogenization and spray drying	(Laokuldilok, Thakeow, Kopermsub, & Utama-Ang, 2016)
2.	β- pinene	Sodium caseinate	Complex coacervation	(Koupantsis & Paraskevopoulou, 2017)
3.	Menthol	Gelatin, modified starch and oil-gelatin emulsion	Fluidized bed coating	Sun, Zeng, He, Qin, and Chen (2013)
4.	Caffeine extract	Sodium alginate	Desolvation and spray drying	(Bagheri, Madadlou, Yarmand, & Mousavi, 2014)
5.	Cardamom extract	Whey protein isolate, guar gum and carrageen gum	Emulsification and freeze drying	(Mehyar, Al-Isamil, Al-Ghizzawi, & Holley, 2014)
6.	Vanilla oil	Chitosan	Complex coacervation	(Yang et al., 2014)
7.	Sweet orange extract	β-Cyclodextrin	Molecular inclusion complex	(Zhu, Xiao, Zhou, & Zhu, 2014)

Table 8.4 Nanoencapsulation of flavor extracts from plant.

S. No	Plant extract	Carrier material	Encapsulation technique	References
1.	Coffee oil	Poly lactic acid	Emulsification and solvent evaporation	(Freiberger et al., 2015)
2.	Vanillin	Poly lactic acid	Emulsification and solvent evaporation	(Dalmolin, Khalil, & Mainardes, 2016)
3.	Peppermint oil	Zein and gum arabica	Dispersion and freeze drying	(Chen & Zhong, 2015)
4.	Thymol-eugenol complex	Zein, sodium caseinate	Antisolvent precipitation; self-assembly	(Chen, Zhang, & Zhong, 2015; Wang & Zhang, 2017)
5.	Carvacrol	Corn starch, sodium caseinate	Emulsion electrospinning	(Tampau, González-Martinez, & Chiralt, 2017)
6.	Saffron extract	Whey protein concentrate, pectin, maltodextrin	Multiple emulsification	(Esfanjani, Jafari, & Assadpour, 2017)

unstable against air, moisture, light and elevated temperatures. Encapsulation of this flavor in skimmed milk powder and modified starch using spray drying and freeze-drying techniques was investigated. The research findings concluded that spray-dried microspheres had greater flavor retention as the freeze-dried product lost the 1,8-cineole component rapidly during storage (Najafi, Kadkhodaee, & Mortazavi, 2011). Natural menthol is a very major constituent of peppermint oil and mint flavor extracts. It is a cyclic monoterpene alcohol compound and is prominently used in chocolates, beverages, pharmaceuticals, candies, liquors, chewing gum, and cooling agents. The fluidized bed drying method was used to trap menthol in gelatin and modified starch using fluidized bed drying methods to study the flavor release kinetics by Sun et al. (2013). Gelatin emulsion was found to be potential coating material with an encapsulation efficiency of about 90.4%. It was also found that about 60% of the encapsulated flavor was released within 11 min. Santos et al. (2014) worked on developing chewing gum incorporated with xylitol and menthol in the encapsulated form to study the cooling effect concerning time and intensity. The compounds were double encapsulated using the coacervation method before incorporation in the chewing gum. The research findings state that the microspheres of the encapsulated compounds can gradually release the compounds during chewing, thus increasing the cooling effect by time and intensity.

8.6 Application of natural plant-based extracts as flavoring agents in the food industry

8.6.1 Beverage industry

The beverage sector is one of the most important aspects of the food processing industries. The beverage industries include the processing of fruit beverages, beer, non-alcoholic sugar beverages, carbonated drinks, tea, coffee and alcoholic drinks. The flavor is a very primary aspect of acceptance of any beverage. Most commonly, liquid flavoring agents are used in beverages. For instant coffee powder production, the coffee soluble solids are extracted from the ground roasted beans by washing underwater currents and separating the aroma compounds using distillation. This extracted coffee flavor is added to the powdered or granulated bean powder (free-flowing or freeze-dried or agglomerated) along with 0.2% of cold extracted coffee oil to give profound coffee flavor to the instant powder (Ashurst, 2012a, 2012b). In juice production industries, the juice extracted and reconstituted would be added with 0.5%–1% of trapped aroma volatiles of the same fruit to improve the aroma profile and sensory characteristics of the fruit juices. In citrus fruits, volatile aromatic oil is expelled from the fruits before extracting juice from them. This citrus oil is clarified and readded to the citrus juices for greater flavor and aroma (Ashurst, 2012a, 2012b). Natural oleoresin exudates from different plants like pepper, ginger, etc., are widely used as flavoring agents in different beverages like soft drinks due to their rich aroma and are widely used for medicinal value in households. Also, certain flavors like cocoa, coffee and tea are characteristically developed during roasting and are not originally occurred in the raw material.

8.6.2 Savory foods

For the flavoring of savory products, the base flavors used are spices and spice blends, whereas the top-notch flavors like caramelized or smoky flavors are formed during the processing and production steps. The obvious loss of the aroma components during the thermal processing of savories is complemented by flavor-rich oleoresins and essential oils from spices and herbs. The hydrolyzed vegetable protein (HYP) can be added to impart the meaty flavor to the products. Soy protein and okara are most widely used for the production of hydrolyzed protein from vegetable sources. An increase in the vegan and vegetarian market share has benefitted from the bulk production of HYP. Top-notch flavors like smoky flavor have been commercially added by using dry powdered flavor composites like smoked paprika, pepper and certain salts (Hall, 2012).

8.6.3 Bakery and confectionary industry

The bakery and confectionery industries are among the vastest food processing industries with a great scope of use of flavorings and volatile chemicals. Most of the ingredients used for producing baked goods like honey, nuts, caramel, chocolate, licorice, coconut, malt, dried fruits are aroma-rich and directly impart desirable flavor to the finished products. Also, the production and processing conditions like temperature and baking time significantly contribute to the development of flavors in bakery products. The most commonly used flavoring extracts in bakery goods are vanilla, chocolate, caramel, butterscotch, etc., essential oils like lavender, peppermint, rose, citrus and citrus peel, rosemary spices like cinnamon, clove, sage, etc., are added to the ingredients. One major problem in flavor induction in baked goods is that there is a possibility of losing the flavors and volatiles during processing due to the high baking temperatures. Encapsulated and dry flavors are most suitable for bakery goods to overcome this problem. In confectionery like soft boiled and hard-boiled candies, the flavoring agents need to be more stable and concentrated on being retained after processing. Most of the manufacturers use synthetic flavorings for this purpose.

Similarly, in jellies and gums, essential oils like citrus are used. In such products, the top notes of the flavors tend to get lost or degraded during storage. Oleoresins of vanilla, cinnamon, nutmeg, rosemary, thyme and ginger are used to impart more profound flavor in these products. The volatile flavors in bakery and confectionery products are generally added into the fat or lipid ingredients during production as they tend to trap the flavor and retain the aroma for longer periods. Also, the sugar in the bakery goods tend to undergo Maillard's reaction during baking, thus imparting its flavor to the product (Armstrong, Luecke, & Bell, 2009; Armstrong & Yamazaki, 1986; Dusterhoft et al., 2006; Lawrence & Ashwood, 1991; Paterson and Piggott, 2006).

8.6.4 Alcoholic beverages

Alcoholic beverages are among the most varied food processing sectors, comprising beer, wine, distilled spirits, cocktails, flavored alcohols, and liqueurs. Beer is one of the most popular alcoholic drinks. It is consumed in both unflavored and flavored forms, making the flavored forms added with added flavor extracts. Also, in unflavored beer, hops are the most important additive that gives beer its characteristic flavor. Hops, a perennial plant, are generally added as an extract in essential oil or oleoresin. The extract of hops consists of terpenoids and nor-isoterpenoids abundantly, of which the terpenoids (myrcene, myrcene acid, myrcene, linalool, geraniol etc.,) are the primary aromatic compounds. Most beer manufacturers use various hops to give a consistent and constant flavor to their beer. However, a few manufacturers use two or three varieties of hop extracts to obtain a desirable aroma note to the beer. A large portion of hop extracts used currently in beer production is produced by supercritical fluid extraction. In naturally flavored beers, the flavors are added through whole fruits or concentrate, spices, seeds, oak, herbs, etc. The type of wheat, method of malting, time for fermentation also contribute to the development of the characteristic flavor of beer and are consistently maintained by the brewers for constant taste and flavor of beer produced in different batches (Buglass & Caven-Quantrill, 2012; Buglass, 2011; Tonutti & Liddle, 2010).

Wine is another alcoholic drink that is widely consumed in popular culture. It is produced by fermenting fruit juices, particularly grapes. The characteristic flavor of wine arises from the type of grapes and depends on the anthocyanin content. The traditional wine flavors are the result of the process of maturation done in old oak barrels. However, in current days, the oak flavor is imparted to the wines by adding small oak chips in the liquor. Herb and spice extracts have been employed in flavoring several wines and are maintained as a manufacturer's secret by several brewers. The typical wine called Vermont originated in Piedmont, Italy, is now produced in several parts of the world. The typical property of this wine is that it uses the extracts of about 30–50 botanicals for its specific wine aroma and taste. The most prominently used flavoring extracts in wine are floral extracts like lavender, rose; spices like clove, star anise, oregano; herbs like chamomile, basil, ginger, sage and unconventional flavors like acacia, wormwood, juniper, etc. The other alcoholic beverages like rum, vodka, gin, brandy, soju, tequila, and whiskey are neutral aroma-less liquors and can be used to produced cocktails, and flavored alcoholic drinks using fruits concentrates, extracts of several herbs, spices and beans. The essential oils, oleoresins, and extracts of citrus fruits and peels, rose, lavender, ginger, basil, etc., are widely used to impart aroma to the beverages (Buglass & Caven-Quantrill, 2012; McKay, Buglass, & Lee, 2011; Proestos et al., 2005; Takoi et al., 2009) (Table 8.5).

Table 8.5 Use of encapsulated plant flavors in the production of different processed food products.

S. No	Type of food product	Encapsulated flavor component	Type of encapsulation	Wall material	Remarks	References
1.	Biscuits	Shrimp oil	Microencapsulation	Sodium caseinate, fish gelatin and glucose syrup (1:1:4 w/v)	The encapsulated product can be incorporated in biscuits up to 6% without affecting sensory properties due to oxidation on light exposure.	(Takeungwongtrakul & Benjakul, 2017)
2.	Biscuits	Cinnamon essential oil	Nanoencapsulation	Maltodextrin	The results revealed the promising benefit of using the encapsulated cinnamon essential oil, as exogenous flavoring, on improving and keeping the overall qualities of biscuits.	(Fadel, Hassan, Ibraheim, Abd El Mageed, & Saad, 2019)
3.	Canned food	Wasabi extract	Microencapsulation	Octenyl succinylated waxy maize starch and maltodextrin	The findings showed that the encapsulated wasabi flavor produced under the studied conditions could be used in canned foods to protect degradation during storage.	(Ratanasiriwat, Worawattanamateekul, & Klaypradit, 2013)
4.	Chocolate-Vanilla dairy drink	*Terminalia arjuna* extract	Microencapsulation	Gum arabica	The final product properties indicated that encapsulation might be effective for incorporating herbal extract into dairy drinks intended for long-duration storage.	(Sawale, Patil, Hussain, Singh, & Singh, 2017)
5.	Extruded product	Lemon flavor extract	Microencapsulation	Sodium caseinate	Sodium caseinate capsules provided flavor protection against harsh extrusion condition (average flavor retention of 67.5%). The extrusion parameters have caused deterioration of the flavor and aroma.	(Yuliani, Torley, D'Arcy, Nicholson, & Bhandari, 2006)

(*Continued*)

Table 8.5 Use of encapsulated plant flavors in the production of different processed food products. *Continued*

S. No	Type of food product	Encapsulated flavor component	Type of encapsulation	Wall material	Remarks	References
6.	Flavored iced tea premix (instant)	Lemon essential oil	Microencapsulation	Gum arabica, modified starch and maltodextrin	Encapsulated lemon oil showed better results in instant ice tea premix for a beverage with a stability of 6 months.	(Sachin, Gadhave, & Jyotsna, 2015)
7.	Hamburger like meat products	Thyme essential oil	Microencapsulation	Casein and maltodextrin	The encapsulated thyme oil exerted antimicrobial and antioxidant properties along with incorporating flavoring into the product.	(Radünz et al., 2020)
8.	Hydroxypropyl β-dextrin (model food system)	Black pepper essential oil	Microencapsulation	Hydroxypropyl β-dextrin	Enhanced antibacterial activity of black pepper oil was improved by 4 times against both *S. aureus* and *Escherichia coli*.	(Rakmai, Cheirsilp, Mejuto, Torrado-Agrasar, & Simal-Gándara, 2017)
9.	Instant tea	Champaca flavor extract	Microencapsulation	Maltodextrin 20% (w/v) and trehalose 0.5% (w/v)	Instant green tea with 0.3% microencapsulated Champaca flavor has shown 96.7% product acceptance.	(Utama-Ang, Kopermsub, Thakeow, & Samakradhamrongthai, 2017)
10.	Milk-based matrices	Extracts of oregano (*Origanum vulgare* L.), citronella (*Cymbopogon nardus* G.) and marjoram (*Majorana hortensis* L.)	Nanoencapsulation	Skimmed milk powder and whey protein concentrate	The product has shown delayed release of flavor over storage time.	(Baranauskiene, Zukauskaite, Bylaite, & Venskutonis, 2006)
11.	Red wine (dealcoholized)	Grape pomace extract	Nanoencapsulation	Zein and L-lysine	The incorporated red wine was studied for pharmacogenetic studies using human subjects.	(Motilva et al., 2016)

12.	Soybean oil	Clove extract	Microencapsulation	Maltodextrin and gum arabica (4.8:2.4)	The soybean oil with encapsulated clove powder allowed the controlled release of the antioxidant, thus retarding oxidation and imparting flavor.	(Chatterjee & Bhattacharjee, 2013)
13.	Starch-based matrices (model food systems)	Peppermint essential oil	Microencapsulation	Chemically modified food starch from various sources	The subsequent retention and release of the flavor and aroma of starch matrices during encapsulation and storage has been studied.	(Baranauskienė, Venskutonis, Dewettinck, & Verhé, 2006)
14.	Yogurt	Red pepper waste extract	Microencapsulation	Whey protein	The final product (yogurt) showed good retention for polyphenol retention and carotenoids and flavor incorporation.	(Šeregelj et al., 2019)
15.	Yogurt	Grape seed extract	Microencapsulation	Whey protein concentrate and gum arabica (3:2)	The resultant product has shown enhanced functional activity and flavor.	(Yadav, Bajaj, Mandal, Saha, & Mann, 2018)

8.7 Safety evaluation and legislation for food flavorings

Food flavorings have been considered as unique food additives and have been excluded by legislation safeguarding the usage of food additives. However, most of the legislation systems have developed a separate protocol for the addition of flavoring agents (natural or synthetic) into food systems. The increase in awareness among the consumers about food labeling and safety has made it mandatory to declare all the additives in food products and the concentration. Most food industries declare the food flavorings generically as "contains flavors" or "contains natural-identical flavors." One advantage of flavoring substances is that they can be self-limiting and can restrict their addition due to their overpowering flavor properties even at low concentrations (Schrankel, 2004). The US-FDA has brought a Food Additives Amendment to the Food, Drug, and Cosmetic Act, 1938 in 1958. This amendment has brought into force to act on the safety of all food additives, including food flavorings that are permitted in the US (Hallagan & Hall, 1995). With the introduction of the concept of "Generally recognized as safe (GRAS)," US-FDA established a scientific panel FEMA, to determine standards for food flavors in food products. The FEMA evaluated the flavoring compounds based on the chemical structure, activity, natural occurrence, sources, rate of metabolic activity and level of toxicity of the particular extract (Woods & Doull, 1991). The FEMA-GRAS list of safe and permitted flavorings have been accepted by US-FDA ever since. The FEMA expert panel committee not only analyses new compounds for their safety and permissible limits, but it also revises the previously declared safe flavorings periodically (Smith et al., 2003).

The JECFA is an international scientific expert panel committee administered cooperatively by the Food and Agriculture Organization (FAO) and the WHO. The JECFA expert panel analyses the risk assessments of different chemical compounds in food and advice several countries to revise their safety guidelines in accordance. Though JECFA was initially concerned only about chemicals and food additives, it slowly expanded its assessment towards food flavorings as it is a very diverse category and needs attention for safe utilization. JECFA has considered the chemically defined flavoring compounds and flavor complexes obtained by physical processes like distillation or extraction using solvents as an important separate category. These complexes may contain oleoresins, esters, essential oils, and plant extracts along with the core flavoring compounds that may or may not have been chemically characterized (Schrankel, 2004). JECFA has adopted a flavor complex scheming protocol given by Feron et al. (2003) based on the safety evaluation procedure, only for the natural flavoring complexes (Smith et al., 2004).

8.8 Conclusion

Flavor is the amalgamation of aroma and taste. The uniqueness of any food substances is directly influenced by its texture and flavor. The flavor of any food matrix is influenced by the sensations like pain, heat, cold, and tactile because of volatile components. At the time of thermal processing, most of the food components lose their flavor or aroma. This can be recovered by the addition of flavoring agents that are synthesized either artificially or naturally. Synthetic flavoring components are carcinogenic and health-threatening due to the presence of chemicals. To overcome such health-threatening risks from synthetic flavors, plant-based flavoring agents are considered as an alternative and boon to humankind because of their health-benefiting properties. Plant-based flavoring compounds are naturally occurring in fruits, leaves, roots, spices, herbs, etc., enhancing the esthetic and organoleptic properties of the foods and improving the shelf-stability and quality of the foodstuffs by fighting against food-spoilage microbes. The extraction of flavoring substances has been practised from ancient times by humankind, and natural flavors like vanilla, banana (ethyl acetate), cocoa, coffee, pineapple, berries, orange, mango, mint, cardamom, etc., have been excessively used in traditional as well as in contemporary foods. The industrial production of flavor matrix from the spices like cinnamon oil from cinnamaldehyde was done during the 19th century. Various extraction methods have been practised so far to recover the flavor matrix from the plant-based substances like direct isolation to separate the essential oils from the plant sources like spices, in which the essential oils are the aromatic compounds followed by SLE, which works based on the osmosis and diffusion of aromatic compounds, super-critical fluid extraction of aromatic compounds by the utilization of liquid CO_2 at atmospheric pressure of >74 bars at 31°C, ultrasound-assisted SLE at lower frequencies from 18 to 40 MHz for the isolation of volatile components like cineole, thujone, and borneol from sage and microwave-assisted extraction by the generation of electromagnetic radiation with frequency ranging from 0.3 to 300 GHz for the recovery of volatile components from leafy herbs like cinnamon, garden mint, basil, and thyme.

Various other novel technologies like isotopic ratio mass spectrometry, radio dating, and microencapsulation have been adopted for the isolation of flavoring matrix from plant sources like vanillin from vanilla. The encapsulated powder flavors can be used in several industries like chewing gum, powdered drinks, instant beverages, and bakery and confectionery products. Flavors obtained from the plant sources are generally recognized as safe, and as per the food regulation acts, it is mandatory to label clearly on the packages the details of the flavoring agents used, whether it is "synthetic" or "natural," the amount and permissible limits of the flavoring substances added in food substances like bakery, confectionery to create the consumer awareness and as a part of food safety.

References

Ahmad, A., & Viljoen, A. (2015). The in vitro antimicrobial activity of Cymbopogon essential oil (lemon grass) and its interaction with silver ions. *Phytomedicine*, *22*(6), 657–665.

Akowuah, G. A., & Htar, T. T. (2014). Therapeutic properties of saffron and its chemical constituents. *Journal of Natural Products*, *7*, 5–13.

Angel, G. R., Menon, N., Vimala, B., & Nambisan, B. (2014). Essential oil composition of eight starchy Curcuma species. *Industrial Crops and Products*, *60*, 233–238.

Armstrong, D. W., & Yamazaki, H. (1986). Natural flavors production: A biotechnological approach. *Trends in Biotechnology*, *4*(10), 264–268. Available from https://doi.org/10.1016/0167-7799(86)90190-3.

Armstrong, L. M., Luecke, K. J., & Bell, L. N. (2009). Consumer evaluation of bakery product flavor as affected by incorporating the prebiotic tagatose. *International Journal of Food Science & Technology*, *44* (4), 815–819.

Asche, S., Michaud, A. L., & Brenna, J. T. (2003). Sourcing organic compounds based on natural isotopic variations measured by high precision isotope ratio mass spectrometry. *Current Organic Chemistry*, *7*(15), 1527–1543.

Ashurst, P. (2012a). *Applications of natural plant extracts in soft drinks. Natural food additives, ingredients and flavourings* (pp. 333–357). Elsevier.

Ashurst, P. R. (2012b). *Food flavorings*. Springer Science & Business Media.

Ayala-Zavala, J. F., González-Aguilar, G. A., & Del-Toro-Sánchez, L. (2009). Enhancing safety and aroma appealing of fresh-cut fruits and vegetables using the antimicrobial and aromatic power of essential oils. *Journal of Food Science*, *74*(7), R84–R91.

Ayseli, M. T., & Ipek Ayseli, Y. (2016). Flavors of the future: Health benefits of flavor precursors and volatile compounds in plant foods. *Trends in Food Science and Technology*, *48*, 69–77. Available from https://doi.org/10.1016/j.tifs.2015.11.005.

Bagheri, L., Madadlou, A., Yarmand, M., & Mousavi, M. E. (2014). Spray-dried alginate microparticles carrying caffeine-loaded and potentially bioactive nanoparticles. *Food Research International*, *62*, 1113–1119.

Baranauskienė, R., Venskutonis, P. R., Dewettinck, K., & Verhé, R. (2006). Properties of oregano (*Origanum vulgare L.*), citronella (*Cymbopogon nardus G.*) and marjoram (*Majorana hortensis L.*) flavors encapsulated into milk protein-based matrices. *Food Research International*, *39*(4), 413–425.

Baranauskiene, R., Zukauskaite, J., Bylaite, E., & Venskutonis, P. R. (2006, October). Aroma retention and flavour release of peppermint essential oil encapsulated by spray-drying into food starch based matrices. In *Proceedings of XIVth International Workshop on Bioencapsulation & COST 865 Meeting*, Lausanne, Switzerland.

Berger, R. G. (2015). Biotechnology as a source of natural volatile flavors. *Current Opinion in Food Science*, *1*(1), 38–43. Available from https://doi.org/10.1016/j.cofs.2014.09.003.

Boulogne, I., Petit, P., Ozier-Lafontaine, H., Desfontaines, L., & Loranger-Merciris, G. (2012). Insecticidal and antifungal chemicals produced by plants: A review. *Environmental Chemistry Letters*, *10*(4), 325–347.

Buglass, A. J. (2011). *Liqueurs and their flavorings, . Handbook of alcoholic beverages: Technical, analytical and nutritional aspects* (Vol. 1, pp. 615–627). Wiley.

Buglass, A. J., & Caven-Quantrill, D. J. (2012). *Applications of natural ingredients in alcoholic drinks. Natural food additives, ingredients and flavorings* (pp. 358–416). Elsevier.

Burdock, G. A. (2019). *Fenaroli's handbook of flavor ingredients* (Vol. 2). CRC press.

Chatterjee, D., & Bhattacharjee, P. (2013). Comparative evaluation of the antioxidant efficacy of encapsulated and un-encapsulated eugenol-rich clove extracts in soybean oil: Shelf-life and frying stability of soybean oil. *Journal of Food Engineering*, *117*(4), 545–550.

Chatterjee, D., & Bhattacharjee, P. (2015). Use of eugenol-lean clove extract as a flavoring agent and natural antioxidant in mayonnaise: product characterization and storage study. *Journal of Food Science and Technology*, *52*(8), 4945–4954.

Chen, H., Zhang, Y., & Zhong, Q. (2015). Physical and antimicrobial properties of spray-dried zein–casein nanocapsules with co-encapsulated eugenol and thymol. *Journal of Food Engineering*, *144*, 93–102.

Chen, H., & Zhong, Q. (2015). A novel method of preparing stable zein nanoparticle dispersions for encapsulation of peppermint oil. *Food Hydrocolloids*, *43*, 593–602.

Cravotto, G., & Cintas, P. (2007). *Extraction of flavorings from natural sources. Modifying flavor in food* (pp. 41–63). Elsevier.

Dusterhoft, E. -M., Minor, M., Nikolai, K., Hargreaves, N., Huscroft, S., & Scharf, U. (2006). *Encapsulated functional bakery ingredients*. Google Patents.

Dalmolin, L. F., Khalil, N. M., & Mainardes, R. M. (2016). Delivery of vanillin by poly (lactic-acid) nanoparticles: Development, characterization and in vitro evaluation of antioxidant activity. *Materials Science and Engineering: C*, *62*, 1–8.

de Sousa Barros, A., De Morais., Ferreira, P. A. T., & Vieira, Í. (2015). HA Chemical composition and functional properties of essential oils from Mentha species. *Industrial Crops and Products*, *76*, 557–564.

Drew, K. (1994). Consumer perceptions of natural foods. In Understanding Natural Flavors, (pp. 164–177). Boston, MA: Springer.

Esfanjani, A. F., Jafari, S. M., & Assadpour, E. (2017). Preparation of a multiple emulsion based on pectin-whey protein complex for encapsulation of saffron extract nanodroplets. *Food Chemistry, 221*, 1962–1969.

Fadel, H. H., Hassan, I. M., Ibraheim, M. T., Abd El Mageed., & Saad, R. (2019). Effect of using cinnamon oil encapsulated in maltodextrin as exogenous flavouring on flavour quality and stability of biscuits. *Journal of Food Science and Technology, 56*(10), 4565–4574.

Fan, J., Li, J., & Fan, Q. (2015). Naringin promotes differentiation of bone marrow stem cells into osteoblasts by upregulating the expression levels of microRNA-20a and downregulating the expression levels of PPARγ. *Molecular Medicine Reports, 12*(3), 4759–4765.

Feron, V. J., Adams, T. B., Cohen, S., Doull, J., Goodman, J. I., Hall, R. L., ... Wagner, B. M. (2003). 51 Safety evaluation of natural flavor complexes. *Toxicology Letters, 144*, s16–s17.

Freiberger, E. B., Kaufmann, K. C., Bona, E., de Araújo., Sayer, C., ... Goncalves, O. H. (2015). Encapsulation of roasted coffee oil in biocompatible nanoparticles. *LWT-Food Science and Technology, 64* (1), 381–389.

Gharsallaoui, A., Roudaut, G., Chambin, O., Voilley, A., & Saurel, R. (2007). Applications of spray-drying in microencapsulation of food ingredients: An overview. *Food Research International, 40*(9), 1107–1121.

Grumezescu, A. M., & Holban, A. M. (2017). *Natural and artificial flavoring agents and food dyes* (Vol. 7). Academic Press.

Hall, R. H. (2012). *Applications of natural ingredients in savory food products. Natural food additives, ingredients and flavourings* (pp. 281–317). Elsevier.

Hallagan, J. B., & Hall, R. L. (1995). FEMA GRAS-A GRAS assessment program for flavor ingredients. *Regulatory Toxicology and Pharmacology, 21*(3), 422–430.

Karathanos, V. T., Mourtzinos, I., Yannakopoulou, K., & Andrikopoulos, N. K. (2007). Study of the solubility, antioxidant activity and structure of inclusion complex of vanillin with β-cyclodextrin. *Food Chemistry, 101*(2), 652–658.

Klinjapo, R., & Krasaekoopt, W. (2018). *Microencapsulation of color and flavor in confectionery products. Natural and artificial flavoring agents and food dyes* (pp. 457–494). Elsevier.

Koupantsis, T., & Paraskevopoulou, A. (2017). Flavour retention in sodium caseinate–Carboxymethylcellulose complex coavervates as a function of storage conditions. *Food Hydrocolloids, 69*, 459–465.

Laokuldilok, N., Thakeow, P., Kopermsub, P., & Utama-Ang, N. (2016). Optimisation of microencapsulation of turmeric extract for masking flavour. *Food Chemistry, 194*, 695–704.

Lawrence, D. V., & Ashwood, D. G. (1991). *The flavoring of confectionery and bakery products. Food flavorings* (pp. 187–223). Springer.

Lesmes, U., & McClements, D. J. (2009). Structure–function relationships to guide rational design and fabrication of particulate food delivery systems. *Trends in Food Science & Technology, 20*(10), 448–457.

Martins, N., Roriz, C. L., Morales, P., Barros, L., & Ferreira, I. C. (2016). Food colorants: Challenges, opportunities and current desires of agro-industries to ensure consumer expectations and regulatory practices. *Trends in Food Science & Technology, 52*, 1–15.

Matasyoh, J. C., Maiyo, Z. C., Ngure, R. M., & Chepkorir, R. (2009). Chemical composition and antimicrobial activity of the essential oil of Coriandrum sativum. *Food Chemistry, 113*(2), 526–529.

McKay, M. A., Buglass, A. J., & Lee, C. G. (2011). *Fermented beverages: Beers, cidars, wines and related drinks*.

Mehyar, G. F., Al-Isamil, K. M., Al-Ghizzawi, H. A. M., & Holley, R. A. (2014). Stability of cardamom (*Elettaria cardamomum*) essential oil in microcapsules made of whey protein isolate, guar gum, and carrageenan. *Journal of Food Science, 79*(10), C1939–C1949.

Menis-Henrique, M. E. C. (2020). Methodologies to advance the understanding of flavor chemistry. *Current Opinion in Food Science, 33*, 131–135. Available from https://doi.org/10.1016/j.cofs.2020.04.005.

Motilva, M. J., Macià, A., Romero, M. P., Rubió, L., Mercader, M., & González-Ferrero, C. (2016). Human bioavailability and metabolism of phenolic compounds from red wine enriched with free or nano-encapsulated phenolic extract. *Journal of Functional Foods*, *25*, 80–93.

Murlidhar, M., & Goswami, T. K. (2012). A review on the functional properties, nutritional content, medicinal utilization and potential application of fenugreek. *Journal of Food Processing and Technology*, *3*(9), 1–10.

Najafi, M. N., Kadkhodaee, R., & Mortazavi, S. A. (2011). Effect of drying process and wall material on the properties of encapsulated cardamom oil. *Food Biophysics*, *6*(1), 68–76.

Nedovic, V., Kalusevic, A., Manojlovic, V., Levic, S., & Bugarski, B. (2011). An overview of encapsulation technologies for food applications. *Procedia Food Science*, *1*, 1806–1815.

Noshad, M., Mohebbi, M., Koocheki, A., & Shahidi, F. (2015). Microencapsulation of vanillin by spray drying using soy protein isolate–maltodextrin as wall material. *Flavour and Fragrance Journal*, *30*(5), 387–391.

Okuro, P. K., Eustáquio de Matos, F., & Favaro-Trindade, C. S. (2013). Technological challenges for spray chilling encapsulation of functional food ingredients. *Food Technology and Biotechnology*, *51*(2), 171–182.

Park, K. S. (2015). Raspberry ketone, a naturally occurring phenolic compound, inhibits adipogenic and lipogenic gene expression in 3T3-L1 adipocytes. *Pharmaceutical Biology*, *53*(6), 870–875.

Paterson, A., & Piggott, J. R. (2006). Flavour in sourdough breads: A review. *Trends in Food Science & Technology*, *17*(10), 557–566.

Prakash, B., Kedia, A., Mishra, P. K., & Dubey, N. K. (2015). Plant essential oils as food preservatives to control moulds, mycotoxin contamination and oxidative deterioration of agri-food commodities–Potentials and challenges. *Food Control*, *47*, 381–391.

Proestos, C., Bakogiannis, A., Psarianos, C., Koutinas, A. A., Kanellaki, M., & Komaitis, M. (2005). High performance liquid chromatography analysis of phenolic substances in Greek wines. *Food Control*, *16*(4), 319–323.

Radünz, M., dos Santos Hackbart., Camargo, T. M., Nunes, C. F. P., de Barros., ... da Rosa Zavareze, E. (2020). Antimicrobial potential of spray drying encapsulated thyme (Thymus vulgaris) essential oil on the conservation of hamburger-like meat products. *International Journal of Food Microbiology*, *330*, 108696.

Rakmai, J., Cheirsilp, B., Mejuto, J. C., Torrado-Agrasar, A., & Simal-Gándara, J. (2017). Physico-chemical characterization and evaluation of bio-efficacies of black pepper essential oil encapsulated in hydroxypropyl-beta-cyclodextrin. *Food hydrocolloids*, *65*, 157–164.

Ramesh, M., & Muthuraman, A. (2018). Flavoring and coloring agents: health risks and potential problems. In Natural and artificial flavoring agents and food dyes, (pp. 1–28). Academic Press.

Ratanasiriwat, P., Worawattanamateekul, W., & Klaypradit, W. (2013). Properties of encapsulated wasabi flavour and its application in canned food. *International Journal of Food Science & Technology*, *48*(4), 749–757.

Rowe, D. J. (2012). *Natural aroma chemicals for use in foods and beverages. Natural food additives, ingredients and flavourings* (pp. 212–230). Elsevier.

Sachin, K., Gadhave, A. D., & Jyotsna, W. (2015). Microencapsulation of lemon oil by spray drying and its application in flavour tea. *Advances in Applied Science Research*, *6*(4), 69–78.

Saifullah, M., Shishir, M. R. I., Ferdowsi, R., Tanver Rahman, M. R., & Van Vuong, Q. (2019). Micro and nano encapsulation, retention and controlled release of flavor and aroma compounds: A critical review. *Trends in Food Science and Technology*, *86*, 230–251. Available from https://doi.org/10.1016/j.tifs.2019.02.030.

Santos, M. G., Carpinteiro, D. A., Thomazini, M., Rocha-Selmi, G. A., da Cruz, A. G., Rodrigues, C. E. C., & Favaro-Trindade, C. S. (2014). Coencapsulation of xylitol and menthol by double emulsion followed by complex coacervation and microcapsule application in chewing gum. *Food Research International*, *66*, 454–462.

Sawale, P. D., Patil, G. R., Hussain, S. A., Singh, A. K., & Singh, R. R. B. (2017). Effect of incorporation of encapsulated and free Arjuna herb on storage stability of chocolate vanilla dairy drink. *Food Bioscience, 19*, 142–148.

Schrankel, K. R. (2004). Safety evaluation of food flavorings. *Toxicology, 198*(1–3), 203–211. Available from https://doi.org/10.1016/j.tox.2004.01.027.

Schwab, W., Davidovich-Rikanati, R., & Lewinsohn, E. (2008). Biosynthesis of plant-derived flavor compounds. *The Plant Journal, 54*(4), 712–732.

Šeregelj, V., Tumbas Šaponjac, V., Lević, S., Kalušević, A., Ćetković, G., Čanadanović-Brunet, J., ... Vidaković, A. (2019). Application of encapsulated natural bioactive compounds from red pepper waste in yogurt. *Journal of Microencapsulation, 36*(8), 704–714.

Shishir, M. R. I., Xie, L., Sun, C., Zheng, X., & Chen, W. (2018). Advances in micro and nano-encapsulation of bioactive compounds using biopolymer and lipid-based transporters. *Trends in Food Science & Technology, 78*, 34–60.

Singh, C. K., Liu, X., & Ahmad, N. (2015). Resveratrol, in its natural combination in whole grape, for health promotion and disease management. *Annals of the New York Academy of Sciences, 1348*(1), 150.

Smith, R. I., Cohen, S. M., Doull, J., Feron, V. J., Goodman, J. I., Marnett, I. J., ... Adams, T. B. (2003). GRAS flavoring substances 21-The 21st publication by the expert panel of the flavor and extract manufacturers association on recent progress in the consideration of flavoring ingredients generally. *Food Technology-Chicago, 57*(5), 46–59.

Smith, R. L., Adams, T. B., Cohen, S. M., Doull, J., Feron, V. J., Goodman, J. I., ... Wagner, B. M. (2004). Safety evaluation of natural flavor complexes. *Toxicology Letters, 149*(1–3), 197–207.

Sun, P., Zeng, M., He, Z., Qin, F., & Chen, J. (2013). Controlled release of fluidized bed-coated menthol powder with a gelatin coating. *Drying Technology, 31*(13–14), 1619–1626.

Takeungwongtrakul, S., & Benjakul, S. (2017). Biscuits fortified with micro-encapsulated shrimp oil: characteristics and storage stability. *Journal of Food Science and Technology, 54*(5), 1126–1136.

Takoi, K., Degueil, M., Shinkaruk, S., Thibon, C., Maeda, K., Ito, K., ... Tominaga, T. (2009). Identification and characteristics of new volatile thiols derived from the hop (Humulus luplus L.) cultivar Nelson Sauvin. *Journal of Agricultural and Food Chemistry, 57*(6), 2493–2502.

Tampau, A., González-Martinez, C., & Chiralt, A. (2017). Carvacrol encapsulation in starch or PCL based matrices by electrospinning. *Journal of Food Engineering, 214*, 245–256.

Taylor, A. J., & Hort, J. (2007). *Modifying flavor: An introduction. Modifying flavor in food* (pp. 1–9). Elsevier.

Tonutti, I., & Liddle, P. (2010). Aromatic plants in alcoholic beverages. A review. *Flavour and Fragrance Journal, 25*(5), 341–350.

Tzeng, T.-F., Chang, C. J., & Liu, I.-M. (2014). 6-Gingerol inhibits rosiglitazone-induced adipogenesis in 3T3-L1 adipocytes. *Phytotherapy Research, 28*(2), 187–192.

Unlu, M., Ergene, E., Unlu, G. V., Zeytinoglu, H. S., & Vural, N. (2010). Composition, antimicrobial activity and in vitro cytotoxicity of essential oil from Cinnamomum zeylanicum Blume (Lauraceae). *Food and Chemical Toxicology, 48*(11), 3274–3280.

Utama-Ang, N., Kopermsub, P., Thakeow, P., & Samakradhamrongthai, R. (2017). Encapsulation of *Michelia champaca L.* extract and its application in instant tea. *International Journal of Food Engineering, 3*(1), 48–55.

Wang, L., & Zhang, Y. (2017). Eugenol nanoemulsion stabilized with zein and sodium caseinate by self-assembly. *Journal of Agricultural and Food Chemistry, 65*(14), 2990–2998.

Wick, E. L., McCarthy, A. I., Myers, M., Murray, E., Nursten, H., & Issenberg, P. (1966). *Flavor and biochemistry of volatile banana components*. ACS Publications.

Woods, L. A., & Doull, J. (1991). GRAS evaluation of flavoring substances by the expert panel of FEMA. *Regulatory Toxicology and Pharmacology, 14*(1), 48–58.

Yadav, K., Bajaj, R. K., Mandal, S., Saha, P., & Mann, B. (2018). Evaluation of total phenol content and antioxidant properties of encapsulated grape seed extract in yoghurt. *International Journal of Dairy Technology*, *71*(1), 96–104.

Yang, Z., Peng, Z., Li, J., Li, S., Kong, L., Li, P., & Wang, Q. (2014). Development and evaluation of novel flavour microcapsules containing vanilla oil using complex coacervation approach. *Food Chemistry*, *145*, 272–277.

Yousuf, B., Gul, K., Wani, A. A., & Singh, P. (2016). Health benefits of anthocyanins and their encapsulation for potential use in food systems: A review. *Critical Reviews in Food Science and Nutrition*, *56*(13), 2223–2230.

Yuliani, S., Torley, P. J., D'Arcy, B., Nicholson, T., & Bhandari, B. (2006). Effect of extrusion parameters on flavour retention, functional and physical properties of mixtures of starch and d-limonene encapsulated in milk protein. *International Journal of Food Science & Technology*, *41*, 83–94.

Zabka, M., Pavela, R., & Slezakova, L. (2009). Antifungal effect of *Pimenta dioica* essential oil against dangerous pathogenic and toxinogenic fungi. *Industrial Crops and Products*, *30*(2), 250–253.

Zhu, G., Xiao, Z., Zhou, R., & Zhu, Y. (2014). Study of production and pyrolysis characteristics of sweet orange flavor-β-cyclodextrin inclusion complex. *Carbohydrate polymers*, *105*, 75–80.

CHAPTER

Plant extracts as coloring agents

9

Nirmal Kumar Meena[1], Vijay Singh Meena[2], M. Verma[3] and Subhrajyoti Mishra[4]

[1]Agriculture University, Kota, India [2]NBPGR, Regional Station, Jodhpur, India [3]SK Rajasthan Agriculture University, Bikaner, India [4]Junagarh Agriculture University, Junagarh, India

9.1 Introduction

Food, which is characterized by several sensory attributes like texture, taste, flavor, and color, is the basic need of everyone. Color is the most important viewable property, which acts as the focal point of costumer in foodstuffs. Color of any food material is an important sensory attribute, which affects consumer acceptance and around 62%–90% consumer assess the food by its color only (Singh, 2006). Basically, color is a molecule that absorbs certain wavelengths and transmits it or reflects it. As per the International Codex Alimentarius Commission (CAC), colorants are the substances that regulate the color of the food or are added to give color (Hastaoğlu, Can, & Vural, 2018). Food colorants are basically substances either natural or artificial, which retain the color of the food and cannot be eaten in the form of food as such (Batu & Molla, 2008). Color of food materials also imparts flavor, taste, and many nutraceutical properties and taken as an indicator for quality attribute. A strong cross linking of color with food makes it a profitable business and boosts the industries. Coloring of food has emerged during 1500 BCE and Egyptians also described colored drugs and wines (Burrows, 2009). It was reported that use of dyes by Egyptians in colored candy, and wine became popular during 400 BCE. According to historical background, the earliest record of using natural dyes was found in China in 2600 BCE. The process of dyeing was known in the Indus Valley period as early as 2500 BCE. There is a back record of saffron which has mentioned in the Bible and henna even before 2500 BCE (Gulrajani, 2001). In 1856 first organic colorant was discovered with the name of mauveine by William Henry Perkin. These colorants were named as "dye." In the beginning of 19th century, coal tar dye was produced from coal.

It has been reported by Burrows (2009) that food coloring substances were derived from some natural sources like saffron, paprika, turmeric, and flowers. Synthetic colorants gained popularity worldwide due to their long-term chemical stability in a small concentration, easy to use, and low cost (Borcakli, 1999). Synthetic colors like carmicin, amaranth, allura red, sunset yellow, tartrazine, ponceu 4R, indigo carmine, caramel etc. most common artificial food colors and are being used widely in food industry (Hastaoğlu et al., 2018). Wide application of these synthetic dyes raised serious health issues. The school going children are severely affected due to consuming processed food, ice creams and bakery products. In the changing scenario, use of natural healthy food becomes an excellent choice of costumer. Plant-based natural colors are considered to be safe and

Plant Extracts: ***Applications in the Food Industry.*** **DOI: https://doi.org/10.1016/B978-0-12-822475-5.00012-0**

healthy. Therefore, in the recent past couple of years, consumer's mood has been driving a force to replace the synthetic dyes with natural plant-based colors.

9.2 Synthetic colors and health impact

Artificial colorants are used in foodstuffs having different structure and source producing specific color. Synthetic colors like carmicin, amaranth, allura red, sunset yellow, tartrazine, ponceu 4R, indigo carmine, caramel etc. are used in nonalcoholic beverages, cola and fruit flavored drinks and quite common in beverage industry (Borcakli, 1999; Özcan, Artık, & Üner, 1997). Similarly, confectionary and baking industry also usesuch colors abundantly without taking health concern into consideration. Dyes such as carmoicin, allura red, amaranth, sunset yellow, and tartrazine are most common in candies, dairy products, and baking industries (Hastaoğlu et al., 2018). These synthetic dyes have negative impacts on health. These may cause allergic reactions, hyperactivity, and skin burning and may cause cancer like disease in children (McCann et al., 2007). Many artificial colors cause behavioral disorders and chromosomal disorders especially when used in child nutrition (Omaye, 2004). It has been reported that some of synthetic dyes like triphenylmethane (malachite green and crystal violet) are illegally used in food materials for coloring and that is also available in fish tissues as a result of residual veterinary drug (Arnold, LeBizec, & Ellis, 2009). Hyperactivity, learning disabilities and behavioral disorders due to food additives in children was already reported (Bateman et al., 2004; Feingold, 1975; McCann et al., 2007). Besides, there are reports that six most common dyes viz. tartrazine(E102), quinolone yellow (E104), sunset yellow FCF (E110), carmoisine/azorubine (E122), ponceau 4R (E129), and allura Red AC (E129) are more common in sugary and beverages, which may cause negative effect on attention deficit, hyperactivity, cancer, chromosomal aberration, neurotoxicity, developmental toxicity, psycotoxicity, and DNA damage (Martins, Roriz, Morales, Barros, & Ferreiral, 2016; Sabnis, Pfizer, & Madison, 2010). The colorants are also associated with the allergy and asthma in human being and some recent evident suggested that Allura red increases urticaria and asthma (Amchova, Kotolova, & Ruda-Kucerova, 2015; Pandey & Upadhyay, 2012; Pollock & Warner, 1990). Impact of synthetic colorants on health is expressed and correlated in different ways. However, exact mechanism and linkage of such disorders are still unknown (Table 9.1).

9.3 Natural food colors

Natural food colors are mostly produced from plant-based sources like leaves, fruits, vegetables, roots, etc. Plant contains an array of natural colorants, which have nutritional and health protective potentials. Coloring of the food with natural plant based colorants is gaining popularity and remarked in trend line worldwide due to health concern and safety point of view. Msagati (2013) mentioned type of food colorants on the basis of their synthesizing source such as natural colorants, natural like colorants and synthetic colorants. Food colorants may be based on their source of origin like plants, bacteria, fungi or animals; as per their hue such as red, blue, green, purple etc. and their chemical derivatives like anthocyanins responsible for blue-purple color (flavonoids derivatives),

Table 9.1 Major permissible synthetic colors approved by the Food Safety Standard Authority of India.

S. No.	Color	Common color name	Color index	Chemical class
1.	Red	Ponceau 4R	16,255 ×	Azo
		Carmoisine	14,720	Azo
		Erythrosine	45,430	Xanthene
2.	Yellow	Tartrazine	19,140	Pyrazolone
		Sunset Yellow	15,985	Azo
3.	Blue	Indigo	73,015	Indigold
		Carmine Brilliant Blue FCF	42,090	Triarylmethane
4.	Green	Fast green FCF	42,053	Triarylmethane

Adopted from Gazette of India, Food Safety Standard Authority of India (FSSAI), GoI, India released on August 1, 2011. https://fssai.gov.in/upload/uploadfiles/files/FSSAI-regulations.pdf (retrieved on 08.06.2021).

carotenoids for red, yellow/orange color (isoprenoid derivatives) and betalains red–pink (nitrogen-–heterocyclic derivatives) (Mittal, Sharma, & Singh, 2007; Viera et al., 2019). Moreover, these plants based colorants are called as pigments which have several medicinal, nutritional and cosmetic values (Frick & Meggos, 1988; Hari, Patel, & Martin, 1994; Shamina, Shiva, & Parthasarathy, 2007a, 2007b). Plant pigments are popularizing globally as a potential source of food color due to safe nature (Shamina et al., 2007a, 2007b). Major drawback of using natural colors in the food is their low stability and off odor to food. It has been reported that color could be used after removal of strong aroma and flavor from red cabbage and radish (Giusti & Wrolstad, 2003). The natural color of foods due to occurrence of natural pigments like carotenoids, chlorophylls, myoglobins, anthocyanins and their modification during processing such as caramel and color additives (Parkinson & Brown, 1981). Plant based colorants posses certain chemicals which can be used directly as such or modified into other form to impart hue. Certain processes like collection of raw materials, extraction, purification and their stabilization followed by formulation are used. As a result of these natural pigments a range of hues can be obtained from green to yellow, orange, blue, violet, red, purple and many more depending upon the source (Shameena et al., 2007). Some of the most common pigments are and their source are depicted in Table 9.2 and described underneath.

9.4 Anthocyanins

The word "anthocyanin" is derived from two Greek words "Anthos," which means flower and "kyanos" means dark blue (Delgado-Vargas & Paredes-Lopez, 2003). Anthocyanins are most spectacular water soluble plant pigments, which are responsible for red, blue, magenta, and purple color of plant-based food. Anthocyanin imparts different colors to the food and makes it attractive. Apart from this, it has health-promoting potential. Several studies suggested that consumption of anthocyanins reduce the risk of some chronic diseases. It also possesses antioxidant properties and helps in boosting of immunity. It is broad group of polyphenolic compounds that is, flavonoids which is

Table 9.2 Major color and pigments obtained from different plant extracts.

S. No.	Color	Type of pigment	Name of plant	Parts used
1.	Red	Anthocyanins	Elderberry	Fruit
2.	Purple/black		Black currant	Fruit
3.	Red, purple, blue		Grape	Fruit
4.	Red		Strawberry	Fruit
5.	Purple		Plum	Fruit
6.	Red, pink, purple		Raspberry	Fruit
7.	Blue		Blueberry	Fruit
8.	Black/dark purple		Black carrot	Root
9.	Red		Red radish	Root
10.	Blue		Blackberry	Fruit
11.	Red, violet	Betanins	Beet root	Root
12.	Yellow	Carotenoids	Yellow carrot	Root
13.	Yellow-orange, red	Bixin, carotenoids	Annatto	Seeds
14.	Orange-red	Crocin	Saffron	Stigma
15.	Pink, purple	Anthocyanin	Karonda	Fruit
16.	Pink	Betalins	Cactus pear	Fruit
17.	Yellow-orange	Lutein	Marigold	Flower
18.	Yellow	Curcumin	Turmeric	Rhizome
19.	Red-orange	Lawsone	Lawsonia	Leaves
20.	Red	Anthocyanins	Roselle	Calyces
21.	Blue	Indigo compounds	True indigo	Leaves
22.	Yellow–orange	Butein	Dhak	Flower
23.	Yellow	Barberine/Anthocyanins	Barberry	Fruits

a secondary metabolite synthesized by plants. These polyphenolic substances are mainly glycosides, polyhydroxy and polymethoxy derivative of 2-phenylbenzopyrylium or flavylium salts (Jackman et al., 1987). Anthocyanins are glycosides of anthocyanidins and sugars (Mortensen, 2006). Anthocyanidins are mostly glycosylated at 3rd position and many times at other positions. A number of anthocyanins are known due to their diverse glycosylation and acylation (Mortensen, 2006). Presence of double bond and carbonyl group on C ring differentiates them to each other. Aglycones have long chromophore with eight conjugated double bonds carrying cation charge fall between 465 to 550 nm visible range. The hydroxyl group on the aglycone is substituted by sugars with glycosidic bond or acylated with aromatic or aliphatic acids (He & Giusti, 2010). Anthocyanins are highly water soluble pigments therefore usually extracted with water and lower concentration of alcohol. To date, more than 635 anthocyanins have been identified but only few have been reported as most important and abundant (Andersen & Jordheim, 2008). Pelargonidin, cyanidin, delphinidin, peonidin, petunidin and malvidin are most common type of anthocyanins present in plants parts. There are several sources of anthocyanins depending on availability. The most common sources for the production of anthocyanins are grape, black currant, red cabbage, elderberry, rose, strawberry, jamun, karonda, black and purple carrot, and red radish. Besides, many flowers, leaves, and plants species are used as source of color. Anthocyanins are mostly used as natural colorant in beverage industry like wines and fruit juices.

Anthocyanins are highly stable in acidic medium but instable and prone to degradation under normal condition and may form insoluble brown product. Anthocyanins produce a diverse range of colors depending on chemical structure. It has been well established that more number of hydroxyl group produces blue color whereas more methoxy group yields red color. Pelargonidin is associated with orange color while delphinidin and malvidin are purple (Mortensen, 2006). Anthocyanins act on the basis of pH of medium. Four major anthocyanin forms exist in equilibria: the red flavylium cation, the blue quinonoidal base, the colorless carbinol pseudobase, and the colorless chalcone (Brouillard & Delaporte, 1977). When pH goes below 2, anthocyanins exist in red flavylium cation form.

Anthocyanin is the chief component of human diet and chiefly present in many colored plant tissues (Naczk & Shahidi, 2004). Human intake anthocyanins from several fruits like grape, pomegranate, plums, berries etc. The composition of anthocyanin varies across the plant species and even variety to variety. However, it is stated that berries provide most anthocyanins content per serving (He & Giusti, 2010). From the available literatures it is envisaged that more than 90% anthocyanins contain glucose as a glycosylating sugar and Cy-3-glu is most widespread anthocyanin in nature (Andersen & Jordheim, 2008; Kong et al., 2003). Besides fruits and vegetables, many processed products like jam, jelly, wines, and juices are also rich source of anthocyanin which offers a sufficient amount of anthocyanin in human diet. With the advancement consumer is becoming more health conscious and synthetic dye has been replaced with natural anthocyanin rich color. Acylated anthocyanins use is increasing significantly in food industry as a natural colorant due to stability. Elder berry, capsicum and grape, flower petals and karonda like fruits have potential use in food industry due to rich in anthocyanin content. It has been reported that natural anthocyanin didn't harm human health on excess oral consumption (Brouillard, 1982). Its commercial application in food industry has been permitted by many countries like Japan, USA, and Europe including India. Natural colored vegetables like radish, sweet potato, black carrot, red cabbage etc. are also used as anthocyanin basket owing to their richness of acylated anthocyanins. Radish and red fleshed potatoes are using as alternative against Federal Food Drug and Cosmetic Red No. 40 (Allura red) (Shipp & Abdel-Aal, 2010). Recently, natural anthocyanins rich beverage has been prepared from fruits of Karonda (*Carissa carandus*) named as "Lalima" (Krishna et al., 2017). It is reported that colorless lemon based drink enriched with Lalima contains 469.2 μg additional anthocyanin (cyanidin-3-glucoside equivalent). The flavonol glucoside from the rose petals can be extracted and used as natural colorant. This phenolic compound is extracted through distillation from *Rosa damascena* Mill. petal which is enriched with sufficient amount of kaempferol and quercetin and its derivatives. Onion is another rich source of quercetin compound which imparts gold and yellow color. Is reported *Opuntia stricta* as a potential source of anthocyanins having red color. Another species of opuntia contains both betacyanins and betaxanthins whereas *O. stricta* conatins only betacyanins (Castellar, Obon, Alacid, & Fernandez-Lopez, 2003; Shamina et al., 2007a, 2007b).

However, differences in pH level affect its stability; thus, majority of natural anthocyanins are used in acidic food products. Krishna et al. (2017) extracted food colorant from Karonda with ethanol solvent. Acetone can be used as efficient solvent for extraction of anthocyanins from red fruits (Garcia-Viguera, Zafrilla, & Tomas-Barberan, 1998; Gil et al., 2000). Identification and characterization of type of anthocyanin is carried out by UV spectrophotometric analysis. Maximum absorbance of anthocyanins occurs at a wavelength of 520–540 nm in visible spectra of spectrophotometer. Some recent advances like use of NMR is also reliable to identify anthocyanins especially acylated anthocyanin (Castaneda-Ovando, Pacheco-Hernandez, Paez-Hernandez, Rodriguez, & Galan-Vidal, 2009; Kosir & Kidric, 2002). The successful application of anthocyanin in food industry can be only possible after suitable procedure and methods for application.

9.5 Betalains

Betalains are another most important group of plant pigment, which are abundantly present in many plants like beetroot, bougainvillea, portulaca, mirabilis, and opuntia. Betalains are heterocyclic and water soluble nitrogenous compound which can be further divided into two major classes according to their chemical structure: betacyanins (red violet color) such as betanin, prebetanin, isobetanin and neobetanin and betaxanthins, responsible for orange-yellow coloring, comprising vulgaxanthin I and II and indicaxanthin (Azeredo, 2009; Saponjac et al., 2016). Betalains are mainly produced by plants belonging to order Caryophyllales.

Betalins are vacuolar nitrogenous compounds having a core structure (protonated 1,2,4,7,7-pentasubstituted 1,7 diazaheptamethin system) known as betalamic acid (Khan & Giridhar, 2015). Betalamic acid condensed with cyclo-DOPA (L-3,4-dihydroxy-phenylalanine)/its glucosyl derivatives, and amino acids/its derivatives which results to formation of betalains viz. betacyanins (violet) and betaxanthins (yellow), respectively. Peterson and Joslyn (1958) mentioned that before 1957, Betalins were categorized as nitrogenous anthocyanin. Production of betanidin on its hydrolysis and isolation of indicaxanthin evidenced that this contains a separate pigment containing system of 1,7-diazaheptamethin which is responsible for their chroma (Mabry, Wyler, Parkih, & Dreiding, 1967; Piattelli, Minale, & Prota, 1964; Wyler and Dreiding,1957). Khan and Giridhar (2015) have reviewed systemically the various aspects such as production, biosynthesis and eco-physiological factors of betalins in a comprehensive mode. Betalins are hydrophilic in nature and due to this, extraction is carried out in aqueous methanol or methanol at pH 5 supplemented with ascorbic acid (Strack, Vogt, & Schliemann, 2003). This stability could be enhanced at 0.25% (w/v) (Khan & Giridhar, 2014). Piattelli et al. (1964) isolated a compound from cactus pear plant (*Opuntia ficusindica*) and identified as indicaxanthin. Indicaxanthin compound (λ_{max} 260, 305, 485 nm) on alkali fusion in the absence of oxygen yielded proline and 4-methylpyridine-2,6-dicarboxylic acid.

Betalains occur in almost all plant parts including flowers, fruits, bracts, inflorescence, and petioles. Since last few decades, a limited betalins sources have been used widely. The most common source like beet root; dragon fruit, opuntia, and amaranth have been used in most of investigation elucidating pigment extraction and/or purification for food uses, and their biological activities in crude and purified extract material (Gandía-Herrero, Escribano, & García-Carmona, 2014; Moreno, García-Viguera, Gil, & Gil-Izquierdo, 2008). Accumulation betacyanins in betalainoplast in vacuoles of tepal cells firstly in *Rebutia* spp. was confirmed by Iwashina, Ootani, and Hayashi (1988). Commercial production of such pigments is not fully exploited because of the lack of quantification methods, species which contain more tissue and extraction technologies. It is estimated that global production of betalins stands at 96.8 Gt out of 99.99% contribution is from beetroot alone. Remaining contribution is obtained from red pitaya, amaranth seeds and opuntia. It was also stated that low production and extraction of betalains might be due to less stability during processing resulting in significant loss. Apart from this, only betacyanins are considered as economically viable and stable as loss of betaxanthins is higher as compare to betacyanins during extraction (Delgado-Vargas, Jiménez, & Paredes-López, 2000; Herbach, Stintzing, & Carle, 2006). Betanin is only abundant betacyanin which has been approved for commercial use in food and pharmaceutical industries (Silva et al., 2016). Several studies have suggested that purified form of betanin can be stable at low temperature and alkaline pH during storage; thus, it may be useful as food colorant and antioxidant additive in meat industry. High antioxidant potential in betalains is very well known (Esatbeyoglu, Wagner, Schini-Kerth, & Rimbach, 2015). For coloring, beet root power and beet extracts are readily available and chiefly used worldwide. Studies suggested that betalains do not change its hue at different pH in food and beverages (JebaKezi & Judia Harriet Sumathy, 2014). Betalains also have good stability against pH and light (Chethana, Nayak, & Raghavarao, 2007; Gonçalves, Da Silva, DeRose, Ando, & Bastos, 2013). Besides color imparting phenomenon, it has enormous health benefit properties.

9.6 Carotenoids

Carotenoids are most common lipophilic natural pigments responsible for red, orange, and yellow color in fruits, vegetables, and sea foods. Yabuzaki (2017) reported more than 1000 chemically distinctive carotenoids in plants. Most of carotenoids absorb light in a spectral region (400–550 nm), thus increases the light harvesting spectrum (Hashimoto, Uragami, & Cogdell, 2016). Carotenoids have good health promoting potential due to their inherent antioxidant properties. Investigation on beta carotene revealed that it has provitamin-A activity (Grune et al., 2010). Similarly, lycopene also a type of carotenoids which is capable to reduce the risk of prostate cancer and zeaxanthin and lutein reduces age related macular generation and also cognitive functions (Carpentier, Knaus, & Suh, 2009; Giovannucci, 2002; Johnson, 2012). Like other pigments, carotenoids also have stability problems and become instable under adverse conditions. The major factors are processing conditions, duration, adverse storage, packaging materials and exposure to light (Rodriguez-Amaya, 2015b). Carotenoids are commercialized in food industry as a colorant produced by chemical extraction methods. There are several fruits, vegetables, flowers and grasses which are rich sources of carotenoids. It is commercially extracted from Paprika, saffron, tomato, marigold, lycopene, annatto and many other plants (Rodriguez-Amaya, 2015b). Orange fleshed sweet potato is also rich source of α carotene and β-carotene which is used to overcome deficiency of vitamin A (Hermanns et al., 2020). Citrus fruits are also reported to be abundant source of carotenoids. On the basis of carotenoids, citrus cultivars are divided into three groups, as β-cryptoxanthin rich, violaxanthin rich, and low in β-cryptoxanthin and violaxanthin (Ikoma, Matsumoto, & Kato, 2016). It is well explained that there is direct correlation between carotenoids content and color of skin. In citrus peel, composition and ratios of the carotenoids affect the fruit color (Xu, Tao, Liu, & Deng, 2006).

Carotenoids are C40 tetraterpenes/tetraterpenoids formed from eight C5 isoprenoid units. Chemically, it is characterized by centrally conjugated double bond, which serves as the chromophore. This chromophore has light absorbing properties which gives attractive colors to carotenoids and also responsible for their typical functions and properties. In plants carotenoids biosynthesis take place in plastids (Li et al., 2016) including chromoplast in flower, fruits, roots and chloroplast in leafy vegetables. Carotenoids synthesized through methylerythritol phosphate (MEP) pathway by utilizing isoprenoid precursors. In this pathway, the enzyme 1-Deoxy-*D*-xylulose 5-phosphate synthase catalyzes the first reaction of *D*-glyceraldehyde 3-phosphate with pyruvate (Rodriguez-Amaya, 2015b). Further, biosynthesis of carotenoids can be divided into two sub categories: first carotenes (includes different carotenes) and another class is xanthophylls (includes Beta cryptoxanthin, zeaxanthin, antheraxanthin, violaxanthin, neoxanthin, zeinoxanthin, lutein etc.). Phytoene is first carotenoid compound of MEP pathway. Citrus accumulates wide array of carotenoids in both peel and pulp tissues depending upon tissue specific regulation. A number of pathways were reported for synthesis of carotenoids in citrus peel. Abscise of the β-cryptoxanthin and zeaxanthin yield red color β-citraurin in the peel of citrus (Hermanns et al., 2020). Recently, Zheng et al. (2019) discovered a natural variation in CCD4b promoter which was found to be most genetic determinant for natural variation in red coloration of citrus peel. Existence of red orange flesh color of melons (*Cucumis melo*) was reported due to predominant β carotene. CmOr was reported to be chief gene governing fruit orange-flesh color (Gur et al., 2017). Similarly, water melon produces a range of carotenoids which impart an attractive red, yellow and orange flesh. Lycopene was reported to be predominant in red fleshed water melon. It is reported that ClLCYB locates in a major flesh color QTL and strongly associated with red flesh color in most modern watermelon cultivars (Guo et al., 2019). Papaya is also having both red and yellow color flesh. Red flesh of papaya is associated with lycopene content whereas yellow fleshed papaya fruits rich in β-carotene and β-cryptoxanthin. In a discovery by Blas et al. (2010) chromoplast and chloroplast localized CpCYCB and CpLCYB were reported to be critical for lycopene accumulation in red-fleshed papaya. Besides, carotenoid accumulation in papaya was also regulated by ethylene and environment signals (Fabi & Do Prado, 2019). Lycopene is considered as ζ carotene, but it is predominant pigment in most of fruits and vegetables like tomato, watermelon, papaya, guava, grapefruit, and pitanga (Rodriguez-Amaya et al. 2008). An appreciable amount of α-carotene is present in carrot, red palm oil, and squashes and pumpkin, whereas γ-carotene are found in rose hips and pitanga abundantly. Rose hips also contain Rubixanthin, a derivative of γ-carotene, which is major pigment of hips (Hornero-Méndez & Mínguez-Mosquera, 2000). β-Cryptoxanthin, a xanthophyll is chiefly present in many orange-fleshed fruits, such as peach, nectarine, papaya, persimmon, tree tomato, and *Spondias lutea* (Rodríguez-Amaya et al., 2008; Rodriguez-Amaya, 2015a). Lutein is also chiefly present in yellow flowers, green leafy vegetables and some cucurbits. However, zeaxanthin is minor carotenoids and only present in limited form because of checking its biosynthesis at β carotene. Some of species specific carotenoids also detected in little amount. Capsanthin and capsorubin are major example which is abundantly present in red pepper (Rodriguez-Amaya, 2015a). Lactucaxanthin, a carotenoid with two ε-rings is rarely present in lettuce (Rodriguez-Amaya, 2015b).

Astaxanthin is pink color xanthophyll, which is derived from carotenoid. It has very high antioxidant potential and also used as a natural colorant. However, it is found in algae, yeast, fishes, and some crustacean byproducts. This is a chief source of color in marine industry. Astaxanthin is preferably more common as synthetic color and available under the commercial brand name CarophyllPink and canthaxanthin as CarophyllRed (Bowen et al., 2002; Gouveia et al., 2002; Storebakken & No, 1992).

Apocarotenoids are the form in which the carbon skeleton has been shortened by removal of fragments from one or both ends of the usual C40 structure. The most common examples of natural apocarotenoids are bixin mostly found in annatto, and crocetin, the chief constituent of saffron color (Rodriguez-Amaya, 2015b). As already mentioned, carotenoids are lipophilic in nature, majority of carotenes are water insoluble pigments but soluble in organic solvents such as acetone, petroleum ether, hexane, alcohol, ethyl ether, chloroform etc. Xanthophylls are better dissolved in methanol and ethanol (Rodriguez-Amaya, 2015b). β-Carotene and the xanthophyll lutein can be soluble in tetrahydrofuran (Craft & Soares, 1992). All the carotenoids have numerous health benefits and it is recommended to take more carotene rich diet for healthy body.

9.7 Porphyrin pigments (chlorophylls)

Natural porphyrin pigments are chlorophylls and chlorophyllins. The green chlorophylls are found in chloroplast where they furnish the process of photosynthesis and facilitate the assimilation of carbon dioxide. Green chlorophylls consist of 4-pyron core with a centrally located atom of Mg. In general chlorophylls are olive green to dark green in color obtained from extraction of green leaves. It is very less stable and sensitive to high heat and temperature. Chlorophylls are used in different beverages, jam, jellies, syrups, pickles and also as medicinal tonic. They have very good health potential and help in curing many diseases, improve metabolism and enhance amount of hemoglobin in blood.

Green natural colorants are constituent of chlorophyll and its derivatives. European current legislation (Regulation EC No. 1333/2008 and its amendment) has allowed two major natural green colorants E140 and E141 which is related with chlorophylls and its derivative extractions respectively. Green colorants can be extracted from different chlorophyll rich fruits, vegetables and other grasses. Herbs and vegetables especially leafy vegetables are abundant source of chlorophyll. Chlorophyll a and b are most dominant chlorophylls used as natural colorant. Some of researchers explained the natural green color named as chlorophyll or E140i which is directly obtained through solvent extraction from natural green plants and mainly composed of chlorophyll a and b and their pheophytin derivatives. Another colorant named as E140ii or chlorophyllin. However, there is no clear cut definition and chemical structure of chlorophyllin but Willstätter (1915) reported that it is chlorophyll derivatives which produced after saponification of chlorophyll without changing in color. Earlier, it was assumed that chlorophyllin composed of that intact Mg molecule but now this theory has rejected and this term is not so common because colorant green pigment include the chemical structure without magnesium molecule. For fixing of green color copper is used. It is reported that water soluble pigment E141ii is most common in food industry and requires saponification and copper treatment for more transformation into green from chlorophylls (Viera et al., 2019). Chlorophylls are stable in their natural existence but prone to change their structure during processing and handling. Most common factors are pH, temperature and nature of component. Major reaction such as substitution of central Mg ion by H ion leads to drastic changes in color of pigment and possibly this is major lacuna in loss of green color during processing of food materials. It is well explained that Mg derivatives are green in color but Mg free mainly pheophytins and pheorphorbides are brown in color. To fix this green color, several attempts such as can coatings, use of alkali agents etc. have been made by researchers but this was found to be least effective with some other negative effects like off flavor (Schwartz et al., 2017). Subsequently, a successful approach has been developed and patented. In a new approach the H ion has been substituted by addition of zinc and copper to form more stable green color metallochlorophylls which consequently "re-green" the food product. A long term investigation was done to fix this process in vegetables during processing (Jones, White, Gibbs, Butler, & Nelson, 1977; LaBorde & von Elbe, 1990; Schanderl, Marsh, & Chichester, 1965). In the continuation, an another patent was registered by Continental Can Company termed as "Veri-Green" in which vegetables were blanched in a Zn^{+2} or Cu^{+2} enriched brine solution to form Zn and Cu pheophytins to make canned beans more greener (Segner, Ragusa, Nank, & Hoyle, 1984; Von Elbe, Huang, Attoe, & Nank, 1986). However, a large amount of Zn is required for this against the permissible limit of 75 ppm by FDA. Therefore, encapsulation is necessary to improve this procedure more efficient. Some of the vegetables are spray dried and their encapsulated green color is using in food industry. It is very difficult to produce natural green hue. In country like Indonesia, natural chlorophyll is being extracted from Suji (*Dracaena angustifolia Medik) Roxb.* plant and using in food industry (Aryanti, Nafiunisa, & Willis, 2016; Prangdimurti, Muchtadi, Astawan, & Zakaria, 2006). However, chlorophylls from this plant are highly prone to enzymatic and non enzymatic degradation.

9.8 Regulatory mechanism for food colors

Food colors and natural pigments are not classified separately by Codex, but they are kept in the category of food additives. Certain limits have been standardized for the handling of synthetic colors but they should have same safety standards as other natural food colors. Licensing and certification process is mandatory for synthetic food colors (Shamina et al., 2007a, 2007b). Legal advisory and regulated use of food colors is governed jointly by WHO and FAO. The CAC along with Codex Committee on Food Additives and Contaminants-CCFAC and Joint Expert Committee on Food Additives—JECFA prepare documents of regulatory practices for safe production of the food which is accepted by majority of countries in all around the world (Luckey, 1968). Moreover, these regulations vary country to country. European Food Safety Authority (EFSA) and the Food and Drug Administration (FDA) are well known agencies in this line (Martins et al., 2016). The FDA is primarily responsible for safe handling and use of food additives, thus a food manufacturer must first get approval from FDA to use new colorant into food.

FDA issues a certificate to manufacture, sale, and use for synthetic colors, but natural colors are exempt from certification process. In USA certified color additives and color additives color additives exempt from certification terms applied. Shamina et al. (2007a, 2007b) mentioned in a comprehensive review that only 43 colorants are authorized as food additives by Council of the European Union, and an E number has been issued. Out of, 16 colorants are based on plant origin and some extracts from juices, fruits and vegetables are also used as colorants. Lehto et al. (2016) collected regulatory information of food colors and compared these regulations in European Union and United States. Only approved colors can be used in both EU and US. Food colors are used as food additive agents and regulated under the comprehensive set of regulations such as regulation

(EC) No. 1331/2008 (European Commission, 2008a) categories the common authorization processes, regulation (EC) No. 1333/2008 (European Commission, 2008b) deals with food additives and its amendment and regulation (EC) No 1129/2011 (European Commission, 2011) presents the rules for food colors. The use of permitted food colors, maximum limits and instruction for uses are also described under regulation (EC) No. 1333/2008. There is also a provision for new color which was not earlier in the market (before 1997) and to be commercialized needs to be evaluated under Regulation (EC) No 258/97 on novel foods (European Commission, 1997) and under the new Regulation (EU) No 2015/2283 (European Commission, 2015) from 2017 onwards, before commencement in the market.

Similarly, there are regulatory provisions in US for federal colors additive which were enforced under US Federal Food, Drug and Cosmetic Act. There are many amendments such as The Food Additive and the Color Additive Amendment, the Food Safety and Modernization Act-2012, the Fair Packaging and Labeling Act-1966 and the Public Health Security and Bioterrorism Preparedness and Response Act (Bioterrorism Act) which came into existence in 2002 also part of this series (Barrows, Lipman, & Bailey, 2003; Northcutt & Parisi, 2013). The main regulatory body for food colors is US Food and Drug Administration which regulates these set of rules under the Title 21 of the Code of Federal Regulations (CFR, 2016). The definition of these doesn't cover the separate classification of colors in the form of natural or synthetic and also their source of origin. The broad classification of approved colors in US is done into two categories that is, certified colors which include all artificial colors and need to be certified by FDA and colors exempt from certification that contain natural pigments (Lehto et al., 2016). Another important point with regard to natural pigments is that all certification exempted color must have purity, specifications and used limitations prescribed under regulatory standards and these must verify by the users not from the FDA. The substances or food materials which impart their own color (like chocolate in milk) when added to the food cannot be considered as a color (21 CFR Section 70.3). In present time there are total 39 colors are authorized and approved for uses in EU whereas in USA only 27 natural colors

Table 9.3 Common available plant extract-based colorants in India.

S. No.	Name of color group	Color name
1.	Carotenoids	Beta carotene Beta-apo 8′-carotenal Methylester of Beta-apo 1′ carotenoic acid Ethylester of Beta-apo 8′ carotenoic acid Canthaxanthin
2.	Chlorophyll	Chlorophyll
3.	Riboflavin (Lactoflavin)	Riboflavin (Lactoflavin)
4.	Caramel	Caramel
5.	Annatto	Annatto
6.	Saffron	Saffron
7.	Curumin or turmeric	Turmeric

Adopted from Gazette of India, Food Safety Standard Authority of India (FSSAI), GoI, India released on August 1, 2011. https://fssai.gov.in/upload/uploadfiles/files/FSSAI-regulations.pdf (retrieved 08.06.2021).

are approved and 9 synthetic colors approved. In spite of different regulatory procedures the overall aim is ensure to make food safer under all circumstances.

In India, as per regulation of Food Safety Standard Authority of India (FSSAI) Regulation 4.2.1 Food additives, unauthorized coloring material except a specifically permitted by these rules is prohibited. Additional coloring material other than mentioned on label attached with article is also prohibited. However, some of the natural colors either produced naturally or extracted synthetically may be used. Common permitted natural colors and synthetic colors are mentioned in Tables 9.2 and 9.3, respectively.

The maximum limit of permitted synthetic food colors has been decided by the authority. For natural colors it is 200 ppm in candy, crystallized and glazed fruits. Beyond that quantity no one can use synthetic colors. As per regulation of FSSAI "The maximum limit of permitted synthetic food colors or mixture thereof which may be added to any food article enumerated 85 in Regulation (vii) 4.2.1 (1) of these Regulations shall not exceed 100 parts per million of the final food or beverage for consumption, except in case of food articles mentioned in clause (c) Regulation 3.1.7 of these Regulations where the maximum limit of permitted synthetic food colors

Table 9.4 International numbering system for different natural food color under the category food additives governed by Codex Alimentarius.

S. No.	International numbering system	Food color name
1.	100	Curcumins
2.	100 (i)	Turmeric
3.	160a (ii)	Natural extracts
4.	160b	Annatto extracts
5.	160c	Paprika Oleoresins
6.	160d	Lycopene
7.	160e	Beta-apo-carotental
8.	160f	Beta-apo-′-carotenic acid, methyl or ethyl ester
9.	161a	Flavoxanthin
10.	161b	Lutein
11.	161c	Krytoxanthin
12.	161d	Rubixanthin
13.	161e	Violoxanthin
14.	161f	Rhodoxanthin
15.	161g	Canthaxanthin
16.	162	Beet red
17.	163	Anthocyanins
18.	163 (ii)	Grape skin extracts
19.	163 (iii)	Black currant extracts
20.	164	Gardenia yellow

Class names and international numbering system for food additives CAC/GL36–1989 (last amendment 2011). Accessed from http://www.fao.org/fao-who-codexalimentarius/shproxy/en/?lnk = 1&url = https%253A%252F%252Fworkspace.fao.org%252Fsites%252Fcodex%252FStandards%252FCXG%2B36%-1989%252FCXG_036e.pdf (retrieved 08.06.2021).

Table 9.5 List of color additives exempted from certification (natural colors) by Food and Drug Administration.

21 CFR section[a]	Straight color	EEC[b]	Year approved	Uses and restrictions
Section73.30	Annatto extract	E160b	1963	Foods generally.
Section73.40	Dehydrated beets (beet powder)	E162	1967	Foods generally.
Section73.75	Canthaxanthin	E161g	1969	Foods generally, NTE 30 mg/lb of solid or semisolid food or per pint of liquid food; May also be used in broiler chicken feed.
Section73.85	Caramel	E150a-d	1963	Foods generally
Section73.90	β-Apo-8′-carotenal	E160e	1963	Foods generally, NTE15 mg/lb solid, 15 mg/pt liquid.
Section73.95	β-Carotene	E160a	1964	Foods generally
Section73.100	Cochineal extract	E120	1969	Foods generally
			2009	Food label must use common or usual name "cochineal extract"; effective January 5, 2011
	Carmine	E120	1967	Foods generally
			2009	Food label must use common or usual name "carmine"; effective January 5, 2011
Section73.125	Sodium copper chlorophyllin	E141	2002	Citrus-based dry beverage mixes NTE 0.2 percent in dry mix; extracted from alfalfa
Section73.140	Toasted partially defatted cooked cottonseed flour	--	1964	Foods generally
Section73.160	Ferrous gluconate	--	1967	Ripe olives
Section73.165	Ferrous lactate	--	1996	Ripe olives
Section73.169	Grape color extract	E163	1981	Nonbeverage food
Section73.170	Grape skin extract (enocianina)	E163	1966	Still & carbonated drinks & ades; beverage bases; alcoholic beverages (restrict. 27 CFR Parts 4 & 5).
Section73.200	Synthetic iron oxide	E172	1994	Sausage casings NTE 0.1% (by wt.).
			2015	Hard and soft candy, mints and chewing gum
			2015	For allowed human food uses, reduce lead from ≤20 to ≤5 ppm
Section73.250	Fruit juice		1966	Foods generally
			1995	Dried color additive
Section73.260	Vegetable juice		1966	Foods generally
			1995	Dried color additive, water infusion
Section73.300	Carrot oil	--	1967	Foods generally
Section73.340	Paprika	E160c	1966	Foods generally
Section73.345	Paprika oleoresin	E160c	1966	Foods generally

(*Continued*)

Table 9.5 List of color additives exempted from certification (natural colors) by Food and Drug Administration. *Continued*

21 CFR section[a]	Straight color	EEC[b]	Year approved	Uses and restrictions
Section73.350	Mica-based pearlescent pigments		2006	Cereals, confections, and frostings, gelatin desserts, hard and soft candies (including lozenges), nutritional supplement tablets and gelatin capsules, and chewing gum.
			2013	Distilled spirits containing not less than 18% and not more than 23% alcohol by volume but not including distilled spirits mixtures containing more than 5% wine on a proof gallon basis.
			2015	Cordials, liqueurs, flavored alcoholic malt beverages, wine coolers, cocktails, nonalcoholic cocktail mixers and mixes and in egg decorating kits.
Section73.450	Riboflavin	E101	1967	Foods generally
Section73.500	Saffron	E164	1966	Foods generally
Section73.530	Spirulina extract		2013	Candy and chewing gum
			2014	Coloring confections (including candy and chewing gum), frostings, ice cream and frozen desserts, dessert coatings and toppings, beverage mixes and powders, yogurts, custards, puddings, cottage cheese, gelatin, breadcrumbs, and ready-to-eat cereals (excluding extruded cereals)
Section73.575	Titanium dioxide	E171	1966	Foods generally; NTE 1% (by wt.)
Section73.585	Tomato lycopene extract; tomato lycopene concentrate	E160	2006	Foods generally
Section73.600	Turmeric	E100	1966	Foods generally
Section73.615	Turmeric oleoresin	E100	1966	Foods generally

[a]Title 21, Code of Federal Regulations (CFR) by Federal Food Drug and Cosmetic Act, 21 CFR Section73.
[b]International Numbering Sys. (European Commission Regulation), European Union.
Note: Accessed from: https://www.fda.gov/industry/color-additive-inventories/summary-color-additives-use-united-states-foods-drugs-cosmetics-and-medical-devices#ftnote1 (accessed 08.06.21).

shall not exceed 200 parts per million of the final food or beverage for consumption." In addition to that, purity of color should be maintained and according to rule, the colors specified in Regulation (v) of 4.2.1 (1) of these Regulations, when used in the preparation of any article of food shall be pure and free from any harmful impurities. Besides, there are some setups of rules (Part 6.16) for selling of synthetic and natural colors. They says that no person can manufacture and sale the synthetic food colors and mixture directly without having any license or label on container mentioning "Food Colors" or chemical/common name of that mixture.

At international level, a specifically international numbering system for food additives adopted by Codex Committee on Food Additives and Contaminants for the identifying food additive in ingredient lists by a numerical approach (Table 9.4). These numerical series also include several synthetic and natural colors and extracts (Table 9.5).

9.9 Challenges with natural colors

Natural colorants are primarily derived from plants, animals, and microbes. However, colorants extracted from the plants changed the perception of consumers. They become more interesting and preferable because of safety and healthy characteristics. Despite certain advantages, plant-based colors have several production and stability challenges. It is reported that plant colorants have higher production cost in the initial, and they are unstable (Prince, 2017; Sen, Barrow, & Deshmukh, 2019). Other challenges with natural colors are heat and light sensitivity and interaction with other ingredients (Prince, 2017). In addition to that colors behavior is highly dependent on the application and that is a tough task to understand by manufacturer for making formulations. Anthocyanin based plant colorants are reported to be highly unstable and highly susceptible to degradation (Jiménez-Aguilar et al., 2011; Sagdic et al., 2013; Zhu et al., 2015). It is also evident that due to external factors, packaging materials, and interference in chemical structure, natural color loses its attractiveness (Jiménez-Aguilar et al., 2011; Lemos, Aliyu, & Hungerford, 2012; Zhu et al., 2015). The degree of susceptibility also varies from source to source from which colorants extracted. It has reported that anthocyanin extracted from red tulip is less sensitive as compare to violet though the plant belongs to same family (Sagdic et al., 2013). Besides, there is no definite legislation separately for natural colors. They are merged with food additives. Therefore, no certain regulatory standards and norms for uses have been assigned to natural colorants. Now a day, consumer satisfaction not only limited to sensory attributes but also with their health impacts, safety and quality (Giusti &Wrolstad, 2003; Kammerer, Kammerer, Valet, & Carle, 2014; Xi et al., 2007). Thus, food industry must satisfy the costumer with its claim for safe, healthy, and reliable in nature and therefore they need strong collaboration with research institutes and clinical agencies.

9.10 Conclusion

Natural food colors have increased demand because of their safety, biological, and health potential over synthetic colors. Uses of natural colors like curcumin, betalains, anthocyanin, paprika, annatto extracts etc. find wide application in the food industry. Their regulatory mechanism has been determined by FDA and different regulatory bodies countrywide to use them in a safe manner. As compared to synthetic dyes, these natural colors are exempted from certification and perceived safer by the human beings. Despite many advantages, extraction process, stability, and longevity are still challenging for natural colorants and need to be addressed by technological interventions. However, producers are bound to keep all regulatory mechanisms, methods of use, and criteria in controlled and regulatory manners.

References

Amchova, P., Kotolova, H., & Ruda-Kucerova, J. (2015). Health safety issues of synthetic food colorants. *Regulatory Toxicology and Pharmacology*, *73*(3), 914–922.

Andersen, O. M., & Jordheim, M. (2008). Anthocyanins—Food applications. In: *Presented at proceeding of 5th international congress pigments foods: For quality and health*, 14–16 August, Helsinki, Finland.

Arnold, D., LeBizec, B., & Ellis, R. (2009). Malachite green in *Residue evaluation of certainveterinary drugs. FAO JECFA Monographs, 6*.

Aryanti, N., Nafiunisa, A., & Willis, F. M. (2016). Ekstraksi dan Karakterisasi Klorofildari Daun Suji (*Pleomele Angustifolia*) sebagai Pewarna Pangan Alami. *Jurnal Aplikasi Teknologi Pangan*, *5*(4), 129–135. Available from https://doi.org/10.17728/jatp.183129.

Azeredo, M. C. (2009). Betalains: Properties, sources, applications, and stability—A review. *International Journal of Food Science & Technology*, *44*, 2365–2376.

Barrows, J. N., Lipman, A. L., & Bailey, C. J. (2003). *Color additives: FDA's regulatory process and historical perspectives*. Magaz: Food Saf.

Bateman, B., Warner, J. O., Hutchinson, E., Dean, T., Rowlandson, P., Gant, C., . . . Stevenson, J. (2004). The effects of a double blind, placebo controlled, artificial food colorings and benzoate preservative challenge on hyperactivity in a general population sample of preschool children. *Archives of Disease in Childhood*, *89*, 506–511.

Batu, A., & Molla, E. (2008). Lokum Üretiminde Kullanılan. Katkı Maddeleri. *Gıda Teknolojileri Elektronik Dergisi* (1), 33–36.

Blas, A. L., Ming, R., Liu, Z., Veatch, O. J., Paull, R. E., Moore, P. H., & Yu, Q. (2010). Cloning of the papaya chromoplast-specific lycopene β-cyclase, CpCYC-b, controlling fruit flesh color reveals conserved microsynteny and a recombination hot spot. *Plant Physiology*, *152*, 2013–2022.

Borcaklı, M. (1999). The use of antimicrobial substances in food production and providing microbiological security, TMMOB publications, Ankara, March p. 16–21.

Bowen, J., Soutar, C., Serwata, R., Lagocki, S., White, D. A., Davies, S. J., & Young, A. J. (2002). Utilization of (3S, 3_S) astaxanthin acyl esters in pigmentation of rainbow trout (Oncorhynchus mykiss). *Aquaculture Nutrition*, *8*, 59–68.

Brouillard, R. (1982). Chemical structure of anthocyanins. *In P. Markakis (ed.), Anthocyanins as food colors* (pp. 1–40). New York: Academic Press.

Brouillard, R., & Delaporte, B. (1977). Chemistry of anthocyanin pigments. 2. Kinetic and thermodynamic study of proton transfer, hydration, and tautomeric reactions of malvidin 3-glucoside. *Journal of the American Chemical Society*, *99*, 8461–8468.

Burrows, J. D. (2009). Palette of our palates: A brief history of food coloring and its regulation. *Comprehensive Reviews in Food Science and Food Safety*, *8*(2), 394–402.

Carpentier, S., Knaus, M., & Suh, M. (2009). Associations between lutein, zeaxanthin, and age-related macular degeneration. *Critical Reviews in Food Science and Nutrition*, *49*, 313–326.

Castaneda-Ovando, A., Pacheco-Hernandez, L., Paez-Hernandez, E., Rodriguez, J. A., & Galan-Vidal, C. A. (2009). Chemical studies of anthocyanins: A review. *Food Chemistry*, *113*, 859–871.

Castellar, R., Obon, J. ,M., Alacid, M., & Fernandez-Lopez, J. A. (2003). Color properties and stability of betacyanins from Opuntia fruits. *Journal of Agricultural and Food Chemistry.*, *51*(9), 2772–2776.

Chethana, S., Nayak, C. A., & Raghavarao, K. S. M. S. (2007). Aqueous two phase extraction for purification and concentration of betalains. *Journal of Food Engineering*, *81*(4), 679–687.

Craft, N. E., & Soares, J. H. (1992). Relative solubility, stability, and absorptivity of lutein and beta-carotene in organic solvents. *Journal of Agricultural and Food Chemistry*, *40*(3), 431–434.

Delgado-Vargas, F., Jiménez, A. R., & Paredes-López, O. (2000). Natural pigments: Carotenoids, anthocyanins, and betalains–characteristics, biosynthesis, processing, and stability. *Critical Reviews in Food Science and Nutrition*, *40*, 173–289.

Delgado-Vargas, F., & Paredes-López, O. (2003). Anthocyanins and betalains. In *Natural Colorants for Food and Nutraceutical Uses* (pp. 167–219). Boca Raton: CRC Press.

Esatbeyoglu, T., Wagner, A. E., Schini-Kerth, V. B., & Rimbach, G. (2015). Betanin—A food colorant with biological activity. *Molecular Nutrition & Food Research, 59*, 36–47.

European Commission. (1997). Regulation (EC) No 258/97 of the European Parliament and of the Council of 27 January 1997 concerning novel foods and novel food ingredients. *Official Journal of the European Union, L43*, 1–6.

European Commission. (2008a). Regulation (EC) No 1331/2008 of the European Parliament and of the Council of December 16, 2008 establishing a common authorisation procedure for food additives, food enzymes and food flavorings. *Official Journal of the European Union, L354*, 1–6.

European Commission. (2008b). Regulation (EC) No 1333/2008 of the European Parliament and of the Council of 16 December 2008 on food additives. *Official Journal of the European Union, L354*, 16–33.

European Commission. (2011). Commission Regulation (EU) No 1129/2011 of 11 November 2011 amending Annex II to Regulation (EC) No 1333/2008 of the European Parliament and of the Council by establishing a Union list of food additives. *Official Journal of the European Union, L295*, 1–177.

European Commission. (2015). Regulation (EU) 2015/2283 of the European Parliament and of the Council of 25 November 2015 on novel foods, amending Regulation (EU) No 1169/2011 of the European Parliament and of the Council and repealing Regulation (EC) No 258/97 of the European Parliament and of the Council and Commission Regulation (EC) No 1852/2001. *Official Journal of the European Union, L327*, 1–22.

Fabi, J. P., & Do Prado, S. B. R. (2019). Fast and furious: Ethylene-triggered changes in the metabolism of papaya fruit during ripening. *Frontiers in Plant Science, 10*, 535.

Feingold, B. F. (1975). Hiperkinesis and learning disabilities linked to artificial food flavorsand colors. *American Journal of Nursing, 75*, 797–803.

Frick, D., & Meggos, H. (1988). Federal food, drug and cosmetic colors. *Food Technology, 7*, 49–56.

Gandía-Herrero, F., Escribano, J., & García-Carmona, F. (2014). Biological activities of plant pigments betalains. *Critical Reviews in Food Science and Nutrition*. Available from https://doi.org/10.1080/10408398.2012.740103.

Garcia-Viguera, C., Zafrilla, P., & Tomas-Barberan, F. A. (1998). The use of acetone as an extraction solvent for anthocyanins from strawberry fruit. *Phytochemical Analysis: PCA, 9*, 274–277.

Gil, M. I. G., Tomas-Barberan, F. A., Hess-Pierce, B., Holcroft, D. M., & Kader, A. A. (2000). Antioxidant activity of pomegranate juice and its relationship with phenolic composition and processing. *Journal of Agricultural and Food Chemistry, 48*, 4581–4589.

Giovannucci, E. (2002). A review of epidemiologic studies of tomatoes, lycopene and prostate cancer. *Experimental Biology and Medicine, 227*, 852–859.

Giusti, M. M., & Wrolstad, R. E. (2003). Acylated anthocyanins from edible sources and their applications in food systems. *Biochemical Engineering Journal, 14*, 217–225. Available from https://doi.org/10.1016/S1369-703X(02)00221-8.

Gonçalves, L. C. P., Da Silva, S. M., DeRose, P. C., Ando, R. A., & Bastos, E. L. (2013). Beetroot-pigment-derived colorimetric sensor for detection of calcium dipicolinate in bacterial spores. *PLoS One, 8*, 737–740.

Gouveia, L., Choubert, G., Pereira, N., Santinha, J., Empis, J., & Gomes, E. (2002). Pigmentation of gilthead seabream, *Sparus aurata* (L. 1875) using *Chlorella vulgaris*(Chlorophyta, Volvocales) microalga. *Aquaculture Research, 33*, 987–993.

Grune, T., Lietz, G., Palou, A., Ross, A. C., Stahl, W., Tang, G., ... Biesalski, H. K. (2010). β-Carotene is an important vitamin A source for humans. *The Journal of Nutrition, 140*, 2268S–2285S.

Gulrajani, M. L. (2001). Present status of natural dyes. *Indian Journal of Fibre & Textile Research, 26*, 191–201.

Guo, S., Zhao, S., Sun, H., Wang, X., Wu, S., Lin, T., . . . Xu, Y. (2019). Resequencing of 414 cultivated and wild watermelon accessions identifies selection for fruit quality traits. *Nature Genetics*, *51*, 1616–1623.

Gur, A., Tzuri, G., Meir, A., Sa'ar, U., Portnoy, V., Katzir, N., . . . Tadmor, Y. (2017). Genome-wide linkage-disequilib- rium mapping to the candidate gene level in melon (*Cucumis melo*). *Scientific Reports*, *7*, 9770.

Hari, R. K., Patel, T. R., & Martin, A. M. (1994). An overview of pigment production in biological systems: Functions, biosynthesis, and applications in food industry. *Food Reviews International*, *10*(1), 49–70.

Hashimoto, H., Uragami, C., & Cogdell, R. J. (2016). Carotenoids and photosynthesis. *SubcellBiochem*, *79*, 111–139.

Hastaoğlu, E., Can, Ö. P., & Vural, H. (2018). The effects of colorants used in hotel kitchens in terms of child health. *Avrupa Bilim ve Teknoloji Dergisi* (14), 10–16.

He, J., & Giusti, M. M. (2010). Anthocyanins: natural colorants with health-promoting properties. *Annual Review of Food Science and Technology*, *1*, 163–187.

Herbach, K. M., Stintzing, F. C., & Carle, R. (2006). Betalain stability and degradation—Structural and chromatic aspects. *Journal of Food Science*, *71*, R41–R50.

Hermanns, A. S., Zhou, X., Xu, Q., Tadmor, Y., & Li, L. (2020). Carotenoid pigment accumulation in horticultural plants. *Horticultural Plant Journal*, *6*(6), 343–360.

Hornero-Méndez, D., & Mínguez-Mosquera, M. I. (2000). Carotenoid pigments in Rosa mosqueta hips, an alternative carotenoid source for foods. *Journal of Agricultural and Food Chemistry*, *48*(3), 825–828.

Ikoma, Y., Matsumoto, H., & Kato, M. (2016). Diversity in the carotenoid profiles and the expression of genes related to carotenoid accumulation among citrus genotypes. *Breed Science*, *66*, 139–147.

Iwashina, T., Ootani, S., & Hayashi, K. (1988). On the pigmented spherical bodies and crystals in tepals of cactaceous species in reference to the nature of betalains or flavonols. *Botanical Magazine Shokubutsu-Gaku-Zasshi*, *101*, 175–184.

Jackman, R. L., Yada, R. Y., Tung, M. A., & Speers, R. A. (1987). Anthocyanins as food colorants—a review. *Journal of Food Biochemistry*, *11*(3), 201–247.

JebaKezi, J., & Judia Harriet Sumathy, V. (2014). Betalain—A boon to the food industry. *Discovery.*, *20*(63), 51–58.

Jiménez-Aguilar, D. M., Ortega-Regules, A. E., Lozada-Ramírez, J. D., Pérez-Pérez, M. C. I., Vernon-Carter, E. J., & Welti-Chanes, J. (2011). Color and chemical stability of spray-dried blueberry extract using mesquite gum as wall material. *Journal of Food Composition and Analysis.*, *24*(6), 889–894. Available from https://doi.org/10.1016/j.jfca.2011.04.012.

Johnson, E. J. (2012). A possible role for lutein and zeaxanthin in cognitive function in the elderly. *The American Journal of Clinical Nutrition*, *96*(5), 1161S–1165S.

Jones, I., White, R., Gibbs, E., Butler, L., & Nelson, L. (1977). Experimental formation of zinc and copper complexes of chlorophyll derivatives in vegetable tissue by thermal processing. *Journal of Agricultural and Food Chemistry*, *25*, 149–153.

Kammerer, D. R., Kammerer, J., Valet, R., & Carle, R. (2014). Recovery of polyphenols from the by-products of plant food processing and application as valuable food ingredients. *Food Research International*, *65*, 2–12. Available from https://doi.org/10.1016/j.foodres.2014.06.012.

Khan, M. I., & Giridhar, P. (2014). Enhanced chemical stability, chromatic properties and regeneration of betalains in *Rivina humilis* L. berry juice. *LWT – Food Science Technology*, *58*, 649–657.

Khan, M. I., & Giridhar, P. (2015). Plant betalains: Chemistry and biochemistry. *Phytochemistry*, *117*, 267–295.

Kong, J. M., Chia, L. S., Goh, N. K., Chia, T. F., & Brouillard, R. (2003). Analysis and biological activities of anthocyanins. *Phytochemistry*, *64*(5), 923–933.

Kosir, I. J., & Kidric, J. (2002). Use of modern nuclear magnetic resonance spectroscopy in wine analysis: Determination of minor compounds. *Analytica Chimica Acta*, *458*, 77–84.

Krishna, H., Chauhan, N., & Sharma, B. D. (2017). Evaluation of karonda (Carissa carandus L.) derived natural colourant cum nutraceuticals-supplement. *RESEARCH ARTICLE PAGE*, 28.

LaBorde, F. F., & von Elbe, J. H. (1990). Zinc complex formation in heated vegetables purees. *Journal of Agricultural and Food Chemistry*, *38*, 484–487.

Lehto, S., Buchweitz, M., Klimm, A., Strassburger, R., Bechtold, C., & Ulberth, F. (2016). Comparison of food color regulations in the EU and the US—Review of current provisions. *Food Additives & Contaminants: Part A*. Available from https://doi.org/10.1080/19440049.2016.1274431.

Lemos, M. A., Aliyu, M. M., & Hungerford, G. (2012). Observation of the location and form of anthocyanin in purple potato using time-resolved fluorescence. *Innovative Food Science and Emerging Technologies*, *16*, 61–68. Available from https://doi.org/10.1016/j.ifset.2012.04.008.

Li, L., Yuan, H., Zeng, Y., & Xu, Q. (2016). Plastids and carotenoid accumulation. *Carotenoids in Nature*, 273–293.

Luckey, T. D. (1968). Introduction to food additives. In T. E. Furia (Ed.), *Handbook of food additives* (2nd ed., pp. 1–27). United States: CRC Press LLC.

Mabry, T. J., Wyler, H., Parkih, I., & Dreiding, A. S. (1967). The conversion of betanidin and betanin to neobetanidin derivatives. *Tetrahedron*, *23*, 3111–3127.

Martins, N., Roriz, C. L., Morales, P., Barros, L., & Ferreiral, I. C. F. R. (2016). Food colorants: Challenges, opportunities and current desires of agro-industries to ensure consumer expectations and regulatory practices. *Trends in Food Science*, 1–72.

McCann, D., Barrett, A., Cooper, A., Crumpler, D., Dalen, L., Grimshaw, K., ... Prince, E. (2007). Food additives and hyperactive behavior in 3-year-old and 8/9-year-old children in the community: A randomised, double-blinded, placebo controlled trial. *Lancet*, *370*, 1560–1567.

Mittal, R., Sharma, A., & Singh, G. (2007). Food colors from plants: Patenting scenario. *Indian Food Industry*, *26*(3), 52–58.

Moreno, D. A., García-Viguera, C., Gil, J. I., & Gil-Izquierdo, A. (2008). Betalains in the era of global agri-food science, technology and nutritional health. *Phytochemistry Reviews.*, *7*, 261–280.

Mortensen, A. (2006). Carotenoids and other pigments as natural colorants. *Pure and Applied Chemistry*, *78* (8), 1477–1491.

Msagati, T. A. M. (2013). *Chemistry of food additives and preservatives* (7th ed., p. 148)New York: John Wiley and Sons publishers.

Naczk, M., & Shahidi, F. (2004). Extraction and analysis of phenolics in food. *Journal of Chromatography A*, *1054*(1–2), 95–111.

Northcutt, J. K., & Parisi, M. A. (2013). Major food laws and regulations. In P. Curtis (Ed.), *Guide to US food laws and regulations* (pp. 73–96). Chishester: Wiley Blackwell.

Omaye, S. (2004). *Food and nutritional toxicology*. New York. New York: CNC Press.

Özcan, G., Artık, N. ve, & Üner, Y. (1997). Gıda katkı maddelerinin tü-ketici bilinci ve insan sağlığı açısından irdelenmesi. *TMMOB, Eylül*, 1–31.

Pandey, R. M., & Upadhyay, S. K. (2012). *Food additive* (5, pp. 1–31). İndia: InTech.

Parkinson, T. M., & Brown, J. P. (1981). Metabolic fate of food colorants. *Annual Review of Nutrition*, *1*, 175–205.

Peterson, R. G., & Joslyn, M. A. (1958). Nature of betanin, the pigment of red beet. *Nature*, *182*, 45–46.

Piattelli, M., Minale, L., & Prota, G. (1964). Isolation, structure and absolute configuration of indicaxanthin. *Tetrahedron*, *20*, 2325–2329.

Pollock, I., & Warner, J. O. (1990). Effect of artificial food colors on childhood behavior archives of disease in childhood. *Archives of Disease in Childhood*, 65–74.

Prangdimurti, E., Muchtadi, D., Astawan, M., & Zakaria, F. R. (2006). Aktivitasantioksidandaun suji.pdf. *Jurnal Teknologi dan Industri Pangan*, *17*(2), 79–86.

Prince, J. (2017). Natural shows its true color. *Nutritional Outlook, 20*(9), 66.

Rodriguez-Amaya, D. B. (2015a). Carotenes and xanthophylls as antioxidants. In *Handbook of antioxidants for food preservation* (pp. 17–50). Woodhead Publishing.

Rodriguez-Amaya, D. B. (2015b). *Food carotenoids: Chemistry, biology and technology*. Oxford: IFT Press-Wiley.

Rodríguez-Amaya, D. B., Kimura, M., Godoy, H. T., & Amaya-Farfan, J. (2008). Updated Brazilian database on food carotenoids: Factors affecting carotenoid composition. *Journal of Food Composition and Analysis, 21*(6), 445–463.

Sabnis, R. W., Pfizer, I., & Madison, N. J. (2010). *Biolojical dyes and stains, synthesis and industrial applications* (pp. 1–521). Canada: Wiley Publication.

Sagdic, O., Ekici, L., Ozturk, I., Tekinay, T., Polat, B., Tastemur, B., ... Senturk, B. (2013). Cytotoxic and bioactive properties of different color tulip flowers and degradation kinetic of tulip flower anthocyanins. *Food Chemical Toxicol, 58*, 432–439.

Saponjac, V. T., Canadanovic-Brunet, J., Cetkovic, G., Jakisic, M., Djilas, S., Vulic, J., & Stajcic, S. (2016). Encapisulation of beetroot pomace extract: Rsm optimization, storage and gastrointestinal stability. *Molecules (Basel, Switzerland), 21*, 584.

Schanderl, S. H., Marsh, G., & Chichester, C. (1965). Colour reversion in processed I. Studies on regreened vegetables pea purges. *Journal of Food Science, 30*, 312–316.

Schwartz, S. J., Cooperstone, J. L., Cichon, M. J., Von Elbe, J. H., & Giusti, M. M. (2017). *Colorants: In Fennema's food chemistry* (2017, pp. 681–752). Boca Raton, FL: CRC Press.

Segner,W. P., Ragusa, T. J., Nank,W. K., & Hoyle, W. C. (1984, September 25). Process for the preservation of green color in canned vegetables. U.S. Patent No. 4473591.

Sen, T., Barrow, C. J., & Deshmukh, S. K. (2019). Microbial pigments in the food industry—Challenges and the way forward. *Frontiers in Nutrition, 6*, 7. Available from https://doi.org/10.3389/fnut.2019.00007.

Shamina, A., Shiva, K. N., & Parthasarathy, V. A. (2007a). Food colors of plant origin. *CAB Reviews: Perspectives in Agriculture, Veterinary Science, Nutrition and Natural Resources*, 2(087). Available from https://doi.org/10.1079/PAVSNNR20072087.

Shamina, A., Shiva, K. N., & Parthasarathy, V. A. (2007b). Food colors of plant origin. *CAB Reviews: Perspectives in Agriculture, Veterinary Science, Nutrition and Natural Resources, 12*, 24.

Shipp, J., & Abdel-Aal, E. S. M. (2010). Food applications and physiological effects of anthocyanins as functional food ingredients. *The Open Food Science Journal, 4*(1).

Silva, D. V., Silva, F. O., Perrone, D., Pierucci, A. P. T. R., Conte-Junior, C. A., Alvares, T. S., ... Paschoalin, V. M. F. (2016). Physicochemical, nutritional, and sensory analyses of a nitrate-enriched beetroot gel and its effects on plasmatic nitric oxide and blood pressure. *Food & Nutrition Research., 60*, 29909.

Singh, S. (2006). Impact of color on marketing. *Management Decision, 44*, 783–789.

Storebakken, T., & No, H. K. (1992). Pigmentation of rainbow trout. *Aquaculture (Amsterdam, Netherlands), 100*, 209–229.

Strack, D., Vogt, T., & Schliemann, W. (2003). Recent advances in betalain research. *Phytochemistry, 62*, 247–269.

Viera, I., Pérez-Gálvez, A., & Roca, M. (2019). Green natural colorants. *Molecules, 24*(1), 154.

Von Elbe, J. H., Huang, A. S., Attoe, E. L., & Nank, W. K. (1986). Pigment composition and color of conventional and Veri-Green canned beans. *Journal of Agricultural and Food Chemistry, 34*, 52–54.

Willstätter, R. (1915). Chlorophyll. *Journal of the American Chemical Society, 37*, 323–345.

Wyler, H., & Dreiding, A. S. (1957). Kristallisiertes betanin. *Vorlaufige Mitteilung Helvetica Chimica Acta, 40*, 191–192.

Xi, L., Qian, Z., Xu, G., Zheng, S., Sun, S., Wen, N., ... Zhang, Y. (2007). Beneficial impact of crocetin, a carotenoid from saffron, on insulin sensitivity in fructosefed rats. *The Journal of Nutritional Biochemistry.*, *18*(1), 64–72. Available from https://doi.org/10.1016/j.jnutbio.2006.03.010.

Xu, J., Tao, N., Liu, Q., & Deng, X. (2006). Presence of diverse ratios of ly- copene/ ß-carotene in five pink or red-fleshed citrus cultivars. *Science Horticulture*, *108*, 181–184.

Yabuzaki, J. (2017). Carotenoids database: Structures, chemical finger-prints and distribution among organisms. *Database*. Available from https://doi.org/10.1093/database/bax004.

Zheng, X., Zhu, K., Sun, Q., Zhang, W., Wang, X., Cao, H., ... Deng, X. (2019). Natural variation in CCD4 promoter underpins species-specific evolution of red coloration in citrus peel. *Molecular Plant*, *12*, 1294–1307.

Zhu, Z., Wu, N., Kuang, M., Lamikanra, O., Liu, G., Li, S., & He, J. (2015). Preparation and toxicological evaluation of methyl pyranoanthocyanin. *Food Chemical Toxicology*, *83*, 125–132. Available from https://doi.org/10.1016/j.fct.2015.05.004.

CHAPTER

Plant extracts as enzymes

10

Vartika Verma[1], Gauri Singhal[2], Sunanda Joshi[1], Monika Choudhary[1] and Nidhi Srivastava[3]

[1]*Department of Bioscience and Biotechnology, Banasthali Vidyapith, Jaipur, India* [2]*Department of Biotechnology, School of Medical and Allied Sciences, Sanskriti University, Mathura, India* [3]*Department of Biotechnology, National Institute of Pharmaceutical Education and Research, Raebareli (NIPER-R), Lucknow, India*

10.1 Introduction

Enzymes are the biological catalysts that enhance the speed of biochemical reactions. In nature, enzymes are specific and effective in less amounts as well as they react under the mild conditions of temperature and pH. All enzymes have been obtained from certain natural sources and get inactivated after the desired transformation of any substrate (Dziezak, 1991). Enzymes, unlike any other inorganic catalysts, are more specific and catalyze only one substrate or a group of closely attached compounds by breaking the specific bond. Owing to this specific nature of enzymes, the formation of by-products has been reduced in high-amount reactions. As enzymes react under optimum conditions of temperature and pH, reduced energy cost has been achieved. The low utility amount of enzymes makes them practical and economical candidates for commercial applications (Dziezak, 1991). Enzymes have been acknowledged as non-toxic, natural food components and have been chosen over chemical products in food industry as they have isolated from the plant and other microbial sources (Simpson & Haard, 1987). In food industry, processing of food can be defined as the methods and practices performed in the food and beverages to alter the raw food material and make it adequate for consumption (Heldman & Hartel, 2007; Monteiro & Levy, 2010).

10.2 History of enzyme use in food production

In food manufacturing, enzymes extracted from different sources such as plants, tissues of animals, and microbes have been used for the centuries. Rennet, a natural enzyme, is an enzyme mixture isolated from the stomach of domestic animals (calves) and used in cheese production. Rennet has a protease enzyme which coagulates the milk and separated the solid curd part from the liquid whey. Some other enzymes have also been produced by the yeast that is used for the fermentation of grape juice to make wine (Shinde, Deshmukh, & Bhoyar, 2015).

Plant Extracts: *Applications in the Food Industry*. DOI: https://doi.org/10.1016/B978-0-12-822475-5.00009-0

10.3 Plant extracts as enzymes

Mainly four different types of plant extracts namely proteases, amylases, lipases, and cellulases have been used as the enzymes in food industry

10.3.1 Protease

Milk coagulation is the main step in cheese production procedure, and some coagulating enzymes like proteases play an important role in cheese manufacturing since thousands of years. This process of cheese manufacturing has seemed to be the oldest known function of protease enzyme. Plant proteases in the crude form or purified form, have been act as a milk coagulant in cheese production process (Harboe, Broe, & Qvist, 2010; Jacob, Jaros, & Rohm, 2011).

10.3.1.1 Source

The crude plant extract can also be further purified to get purified enzyme depending on the degree of purification. Precipitation technique with ammonium sulfate has been an efficient method to yield significant amount of active proteases from *Cynara cardunculus* flowers (Barros et al., 2003). Cardosin A and B, a type of protease enzyme has been extracted from the stigma and stylet of *C. cardunculus* flower (Silva, Allmere, Malcata, & Andrén, 2003). Proteases have been extensively isolated from the stigma of *Cynara scolymus*, dried flowers of *Moringa oleifera* and fresh flowers of *Silybum marianum* (Pontual et al., 2012; Sidrach, García-Cánovas, Tudela, & Rodríguez-López, 2005; Vairo-Cavalli, Claver, Priolo, & Natalucci, 2005). Partially purified protease enzyme extract (onopordosin) has been extracted from the stigma and style of *Onopordum acanthium* flowers (Brutti, Pardo, Caffini, & Natalucci, 2012). Hieronymain, a protein extract was obtained from the fruits of *Bromelia hieronymi* (Bruno et al., 2010). Protease enzyme extract was also obtained from peeled ginger rhizomes (Hashim, Mingsheng, Iqbal, & Xiaohong, 2011). Various plant seeds have also been used to prepare the plant extract as a protease source for cheese making process. Latex of fig tree has also been used for the extraction of protease enzyme and assessed for milk clotting properties (Kumari, Sharma, & Jagannadham, 2012; Sharma, Kumari, & Jagannadham, 2012).

10.3.1.2 Extraction procedure

Proteases have been present in almost all plant tissues and extracted from their natural sources or by in vitro culture techniques to get continuous supply of plant proteases. Protease enzyme have been isolated from a variety of plant parts including seed, latex, roots, leaves as well as flower and extensively studied for its role in milk coagulation (González-Rábade, Badillo-Corona, Aranda-Barradas, & del Carmen Oliver-Salvador, 2011; Shah & Mir, 2014; Shah, Mir, & Paray, 2014). Generally, protease enzyme has been extracted from plant parts using aqueous maceration process. The aqueous extract of these plant parts can be prepared by several ways. For this process, the whole or crushed dried plant parts are soaked in water for variable time period at room temperature. After soaking, filtrate or crude extract is collected and used as an enzyme (milk coagulant) (Roseiro, Barbosa, Ames, & Wilbey, 2003). An alternative process has included grinding of dried plant part with crude salt in presence of warm milk, stain the filtrate and solubilization of enzyme (Sousa & Malcata, 2002). The detailed procedure of enzyme extraction has been demonstrated in Fig. 10.1.

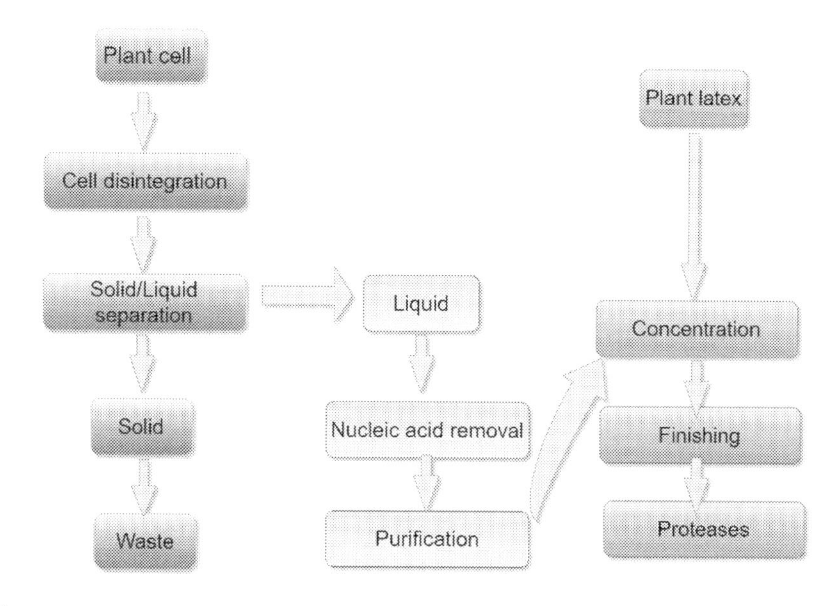

FIGURE 10.1

Extraction procedure of protease enzyme.

10.3.1.3 Types

Mostly, milk coagulant belongs to the aspartic protease group; however, some milk coagulants have also reported from other protease group like serine and cysteine proteases under proper conditions.

10.3.1.3.1 Aspartic proteases

Aspartic proteases possess two aspartic residues at the catalytic site of the enzyme. At the acidic pH level, aspartic proteases have been the most active and demonstrate special cleavage specificity for peptide bonds present between hydrophobic amino acid residues (Domingos et al., 2000). Some aspartic proteases with milk coagulant activity reported from various plant sources including *C. scolymus*, *Onopordum turcicum*, milk thistle, *Centaurea calcitrapa* and rice kernels (Asakura, Watanabe, Abe, & Arai, 1997; Domingos et al., 2000; Llorente, Brulti, & Natalucci, 1997; Tamer, 1993; Vairo-Cavalli et al., 2005). In Mediterranean region, flowers of *C. cardunculus* have been traditionally used for the cheese manufacturing as a source of aspartic proteases (Barros et al., 2003). The *C. cardunculus* have produced cardosins and cyprosins, a type of aspartic proteases that only get accumulated in mature flower parts including petals and pistils however, this enzyme has not accumulated in leaves and seeds (Cordeiro, Pais, & Brodelius, 1998). Cardosin A, an aspartic protease has also been isolated from pistils of *C. cardunculus* abundantly. Three cyprosins (aspartic proteases) that have milk clotting activities were also isolated, purified and characterized from dried flowers of *C. cardunculus* (Heimgartner et al., 1990). The specificity and kinetic parameters of these aspartic proteases are similar to chymosin and pepsin (Veríssimo et al., 1996; Veríssimo, Esteves, Faro, & Pires, 1995).

10.3.1.3.2 Papain

Papain has been classified in the cysteine proteases that processes the protein more broadly than pancreatic protease compounds. Papain mostly comprises a single peptide chain with a sulfhydryl gathering and three sulfide spans. Papain has been isolated from the papaya latex that assembled from the dried unripen papaya. The movement of protein relied upon the unripen papaya natural product. Fundamentally, papain has been settled by disulfide connects and collapsed around these scaffolds. This structure made a functioning site accessible for the cooperation of new particles. 3D structure of papain has principally comprising of two unmistakable complex particles with a cleavage between them that conveyed a functioning site for the catalysis. Papain has a sub-atomic load of 23,406 Da and globular structure with 212 amino acids. It has been steady and dynamic under a wide scope of pH, temperature and fixation. Papain has stayed dynamic at high temperature go, 3–9 pH go and has protection from higher grouping of denaturing substances (Edwin & Jagannadham, 2000). As a crystalline suspension, papain can be steady for 6 years at 50C utilizing settling operators like EDTA, cysteine and dimercaptopropanol. Papain assumed different jobs in food industry be that as it may; enzymatic combination of peptides, amino acids and different atoms has been the most widely recognized employments of papain (Esti, Benucci, Lombardelli, Liburdi, & Garzillo, 2013). Papain has additionally assumed pivotal jobs in food, organic procedures and medication industry. Papain has a major proteolytic action toward amino acid esters, amide joins, short chain peptides, and proteins (Mamboya, 2012).

10.3.1.3.3 Bromelain

Bromelain belongs to the family of sulfhydryl proteolytic enzymes and mainly obtained from the pineapples. Bromelain is a mixture of enzyme that has been used for the digestion of proteins. It has been mainly extracted from the pineapple, some fruits and stem as well. The one extracted from the fruits has known as fruit Bromelain and from the stem has known as stem Bromelain. It has been consisting of 212 amino acids and has molecular weight of 33 kDa (Babu, Rastogi, & Raghavarao, 2008). During protein breakage, it has remained stable at temperature range of 40°C–60°C in 3–7 pH range (Mohapatra, Rao, & Ranjan, 2013; Srujana & Narayana, 2017). This enzyme showed the maximum activity at 50°C and pH 7 at simple extraction and higher proteolytic activity at 60°C and pH 8 (Martins et al., 2014). Similar to papain, Bromelain also played major role in pharmaceutical, food, cosmetics and other industries. Bromelain has also been used for the tenderization of meat, solubilization of grain proteins, clarification of beer and cookies baking. In fresh apple juices, Bromelain has acted as a enzymatic browning inhibitor (Mohan, Sivakumar, Rangasamy, & Muralidharan, 2016).

10.3.1.3.4 Ficain

Ficain, a proteolytic enzyme is also known as Ficin. It has been extracted from the fig tree and belonging to the family of sulfhydryl or proteinases enzymes. It has been isolated from the clarified latex of fig tree. As an specific enzyme, Ficain can hydrolyzed the chemical bonds of natural proteins that helps on the proper digestion of protein. The structure and hydrolysis mechanism of Ficain has been much similar to the papain. It has a good stability and wide applications in the different industries including healthcare and food industry. The separation and purification of raw

Ficain can be processed via various methods like precipitation, chromatography and electrophoresis (Arribére, Caffini, & Priolo, 2000).

10.3.1.4 Role in food industry

In cheese making process, plant extracts have been used since ancient times as milk coagulants. Enzyme extract isolated from vegetables as a coagulant has also been used for the cheese manufacturing in areas like Mediterranean, south European countries and West Africa. In Spain and Portugal, largest variety of cheese has been produced through vegetable coagulant using *Cynara* spp. (Roseiro et al., 2003). The vegetable coagulant of *Cynara* spp. has been used for the production of Portuguese Serra cheese, Serpa cheeses, Spanish Los Pedroches, La Serena, Torta del Casar cheeses (from ewes' milk), Los Ibores cheese (from goats' milk) and Flor de Guía cheese (mixture of ewes' and cows' milk) (Fernández-Salguero & Sanjuán, 1999; Fernández-Salguero, Sanjuán, & Montero, 1991; Macedo, Faro, & Pires, 1993; Roa, López, & Mendiola, 1999; Sanjuán et al., 2002). Traditionally, in Nigeria and Republic of Benin, the extract of *Calotropis procera* has been used in cheese production processes (Roseiro et al., 2003). However, most vegetable coagulants have excessive proteolytic nature that limited the use of coagulant in cheese production as it lowered the texture, flavor and yield of cheese (Lo Piero, Puglisi, & Petrone, 2002). The search of new prospective milk coagulant enzyme from plant sources is in continuous process to make the enzymes more useful in industries and complete the increasing global demand for high quality and diversified cheese production (Hashim et al., 2011).

10.3.2 Lipases

Lipases have been the most extensively used class of enzymes (Hasan, Shah, & Hameed, 2006; Schmid & Verger, 1998). Lipases are found in almost all unicellular and multicellular organisms and are ubiquitous in nature. In industrial applications, mostly lipases have been obtained from yeast and fungi (Sharma & Kanwar, 2014). However, lipases can also obtained from different sources like bacterial, fungal, animal, plant and algal (Patil, Chopda, & Mahajan, 2011). Plant can be used as a novel source of lipase enzyme because of their advantages including easy acceptability, low cost source, specific application of plant enzyme and their direct applications as biocatalyst. Lipases extracted from plant tissues have included different types of non-specific monoacylglycerol, lipid acylhydrolases, triacylglycerol lipases and phospholipases A1, A2, B, C, D. the type TAG1 lipase have been mainly present in the plant seeds as energy reservoir (Beevers, 1969; Hutton & Stumpf, 1969; Lin et al., 1982). Now-a-days many lipases have been isolated and purified from different plant sources including latex of *Carica papaya* and scutella of *Z. mays*. In oilseeds, lipases have been localized in the oil bodies and glyoxysomes (Lin & Huang, 1983; Rosnitschek & Theimer, 1980). The extraction procedure of lipases from seeds is given in Fig. 10.2.

10.3.2.1 Role in food industry

In food industry, plant lipases have focal points over different sources including microbial lipases as a result of their adequacy. Plant lipases exhibit extraordinary soundness in solvent-catalyzed responses like interesterification. Plant lipases likewise have an extra significance as these lipases have minimal effort of handling and creation. Lipase extracted from *C. papaya* (CPL) has been viably utilized for the

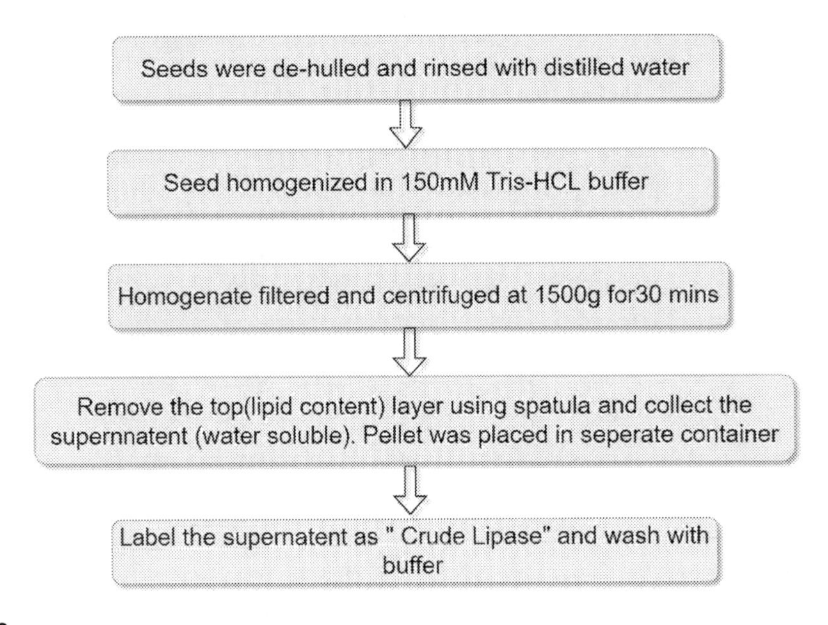

FIGURE 10.2

Flow sheet for extraction of lipase from seeds.

low calorie short and long chain triacylglycerols union for the newborn child formulae. These papaya lipases have likewise been utilized for the creation of triacylglycerols utilizing ethyl esters interesterification with the assistance of tri palmitin (Gandhi & Mukherjee, 2001). The significant expense and unavailability of human milk fat can be overwhelmed by union of human took after milk fat. The human took after milk fat has been done by trans-esterification of unsaturated fat of rapeseed oil with tri palmitin utilizing papaya latex (Mukherjee & Kiewitt, 1998).

Nowadays, CPL has been self-immobilized in papaya latex and servesas a substitute for human milk fat. This CPL has go about as a biocatalyst and utilized for the creation of business lipases as a minimal effort substitute (Tecelão, Rivera, Sandoval, & Ferreira-Dias, 2012). CPL has likewise been utilized for the union of cocoa margarine that can be utilized for the chocolate creation requiring little to no effort. CPL has additionally created some side-effects when contrasted with the synthetic amalgamation (Pinyaphong & Phutrakul, 2009). In TAG interesterification, fat and oil adjustment have additionally been finished utilizing CPL dependent on sound system selectivity (Villeneuve, Pina, Skarbek, Graille, & Foglia, 1997). In this way, CPL has been end up being a savvy plant catalyst arranged from the unrefined papaya latex and had different business applications (Lin & Huang, 1983).

10.3.3 Cellulase

In various industries, cellulase is the most commonly used enzyme after proteases and has a variety of applications in industrial biotechnology. In the processing of lignocellulosic materials, cellulase has played the important role along with hemicellulase and pectinase for the production of fuel and feedstock's.

10.3.3.1 Role in food industry

In today's context, during the time of urbanization, climate change, and increase in population, food production has become a major concern for the human beings. In agriculture practices, food production has been increased with the technological advances and required vigorous food and beverage industries for the processing, preservation and value addition of raw food materials. Across the globe, these industries categorized as major industries and significantly contribute toward the catering to the people's needs. The food industry has mainly processed the food materials by improving its nutritional quality and reduced the concerns of human health with extending shelf life of the food products. Food industries have also improved the flavor, texture, color, packaging and consumption of the food products (Chandrasekaran, 2012). In food industry, many of these roles have been played by the enzymes to achieve the specific target.

Cellulases have been used in the food industry to increase the nutritional value of the food products. Generally, cellulases have been used in combination with hemicellulase and pectinases in food industries (Bhat, 2000; Kumar, 2015). The macerating enzymes have been made up of the combination of these three enzymes and used for the extraction of olive oil. They have been used to increase the extraction yield, lower rancidity, more antioxidants, better quality and reduced wastage (Galante, DE Conti, & Monteverdi, 1998). Fruits in their natural state have tend to rot fast hence enzymes can be used to convert them to the juice or puree can extend the shelf life of fruits. The addition of maceration enzyme in this process has yield better product and less browning of product (Sims & Bates, 1994).

10.3.4 Amylase

Commercially, amylase has been considered as an interesting enzyme as it hydrolyzes the starch material present in food products. Amylolytic (amylase) enzymes have been widely spread among the plant tissues (storage tissues) and vegetative organs including seeds, tubers and leaves. α-Amylase has been the major starch hydrolyzing enzyme in plant parts but in leaves, sometimes, the activity of amylase has been substantial (Dreier, Schnarrenberger, & Börner, 1995). Previously, amylase has been isolated and purified from different plant sources including soybean, sweet potato, pea, barley and rye apart from other microbial sources (Yamamoto, 1988).

10.3.4.1 Role in food industry

Starch is the main source of carbohydrate that originates from the plant sources. Starch derivatives like glucose syrup, maltodextrin, cyclodextrin, hydrolysates, and other modified starch have significant roles in different food, feed, and beverage industries. Starch modified enzymes have important role in production of starch derivatives and it has been a growing industries. According to updated information, around 11 amylases have been used in food industries, 215 amylases have applied in food processing industries like fruit juice preparation, brewing, baking, starch syrup etc. (Mobini-Dehkordi & Javan, 2012). In dough, starch has been broken down to the limited dextrins, an intermediate product of starch hydrolysis along with fermentable sugars during bread making process. This dextrin has been further fermented to yield the alcohol and carbon dioxide (Prakash &

Jaiswal, 2010). The formation of low molecular weight dextrins has reduced the hardness in bread. In wheat flour, β-amylase has present in abundance however; α-amylase has been absent. This β-amylase has catalyzed the undamaged native starch granules present in wheat flour. In dough, starch hydrolysis has been achieved by heterogeneous action of α-amylase and β-amylase. In the milling process, starch granules present in the flour have been broken down and made them more susceptible to amylase activity. In the baking process, these starch granules get gelatinized and their liquefaction occurred with the help of α-amylase. In flour, β-amylase has also converted the dextrin into maltose and later maltose has been fermented with baker's yeast. This enzymatic hydrolysis has maintained vigorous yeast fermentation so that lively dough has been produced with large loaf volumes (Dekker, 1994). The diversified role of amylase has been demonstrated in Fig. 10.3.

During the storage of baked products, properties of the products may be affected by the staling effect that causes distasteful changes in the products including reduced crust crisp, crumb firmness, loss of flavor, and moisture content. The short amylopectin side chains present in the fresh, soft bread has been crystalized gradually during storage. Owing to starch crystallization, moisture has been migrated within the crumb that increased the crumb firmness and reduced the crumb resilience. To stop this changes during storage, thermostable α-amylase have been used as anti-staling agents. α-amylase has limited the recrystallization, network formation of amylopectin and reduced the water immobilization which helped in the softness retention and shelf life improvement in the baked products (Jana et al., 2013).

10.3.5 Lipoxygenase

Lipoxygenase also known as lipoxidase has been widely distributed among the plants, animals, and fungi. It has been used to catalyze the oxidation reaction of cis, cis-diene units of fatty acids, and converted them to hydroperoxidienoic compounds. Recently, the pH effect of Lipoxygenase has been identified on ionic strength of substrate. It has been abundantly present in the plant sources

FIGURE 10.3

Role of amylase in food industry.

like legumes and potato tubers. Generally, Lipoxygenase has catalyzed the main polyunsaturated fatty acids (linoleic and linoleic acids) of the plant tissues. This enzyme has found in the vegetative tissues and played crucial role in plant defense system however, its amount in plant vegetative tissues is very less. It has been reported that Lipoxygenase used to eliminate the production of assonate and proteinase inhibitors from potato leaves and reduced the susceptibility of insect attack (Royo et al., 1999).

Lipoxygenase has been widely distributed in plants, and thus different methods have been used to isolate and purify it. The main source of Lipoxygenase enzyme is wheat. The extraction process involved isolation and purification of enzyme through chromatography. The purified enzyme has been characterized on the basis of various enzymatic parameters including thermal sensitivity, pH and amino acid composition (Shiiba, Negishi, Okada, & Nagao, 1991).

10.3.5.1 Role in food industry

Lipoxygenase plays both positive and negative roles in food industry. In a positive manner, it acts as an ingredient in bread production and also as an aroma enhancer. However, it has also affect flavor, color and anti-oxidant properties of food in a negative manner (Barrett, 1975; Baysal & Demirdöven, 2007).

10.3.6 Pectinases

Pectinases have been used for the hydrolysis of glycosidic bonds present in the pectic polymers. These pectic polymers have been generally found in the citrus fruits like pineapple, apple, tomato, orange, lemon pulp and act as the natural substrate for the pectinase. On the basis of enzyme functions, pectinase can be categorized as pectin esterases (remove acetyl and methoxyl groups from pectin), polygalacturonases (hydrolyze glycosidic α-(1−4) bonds), pectin lyase, and pectate lyase (Saadoun, Dawagreh, Jaradat, & Ababneh, 2013). Apart from plant sources, pectinase can also be produced from the natural and recombinant microbial sources with increased yield and thermostability (Rebello et al., 2017). Pectinases can work on both smooth and hairy regions of the pectin. There have been acidic and alkaline pectinases based on their pH and categorized in endo and exo-pectinases. When this enzyme has cleaved any substrate randomly, it called as endopectinases however; when terminal ends have been targeted, it called as exopectinases (Pedrolli, Monteiro, Gomes, & Carmona, 2009).

10.3.6.1 Role in food industry

Pectinase enzyme has been used in various industries including food industry, paper bleaching, and remediation (Pasha, Anuradha, & Subbarao, 2013). Pectinase-added juices have clearer appearance and filterability than enzyme-less juices (Saadoun et al., 2013). Pectinase have also been used to improve the flavor and color of drinks apart from reducing turbidity and haze generation from natural fruit juices like banana and apple. Haze removal has been the most costly part of juice production. To overcome this problem, pectinase has been added in the fruit juices along with gelation that increased the viscosity and turbidity of the juices and reduced the haze. Biogenic enzymes like pectinases have acted nearly nine times better than mechanical maceration in juice production to get good results (Rebello et al., 2017).

10.3.7 Peroxidase

Plants are the main source of peroxidase enzyme, and peroxidase is mainly located in the roots and sprouts of the higher plants including beetroot, potato tuber, horse radish, soybean, banana, carrot, tomato, papaya, wheat, turnip, beats, dates, and strawberries (Ambreen, Rehman, Zia, & Habib, 2000; Reed, 1975). The plant peroxidase superfamily has comprised of heme containing glycoproteins that vary in their catalytic properties and structure. It has been isolated and characterized from different plant sources such as tubers, fruits, grains and leaves. The availability of peroxidase with high specificity and stability has been used for the improvement of immune enzymatic analytical kit, development of new analytical methods and potential industrial processes (Idesa, 2018; Rosa et al., 2020).

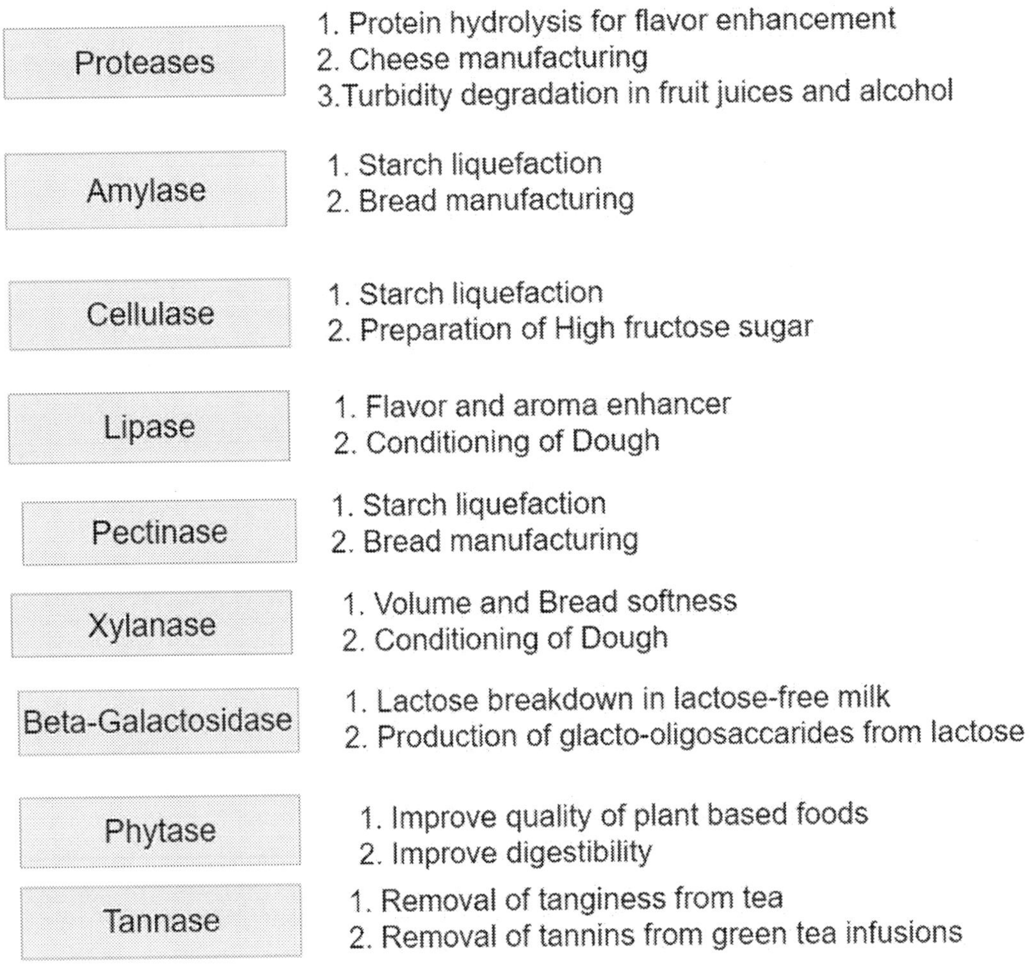

Proteases
1. Protein hydrolysis for flavor enhancement
2. Cheese manufacturing
3.Turbidity degradation in fruit juices and alcohol

Amylase
1. Starch liquefaction
2. Bread manufacturing

Cellulase
1. Starch liquefaction
2. Preparation of High fructose sugar

Lipase
1. Flavor and aroma enhancer
2. Conditioning of Dough

Pectinase
1. Starch liquefaction
2. Bread manufacturing

Xylanase
1. Volume and Bread softness
2. Conditioning of Dough

Beta-Galactosidase
1. Lactose breakdown in lactose-free milk
2. Production of glacto-oligosaccarides from lactose

Phytase
1. Improve quality of plant based foods
2. Improve digestibility

Tannase
1. Removal of tanginess from tea
2. Removal of tannins from green tea infusions

FIGURE 10.4

Major applications of plant enzymes in food industry.

This enzyme has been used to catalyze the wide range of substrate using different peroxide such as hydrogen peroxide (Holm, 1995). In food industry, peroxidase has been used for flavor, texture, and color production as well as for the improvement of food nutritional quality (Bansal & Kanwar, 2013).

10.4 Applications of plant enzymes in food industry

Enzymes have been used for the production and quality enhancement of different types of food in the food industry including yoghurt, cheese, and bread syrup. Prior to our knowledge of enzymes, some food processes were done traditionally like cheese manufacturing, brewing meat tenderization, and condiment preparation using papaya leaves based on proteolysis. The major applications of plant extract as enzyme are mentioned in Fig. 10.4.

10.5 Conclusion

The magnificence and enchantment of plant extricates as enzymes utilized in food ventures have been broadened. In not so distant future, the worldwide interest of enzymes has been anticipated to be ascending at a quick pace. Catalysts are sensitive to obtain under extraordinary ecological conditions in the past decades, and in this way various businesses were held to grasp enzyme innovation. In this field, a few methodologies were utilized to produce novel compounds from the nature for upgrading the synergist properties, catalyst specialization to serve new capacities, again structuring of biocatalysts, and advancement of protein planning. The advances in plant biotechnology have proposed various techniques to manufacture the enzymes that can be utilized in food industry, and their applications will improve the nature of human life. These exercises indicated creative methodologies for the planning of improved biocatalysts with greater steadiness (pH, temperature), less reliance on metal particles, and diminished weakness to inhibitory specialists while keeping up focused action as well as advancing novel exercises. This has been important for the utilizations of enzymes in food and feed ventures that permitted upgraded execution under ideal conditions that has diminished the danger of microbial tainting.

References

Ambreen, S., Rehman, K., Zia, M. A., & Habib, F. (2000). Kinetic studies and partial purification of peroxidase in soybean. *Pakistan Journal of Agricultural Sciences*, *37*(3–4), 119–122.

Arribére, M. P. M. C., Caffini, O., & Priolo, S. (2000). Proteolytic enzymes from the latex of Ficus punzila L. (Moraceae). *Acta Farm. Botiueretise*, *19*(4), 257–262.

Asakura, T., Watanabe, H., Abe, K., & Arai, S. (1997). Oryzasin as an aspartic proteinase occurring in rice seeds: Purification, characterization, and application to milk clotting. *Journal of Agricultural and Food Chemistry*, *45*(4), 1070–1075.

Babu, B. R., Rastogi, N. K., & Raghavarao, K. S. M. S. (2008). Liquid–liquid extraction of bromelain and polyphenol oxidase using aqueous two-phase system. *Chemical Engineering and Processing: Process intensification*, *47*(1), 83–89.

Bansal, N., & Kanwar, S. (2013). Peroxidase (s) in environment protection. *Science World Journal*, *714639*, 9.

Barrett, F. F. (1975). Enzyme uses in the milling and baking industries. In Reed, G. (Ed.), *Enzymes in food processing*. Academic Press, Inc., NY.

Barros, R. M., Extremina, C. I., Gonçalves, I. C., Braga, B. O., Balcão, V. M., & Malcata, F. X. (2003). Hydrolysis of α-lactalbumin by cardosin A immobilized on highly activated supports. *Enzyme and Microbial Technology*, *33*(7), 908–916.

Baysal, T., & Demirdöven, A. (2007). Lipoxygenase in fruits and vegetables: A review. *Enzyme and Microbial Technology*, *40*(4), 491–496.

Beevers, H. (1969). Glyoxysomes of castor bean endosperm and their relation to gluconeogenesis. *Annals of the New York Academy of Sciences*, *168*(2), 313–324.

Bhat, M. (2000). Cellulases and related enzymes in biotechnology. *Biotechnology Advances*, *18*(5), 355–383.

Bruno, M. A., Lazza, C. M., Errasti, M. E., López, L. M., Caffini, N. O., & Pardo, M. F. (2010). Milk clotting and proteolytic activity of an enzyme preparation from Bromelia hieronymi fruits. *LWT-Food Science and Technology*, *43*(4), 695–701.

Brutti, C. B., Pardo, M. F., Caffini, N. O., & Natalucci, C. L. (2012). Onopordum acanthium L.(Asteraceae) flowers as coagulating agent for cheesemaking. *LWT-Food Science and Technology*, *45*(2), 172–179.

Chandrasekaran, M. (Ed.), (2012). *Valorization of food processing by-products*. CRC press.

Cordeiro, M. C., Pais, M. S., & Brodelius, P. E. (1998). *Cynara cardunculus subsp. flavescens (Cardoon): In vitro culture, and the production of cyprosins—Milk-clotting enzymes. Medicinal and aromatic plants X* (pp. 132–153). Berlin, Heidelberg: Springer.

Dekker, R. F. H. (1994). Enzymes in food and beverage processing. 1. *Food Australia*, *46*(3), 136–139.

Domingos, A., Cardoso, P. C., Xue, Z. T., Clemente, A., Brodelius, P. E., & Pais, M. S. (2000). Purification, cloning and autoproteolytic processing of an aspartic proteinase from Centaurea calcitrapa. *European Journal of Biochemistry*, *267*(23), 6824–6831.

Dreier, W., Schnarrenberger, C., & Börner, T. (1995). Light-and stress-dependent enhancement of amylolytic activities in white and green barley leaves: β-amylases are stress-induced proteins. *Journal of Plant Physiology*, *145*(3), 342–348.

Dziezak, J. D. (1991). Enzymes-catalysts for food processes. *Food Technology*, *45*(1), 78.

Edwin, F., & Jagannadham, M. V. (2000). Single disulfide bond reduced papain exists in a compact intermediate state. *Biochimica et Biophysica Acta (BBA)-Protein Structure and Molecular Enzymology*, *1479*(1–2), 69–82.

Esti, M., Benucci, I., Lombardelli, C., Liburdi, K., & Garzillo, A. M. V. (2013). Papain from papaya (Carica papaya L.) fruit and latex: Preliminary characterization in alcoholic–acidic buffer for wine application. *Food and Bioproducts Processing*, *91*(4), 595–598.

Fernández-Salguero, J., & Sanjuán, E. (1999). Influence of vegetable and animal rennet on proteolysis during ripening in ewes' milk cheese. *Food Chemistry*, *64*(2), 177–183.

Fernández-Salguero, J., Sanjuán, E., & Montero, E. (1991). A preliminary study of the chemical composition of Guía cheese. *Journal of Food Composition and Analysis*, *4*(3), 262–269.

Galante, Y. M., DE Conti, A. L. B. E. R. T. O., & Monteverdi, R. (1998). Application of trichoderma enzymes. *Trichoderma and Gliocladium, Volume 2: Enzymes, Biological Control and commercial applications* (2), 327.

Gandhi, N. N., & Mukherjee, K. D. (2001). Reactivity of medium-chain substrates in the interesterification of tripalmitin catalyzed by papaya lipase. *Journal of the American Oil Chemists' Society*, *78*(9), 965–968.

González-Rábade, N., Badillo-Corona, J. A., Aranda-Barradas, J. S., & del Carmen Oliver-Salvador, M. (2011). Production of plant proteases in vivo and in vitro—A review. *Biotechnology Advances*, *29*(6), 983–996.

Harboe, M., Broe, M. L., & Qvist, K. B. (2010). The production, action and application of rennet and coagulants. *Technology of Cheesemaking*, *2*.

Hasan, F., Shah, A. A., & Hameed, A. (2006). Industrial applications of microbial lipases. *Enzyme and Microbial Technology*, *39*(2), 235–251.

Hashim, M. M., Mingsheng, D., Iqbal, M. F., & Xiaohong, C. (2011). Ginger rhizome as a potential source of milk coagulating cysteine protease. *Phytochemistry*, *72*(6), 458–464.

Heimgartner, U., Pietrzak, M., Geertsen, R., Brodelius, P., da Silva Figueiredo, A. C., & Pais, M. S. S. (1990). Purification and partial characterization of milk clotting proteases from flowers of Cynara cardunculus. *Phytochemistry*, *29*(5), 1405–1410.

Heldman, D. R., & Hartel, R. W. (2007). *Principles of food processing*. London: Chapman & Hall.

Holm, K. A. (1995). Automated determination of microbial peroxidase activity in fermentation samples using hydrogen peroxide as the substrate and 2, 2′-azino-bis (3-ethylbenzothiazoline-6-sulfonate) as the electron donor in a flow injection system. *Analyst*, *120*(8), 2101–2105.

Hutton, D., & Stumpf, P. K. (1969). Fat metabolism in higher plants. XXXVII. Characterization of the β-oxidation systems from maturing and germinating castor bean seeds. *Plant Physiology*, *44*(4), 508–516.

Idesa, G. D. (2018). *Extraction and partial purification of peroxidase enzyme from plant sources for antibody labeling* (Doctoral dissertation, Mekelle University).

Jacob, M., Jaros, D., & Rohm, H. (2011). Recent advances in milk clotting enzymes. *International Journal of Dairy Technology*, *64*(1), 14–33.

Jana, M., Maity, C., Samanta, S., Pati, B. R., Islam, S. S., Mohapatra, P. K. D., & Mondal, K. C. (2013). Salt-independent thermophilic α-amylase from Bacillus megaterium VUMB109: An efficacy testing for preparation of maltooligosaccharides. *Industrial Crops and Products*, *41*, 386–391.

Kumar, S. (2015). Role of enzymes in fruit juice processing and its quality enhancement. *Advances in Applied Science Research*, *6*(6), 114–124.

Kumari, M., Sharma, A., & Jagannadham, M. V. (2012). Religiosin B, a milk-clotting serine protease from Ficus religiosa. *Food Chemistry*, *131*(4), 1295–1303.

Lin, Y. H., Moreau, R. A., & Huang, A. H. (1982). Involvement of glyoxysomal lipase in the hydrolysis of storage triacylglycerols in the cotyledons of soybean seedlings. *Plant Physiology*, *70*(1), 108–112.

Lin, Y. H., & Huang, A. H. (1983). Lipase in lipid bodies of cotyledons of rape and mustard seedlings. *Archives of Biochemistry and Biophysics*, *225*(1), 360–369.

Llorente, B. E., Brulti, C. B., & Natalucci, C. L. (1997). From Artichoke (Cynara Scolymus L., Asteraceae). *Acta Farmaceutica Bonaerense*, *16*(1), 37–42.

Lo Piero, A. R., Puglisi, I., & Petrone, G. (2002). Characterization of "Lettucine," a serine-like protease from Lactuca sativa leaves, as a novel enzyme for milk clotting. *Journal of Agricultural and Food Chemistry*, *50*(8), 2439–2443.

Macedo, I. Q., Faro, C. J., & Pires, E. M. (1993). Specificity and kinetics of the milk-clotting enzyme from cardoon (Cynara cardunculus L.) toward bovine. kappa.-casein. *Journal of Agricultural and Food Chemistry*, *41*(10), 1537–1540.

Mamboya, E. A. F. (2012). Papain, a plant enzyme of biological importance: A review. *American Journal of Biochemistry and Biotechnology*, *8*(2), 99–104.

Martins, B. C., Rescolino, R., Coelho, D. F., Zanchetta, B., Tambourgi, E. B., & Silveira, E. (2014). Characterization of bromelain from ananas comosus agroindustrial residues purified by ethanol factional precipitation. *Chemical Engineering Transactions*, *37*, 781–786.

Mobini-Dehkordi, M., & Javan, F. A. (2012). Application of alpha-amylase in biotechnology. *Journal of Biology and Today's World*, *1*, 39–50.

Mohan, R., Sivakumar, V., Rangasamy, T., & Muralidharan, C. (2016). Optimisation of bromelain enzyme extraction from pineapple (Ananas comosus) and application in process industry. *American Journal of Biochemistry and Biotechnology*, *12*(3), 188–195.

Mohapatra, A., Rao, V. M., & Ranjan, M. (2013). Comparative study of the increased production and characterization of Bromelain from the peel, pulp and stem pineapple (Anannus commas). *International Journal of Advancements in Research & Technology*, 2(8), 249–277.

Monteiro, C.A., Levy, R.B., Claro, R.M., Castro, I.R.R.D. and Cannon, G., 2010. A new classification of foods based on the extent and purpose of their processing. *Cadernos de saude publica*, 26, pp. 2039–2049.

Mukherjee, K. D., & Kiewitt, I. (1998). Structured triacylglycerols resembling human milk fat by transesterification catalyzed by papaya (Carica papaya) latex. *Biotechnology Letters*, *20*(6), 613–616.

Pasha, K. M., Anuradha, P., & Subbarao, D. (2013). Applications of pectinases in industrial sector. *International Journal of pure and Applied sciences and Technology*, *16*(1), 89.

Patil, K. J., Chopda, M. Z., & Mahajan, R. T. (2011). Lipase biodiversity. *Indian Journal of Science and Technology*, *4*(8), 971–982.

Pedrolli, D. B., Monteiro, A. C., Gomes, E., & Carmona, E. C. (2009). Pectin and pectinases: Production, characterization and industrial application of microbial pectinolytic enzymes. *Open Biotechnology Journal*, 9–18.

Pinyaphong, P., & Phutrakul, S. (2009). Synthesis of cocoa butter equivalent from palm oil by Carica papaya lipase-catalyzed interesterification. *Chiang Mai Journal of Science*, *36*(3), 359–368.

Pontual, E. V., Carvalho, B. E., Bezerra, R. S., Coelho, L. C., Napoleão, T. H., & Paiva, P. M. (2012). Caseinolytic and milk-clotting activities from Moringa oleifera flowers. *Food Chemistry*, *135*(3), 1848–1854.

Prakash, O., & Jaiswal, N. (2010). α-Amylase: An ideal representative of thermostable enzymes. *Applied Biochemistry and Biotechnology*, *160*(8), 2401–2414.

Rebello, S., Anju, M., Aneesh, E. M., Sindhu, R., Binod, P., & Pandey, A. (2017). Recent advancements in the production and application of microbial pectinases: An overview. *Reviews in Environmental Science and Bio/Technology*, *16*(3), 381–394.

Reed, G. (1975). Oxidoreductase. In *Enzymes in food processing*. Academic Press, Inc., NY.

Rosa, G. P., Barreto, M. D. C., Pinto, D. C., & Seca, A. M. (2020). A green and simple protocol for extraction and application of a peroxidase-rich enzymatic extract. *Methods and Protocols*, *3*(2), 25.

Roa, I., López, M. B., & Mendiola, F. J. (1999). Residual clotting activity and ripening properties of vegetable rennet from Cynara cardunculus in La Serena cheese. *Food Research International*, *32*(6), 413–419.

Roseiro, L. B., Barbosa, M., Ames, J. M., & Wilbey, R. A. (2003). Cheesemaking with vegetable coagulants—The use of Cynara L. for the production of ovine milk cheeses. *International Journal of Dairy Technology*, *56*(2), 76–85.

Rosnitschek, I., & Theimer, R. R. (1980). Properties of a membrane-bound triglyceride lipase of rapeseed (Brassica napus L.) cotyledons. *Planta*, *148*(3), 193–198.

Royo, J., León, J., Vancanneyt, G., Albar, J. P., Rosahl, S., Ortego, F., ... Sánchez-Serrano, J. J. (1999). Antisense-mediated depletion of a potato lipoxygenase reduces wound induction of proteinase inhibitors and increases weight gain of insect pests. *Proceedings of the National Academy of Sciences*, *96*(3), 1146–1151.

Saadoun, I., Dawagreh, A., Jaradat, Z., & Ababneh, Q. (2013). Influence of culture conditions on pectinase production by Streptomyces sp.(strain J9). *International Journal of Life Science and Medical Research*, *3* (4), 148.

Sanjuán, E., Millán, R., Saavedra, P., Carmona, M. A., Gómez, R., & Fernández-Salguero, J. (2002). Influence of animal and vegetable rennet on the physicochemical characteristics of Los Pedroches cheese during ripening. *Food Chemistry*, *78*(3), 281–289.

Schmid, R. D., & Verger, R. (1998). Lipases: Interfacial enzymes with attractive applications. *Angewandte Chemie International Edition*, *37*(12), 1608–1633.

Shah, M. A., & Mir, S. A. (2014). Plant proteases in food processing. In J.-M. Mérillon, & K. G. Ramawat (Eds.), *Bioactive molecules in food, reference series in phytochemistry*. Cham: Springer. Available from https://doi.org/10.1007/978-3-319-54528-8_68-1.

Shah, M. A., Mir, S. A., & Paray, M. A. (2014). Plant proteases as milk-clotting enzymes in cheesemaking: A review. *Dairy Science and Technolnology*, *94*, 5–16.

Sharma, S., & Kanwar, S. S. (2014). Organic solvent tolerant lipases and applications. *The Scientific World Journal*, *2014*.

Sharma, A., Kumari, M., & Jagannadham, M. V. (2012). Religiosin C, a cucumisin-like serine protease from Ficus religiosa. *Process Biochemistry*, *47*(6), 914–921.

Shiiba, K., Negishi, Y., Okada, K., & Nagao, S. (1991). Purification and characterization of lipoxygenase isozymes from wheat germ. *Cereal Chemistry*, *68*(2), 115–122.

Shinde, V. B., Deshmukh, S. B., & Bhoyar, M. G. (2015). Applications of major enzymes in food industry. *Indian Farmer*, *2*(6), 497–502.

Sidrach, L., García-Cánovas, F., Tudela, J., & Rodríguez-López, J. N. (2005). Purification of cynarases from artichoke (Cynara scolymus L.): Enzymatic properties of cynarase A. *Phytochemistry*, *66*(1), 41–49.

Silva, S. V., Allmere, T., Malcata, F. X., & Andrén, A. (2003). Comparative studies on the gelling properties of cardosins extracted from Cynara cardunculus and chymosin on cow's skim milk. *International Dairy Journal*, *13*(7), 559–564.

Simpson, B. K., & Haard, N. F. (1987). Cold-adapted enzymes from fish. M. Dekker (Ed.).

Sims, C. A., & Bates, R. P. (1994). Challenges to processing tropical fruitjuices: Banana as an example. In *Proceedings of the Florida State Horticultural Society*, *107*, 315–318.

Sousa, M. J., & Malcata, F. X. (2002). Advances in the role of a plant coagulant (Cynara cardunculus) in vitro and during ripening of cheeses from several milk species. *Le Lait*, *82*(2), 151–170.

Srujana, N. S. V., & Narayana, M. K. (2017). Extraction, purification and characterization of bromelain from pineapple and in silico annotation of the Protein. *HELIX*, *7*(4), 1799–1805.

Tamer, I. M. (1993). Identification and partial purification of a novel milk clotting enzyme from Onopordum turcicum. *Biotechnology Letters*, *15*(4), 427–432.

Tecelão, C., Rivera, I., Sandoval, G., & Ferreira-Dias, S. (2012). Carica papaya latex: A low-cost biocatalyst for human milk fat substitutes production. *European Journal of Lipid Science and Technology*, *114*(3), 266–276.

Vairo-Cavalli, S., Claver, S., Priolo, N., & Natalucci, C. (2005). Extraction and partial characterization of a coagulant preparation from Silybum marianum flowers. Its action on bovine caseinate. *The Journal of Dairy Research*, *72*(3), 271.

Veríssimo, P., Esteves, C., Faro, C., & Pires, E. (1995). The vegetable rennet of Cynara cardunculus L. contains two proteinases with chymosin and pepsin-like specificities. *Biotechnology Letters*, *17*(6), 621–626.

Veríssimo, P., Faro, C., Moir, A. J., Lin, Y., Tang, J., & Pires, E. (1996). Purification, characterization and partial amino acid sequencing of two new aspartic proteinases from fresh flowers of Cynara cardunculus L. *European Journal of Biochemistry*, *235*(3), 762–768.

Villeneuve, P., Pina, M., Skarbek, A., Graille, J., & Foglia, T. A. (1997). Specificity of Carica papaya latex in lipase-catalyzed interesterification reactions. *Biotechnology Techniques*, *11*(2), 91–94.

Yamamoto, T. (1988). *Handbook of amylases and related enzymes. The amylase research society of Japan* (pp. 40–44). *Oxford*: *Pergamon Press*.

CHAPTER

Plant extracts as packaging aids

11

Nazila Oladzadabbasabadi[1], Abdorreza Mohammadi Nafchi[1,2], Fazilah Ariffin[1] and A.A. Karim[1]

[1]*Food Biopolymer Research Group, Food Technology Division, School of Industrial Technology, Universiti Sains Malaysia, Malaysia* [2]*Food Biopolymer Research Group, Food Science and Technology Department, Damghan Branch, Islamic Azad University, Damghan, Iran*

11.1 Introduction

In the last few years, several new ideas have been developed for food packaging. Consumers are demanding more and more mildly preserved convenience foods with better quality. Moreover, there have been changes in retail and distribution practices such as the central implementation of operations, new trends and internationalization of markets. Consequently, a range of different products with different temperature requirements has increased distribution distances and longer storage times and have also placed great demands on the food packing sector (Han, Ruiz-Garcia, Qian, & Yang, 2018; Silberbauer & Schmid, 2017). The ability to extend the shelf life of food products is limited to traditional packaging concepts (Wyrwa & Barska, 2017).

11.1.1 Smart packaging

Smart packaging, which includes active and intelligent packaging, is an enhancement to traditional packaging functions and is anticipated to be the future of food packaging solutions (Lloyd, Mirosa, & Birch, 2019).

11.1.1.1 Active packaging

Active packaging is an innovative approach that can be described as a form of packaging that changes packaging conditions to increase shelf life or enhance safety or sensory properties while maintaining food quality (Wyrwa & Barska, 2017; Yildirim et al., 2018). Nowadays, the development of food packaging with the idea of incorporating active compounds into packaging materials or their circumstances known as "active packaging" has been evolved (Adilah, Jamilah, Noranizan, & Hanani, 2018; Mir et al., 2017). The researchers have gained interest in this innovation because of the need to increase the shelf life of food, preserve food and food quality, and develop organoleptic functionality. In addition, eco-friendly active packaging and natural preservatives could provide better solutions to health and environmental problems (Guillard et al., 2018). The packaging cannot provide information on food storage history, including changes in temperature, the stability of the package, the environmental conditions and food spoil, even though the development of the

Plant Extracts: ***Applications in the Food Industry*****. DOI: https://doi.org/10.1016/B978-0-12-822475-5.00001-6**

packaging industry guaranteed quality of food. Efforts are also being made to build a framework capable of informing customers of food quality and health, and the implementation of intelligent packaging systems is one of the most recent approaches for this purpose (Eskandarabadi et al., 2019; Yousefi et al., 2019).

11.1.1.2 Intelligent packaging

Another modern type of packaging that characterizes the ability to monitor the condition of packaged food or the environment by providing information on various factors during the transport and stocking is intelligent packaging (Pereira de Abreu, Cruz, & Paseiro Losada, 2012).

11.1.2 Plant extract

In recent years, because of the natural compounds they contain, interest in plant extracts has increased. The biological function of the substances has gained economic significance and has contributed to the growing use of these plant extracts in natural therapies and alternative medicines (Yilar, Kadioglu, & Telci, 2018). For many years, plant extracts have been used for various purposes. As a natural source of antioxidant and antimicrobials, they have recently become a widespread interest. Plant extracts are especially valuable because they are relatively safe, improve food shelf-life, are generally have embraced by consumers and can be used for a variety of different applications due to their potential (Rakholiya, Kaneria, & Chanda, 2013). Plant extract can be combined into the films due to their phenolic ingredients with antioxidant and antimicrobial activities (Lee, Yang, Lee, & Song, 2016). Essential oils (EOs) extracted from plants or spices are rich sources of biologically active compounds such as terpenoids and phenolic acids (Ruiz-Navajas, V.-M., Sendra, Perez-Alvarez, & Fernández-López, 2013). EOs extracted from thyme (Karabagias, Badeka, & Kontominas, 2011), cinnamon (Bonilla & Sobral, 2016; Hu, Wang, Xiao, & Bi, 2015), rosemary (Eskandarabadi et al., 2019) and oregano (Lee et al., 2016) showed significant antimicrobial activities versus various microorganisms. It has been long recognized that some of the EOs have antioxidant properties (Alexandre, Lourenço, Bittante, Moraes, & Sobral, 2016; Bhavaniramya, Vishnupriya, Al-Aboody, Vijayakumar, & Baskaran, 2019; Espitia et al., 2014; Ruiz-Navajas et al., 2013). Role of plant extracts on the properties of packaging material including physical, mechanical, barrier, and functional properties will be addressed in this chapter.

11.2 Potential plant extract for packaging

The numerous functionalities and status of herbal extracts for use in foods have recently created great interest in food applications by mixing them into dietary biopolymer-based packaging materials (Wang, Marcone, Barbut, & Lim, 2012b).

Nowadays, significant attention is being paid to emerging innovative ideas of smart/intelligent, active, and eco-friendly food packaging systems. The packaging has been given new functionalities, mainly resulting from the recent consumer's desire for organic and clean-label- high-quality products. The most prevalent approach used to create active packaging is the integration of the active agent into the packaging matrix. The active agents would be classified as direct additives since the functional agent is predestinated to contribute to the profile of the ingredients of the food product. The successful packaging techniques often offer remedial approaches for food safety and preservations, without food additives being involved. For instance, the covalent bond between the active agent and the packaging material enables it to provide the activity without migrating to the structure of the food. Antimicrobial active packaging that belongs to the active packaging family is basically a packaging device that includes antimicrobial agents (AAs). Traditional direct application of AAs on food surfaces (e.g., dipping, spraying or pulverization) may lead to changes in taste due to excessive amounts of active components. Early evaporation and inactivation or denaturation of active agents by food additives, as well as rapid migration into the food mass, can take place using direct application techniques. Edible films give environmental compatibility along with improved food quality and shelf life and reduce potential loss of volatile flavors and aromas (Lim, Jang, & Song, 2010).

Currently, the tendency to decrease the use of synthetic additives in packaging has focused on replacing them with natural antioxidants, in particular tocopherol, plant extracts and herbal essential oils such as rosemary, oregano, and tea, which are safer and, in most cases, support multiple health benefits. Special consideration has been given to active packages with antioxidant properties because they are one of the most promising substitutes to traditional packaging where antioxidants are incorporated into or coated on food packaging materials to prevent oxidation of the product, which is one of the main causes of food spoilage (López-de-Dicastillo, Gómez-Estaca, Catalá, Gavara, & Hernández-Muñoz, 2012).

Grape skin is a by-product of the juicing operation as well as abundant in phenolic compounds including non-flavonoids and flavonoids. Glucosides of cyanidin, malvidin, delphinidin, peonidin, petunidin, and pelargonidin are the most plentiful anthocyanin in grape skin. In fact, the most antioxidant activity is observed in the skin of the grape. Hence, it is potentially valuable to incorporate grape skin extracts into polymers to improve food packaging (Alexandre et al., 2016; Oladzadabbasabadi, Karazhiyan, & Keyhani, 2017).

Plant-based products are superlative substitutes to chemical preservatives, and their use in food meets consumer demands for minimally processed natural products while providing some additional benefits for both food and consumers. Extract of the grape seed comprises fairly high amounts of flavonoids, including monomeric flavanols, trimeric, dimeric, polymeric, and phenolic acids. *Zataria multiflora* essential oil (ZEO) has large amounts of oxygenated phenolic monoterpenes and in in-vitro revealed antifungal, antioxidant, and antimicrobial activity (Moradi et al., 2012).

One of the superlative consumed drinks is tea due to its bioactive compounds related to various health benefits. In most cases, tea extracts are potent antioxidants. Epigallocatechin is 30 times greater than vitamin E and 20 times more active than vitamin C in some studies. Research studies demonstrated that both green tea and black tea extracts (BTE) had the powerful antimicrobial and antioxidant ability (Peng, Wu, & Li, 2013). Phenolic compounds are natural components which are typically extracted from vegetables and fruit. Phenolic ingredients are mostly combined into starch-based films to enhance active packaging due to strong antioxidant and microbial activities. The phenolic-rich starch-based films could be used to prolonging the food shelf life (Qin et al., 2019). The come about proposed that the Nano-composite films containing grapefruit seed extract (GSE) or thymol may amplify the shelf life of food (Lim et al., 2010). Antioxidant active films were successfully developed using a casting method based on an ethylene-vinyl alcohol copolymer and natural antioxidants. Thermal analysis showed that the materials had degradation of the crystallinity structure and about 4% of the remaining solvent that led to a loss of the barrier properties compared with extrusion materials (López-de-Dicastillo et al., 2012). Ma, Ren, and Wang (2017) carried out the study on the formulation of a pH-sensing film based on Tara gum that combines cellulose with grape skin extracts. They found that in the pH range of 1–11 the visual color changed from red to slightly green and the resulting film reacted well to the milk spoiling process. The film may be used as a clear pH- label, and the film's color change offers a simple path to tracking packaged food freshness. Moradi et al. (2012) reported that the addition of grape seed extract and *Z. multiflora* Boiss essential oil into the chitosan film improved surface wettability, total phenol activity and antioxidant activity. The films integrated by neat chitosan and ZEO had a light yellowish color, while the films incorporated by GSE + ZEO were gray. The findings also showed that GSE formulated chitosan film can be used as an active film due to its excellent percentage of % SI and antioxidant properties in medium moisture products, such as muscle food. Comprehensive reviews on inclusion essential oils and nanoparticles into meat and meat products active packaging mechanisms have been published by (Kargozari & Hamedi, 2019). The effects on the physical, structural and antioxidant properties of chitosan films of 0.5%, 1% and 2% green tea extracts (GTE) and BTE were studied (Peng et al., 2013). Results showed that the incorporation of tea extracts considerably reduced the water vapor permeability and improved film antioxidant capability.

In all food simulants ethanol, the 2,2-diphenyl-1-picrylhydrazyl (DPPH) radical scavenging capability of GTE films was stronger than that of BTE films. The equilibration time in different simulants of foods decreased with increased concentration of ethanol. This study found that active chitosan film could be obtained by incorporating tea extracts, which could provide new formulation options for the development of active antioxidant packaging. Active and intelligent packaging films have been successfully developed by incorporating various quantities of *Lycium ruthenicum* anthocyanins (LRA) into the starch matrix (Qin et al., 2019). The incorporation of LRA has significantly enhanced the moisture content, WVP and mechanical strength of starch film owing to the hydrogen bonds formed between starch and LRA. Besides, the addition of LRA greatly improved the light barrier, antioxidant, and pH-sensitive properties due to anthocyanin functionalities. In future, starch-LRA films, besides that they can be used as active packaging films to increase the shelf life of foods. Also, they will be able to use them as smart packaging films to monitor food freshness in real-time. The results of the xyloglucan film from *Tamarindus* revealed that the film containing 4.5% xyloglucan and 1.5% glycerol was considered appropriate for use as a wrapping film in cut-up 'Sunrise Solo' papayas. They displayed a higher mass loss to the PVC film in their usage and were statistically equivalent to the production of ethylene and carbon dioxide. Despite these findings, the characterization and usage data indicated that xyloglucan from *Tamarindus indica* has acceptable properties about food and packaging (Santos et al., 2019). Cross-linked polymers prepared from sodium alginate/pectin (P) with tartaric acid and citric acid by Singh, Baisthakur, and Yemul (2020). Pectin and sodium alginate were extracted from waste pineapple rind and seaweed, respectively. The crosslinked films have been characterized using different analytical methods. Through the edibility test by mice feeding method, they proved that these crosslinked polymers are edible. They claimed that these newly polymeric films as a green food wrapping material can be useful to enhance the shelf life of food such as chocolate and Indian vegetable puff. The waste pineapple shell has successfully been turned into an edible packaging film with value-added application in food packaging. Poly (ethylene terephthalate)/ polypropylene (PET/PP) films containing extract of olive leaves (OLE) were produced in batch (BM) and semi-continuous modes using supercritical solvent impregnation by Cejudo Bastante et al. (2019). The research focused on the effect on the properties of the impregnated films from pressure, temperature, CO_2 flow and OLE. Thermal examination of unimpregnated samples showed a decline in the crystallinity of the treated PP layer at 35°C and a rise in the Tg of PET treated at 55°C due to CO_2 sorption. Conditions of 400 bar and 35°C in BM overall were optimal for the development of highly antioxidant films with slight structural changes. Agar-based bionanocomposite films with ZnO nanoparticles as an active material for shelf life development of green grape packaging were investigated by Kumar, Boro, Ray, Mukherjee, and Dutta (2019). They have synthesized zinc oxide nanoparticles (ZnO NPs) using Mimusops elengi fruit extract as a new resource. Incorporation of ZnO NPs in composite films decreased tensile strength and transparency, whereas enhanced thermal stability, elongation, and film thickness. The films were utilized for packaging of green grape and the forms of the fruits were controlled throughout storage. Grapes packaged in composite films with 4% (w/w) ZnO NPs displayed fresh appearance up to 21 days during storage. The findings demonstrated the ability of the fabricated agar-ZnO nanocomposite film as useful packaging materials to improve the shelf life of fresh fruit, such as green grapes, after harvest. Wang, Marcone, Barbut, and Lim (2012a) demonstrated anthocyanin-rich red raspberry (*Rubusstrigosus*) extract (ARR) can be used to enhance the material properties of soy protein isolate (SPI) films and probably other protein-based films. ARRE resulted in an SPI film having remarkably improved tensile strength ($P < .05$) and % elongation at break ($P < .05$), also increased water swelling ratio ($P < .05$) and in vitro pepsin digestibility ($P < .05$). Besides, ARRE increased darkness, redness, and yellowness film appearance than the control film. The produced films also exhibited greatly reduced water solubility and water vapor permeability. Wang et al. (2012b) have reviewed. This review showed that current evidence that with the incorporation of plant extracts, various physical characteristics of edible biopolymer materials can be modified. Moreover, bioactive ingredients (antioxidant or antimicrobial activities) of plant extract in these materials have also been identified and considered. The interaction between plant extract and biopolymers has a well-recognized. Plant extracts can improve the properties of dietary polymers in food packaging. In overall, it could be claimed that the application of natural plant extracts would be a high potential non-toxic and natural compound to make dietary biopolymer-based materials with different usage.

Song, Shin, and Song (2012) have evaluated the physical properties of a composite film prepared from barley bran protein and gelatin (BBG). With an increase of barley bran protein content, Tensile strength (TS) and elongation at break (E) values of the BBG film were reduced. E values decreased by increasing gelatin content, while TS increased. Also, they investigated a BBG film containing GSE to inhibit the growth of pathogenic bacteria. Populations of *Escherichia coli* O157: H7 and *Listeria monocytogenes* inoculated on salmon packaged decreased after 15 days of storage. Besides, decrease the peroxide value and the thiobarbituric acid value was observed in packing salmon with the BBG films containing GSE. Because of film-forming ability and low cost of BBP, it can be used as an effective edible film source for packing of salmon during storage. In another study, chitosan-based composite films with different amounts of GSE were fabricated via casting method. In addition, the packaging of bread samples containing composite GFSE chitosan-based films prevented fungal growth in comparison with control samples. Therefore, GFSE composite chitosan-based films have the potential to be an important alternative in food technology (Tan, Lim, Tay, Lee, & Thian, 2015).

11.2.1 Antimicrobial activity aids

Nowadays, attention to the production of high-quality safe food and excellent extending shelf life has led to increasing research and efforts to develop new bio-based packaging materials and antimicrobial packaging (Kanmani & Rhim, 2014a; Marvizadeh, Oladzadabbasabadi, Mohammadi Nafchi, & Jokar, 2017; Sganzerla et al., 2020; Treviño-Garza, Yañez-Echeverría, García, Mora-Zúñiga, & Arévalo-Niño, 2020; Wang, Lim, Tong, & Thian, 2019). In total, antimicrobial packaging is accomplished by merged antimicrobial agents directly into packaging films, coating antimicrobial packaging films, and packaging materials made from polymers with natural antimicrobial properties (Wang, Lim, et al., 2019).

In one hand, it is of great importance to use biodegradable nanocomposite films in active packaging because they can provide a controlled release of antimicrobial compounds (Sani, Ehsani, & Hashemi, 2017).

On the other hand, to produce active packaging films, compounds including bacteriocins, enzymes, spice extracts or essential oils, plant seed extract, organic acids, fatty acids, nano-sized metal, and metallic oxides have been used as efficient antimicrobials. Among them, because of their potential antimicrobial activity and adaptability with biopolymer matrices, natural antimicrobials such as essential oils, spice extracts, and fruit seed extracts are extensively used particularly in the food packaging field (Kanmani & Rhim, 2014a). Current knowledge shows that several plant extracts may significantly reduce or inhibit the growth of pathogenic and spoilage microorganisms. Most of the natural options for synthetic food preservatives studied in recent research are raw or pure plant extracts, that is, essential oils or pure ingredients that have become the center of interest for on-site use in food products (Bouarab Chibane, Degraeve, Ferhout, Bouajila, & Oulahal, 2019). Further, delay and inhibition of bacterial growth on food products play an important role in the preparation of active packaging because the attack on causative bacteria and food poisoning is the main cause of food deterioration among inherent and external factors (Nguyen et al., 2020). Therefore, microbial growth regulation in food products has always been the main concern for the various stakeholders in the agri-food industry. Then, a twofold challenge must be taken into consideration: maintaining both food health and reducing food waste. The use of natural antimicrobials to preserve food is a concept pursued by consumers and food producers. Due to the increasing requirement for minimally processed products, especially those containing natural additives, their use is expected to increase progressively in future. Despite significant efforts to develop manufacturing methods, distribution, hygiene practices and public awareness, spoilage and food-borne pathogenic microorganisms still result in tremendous unacceptable human costs and economic losses. Owing to Increased consumption of minimally processed, fresh and ready-to-eat food has resulted in new ecological pathways for microbial growth. Consumers demand "healthier" and more environmentally sustainable food production processes to ensure the microbial protection of their products, which encourage the creation of creative bio preservation technologies focused on the use of natural antimicrobial agents rather than synthetic preservatives (Bouarab Chibane et al., 2019). Thus the application of plant extracts in edible coating films as natural antimicrobials may fulfill the rising consumer requirement for safe, convenient, and fresh food (Ma, Ren, Gu, & Wang, 2017; Nguyen et al., 2020).

This research was conducted to determine the effectiveness of whey protein isolate and cellulose nanofiber nanocomposite films titanium dioxide and rosemary essential oil in maintaining the microbial and sensory consistency of lamb meat during storage at 4 ± 1°C. The best concentration of each compound to be applied to the film was initially calculated by methods of micro-dilution and disk diffusion. During 15 days of storage, the microbial and sensory properties of lamb meat were tested in two groups (control and treatment). For the sensory analysis, microbial analysis, and a hedonic scale of nine points were applied. Tests showed that the application of nanocomposite films greatly decreased the treatment group's bacterial counts. Higher inhibition effects on Gram-positive bacteria were observed compared to Gram-negative bacteria. The microbial and sensory tests also indicated that the nanocomposite films prolonged the shelf-life of treated meat considerably (15 days) comparison to the control meat (6 days). Edible nanocomposite films have been efficient in protecting the microbial and sensory qualities of lamb meat; thus, this application is especially suggested in red meat by (Sani et al., 2017).

Chen et al. (2020) developed antimicrobial packaging film based on cellulose nanofiber and a pH indicator with incorporation of oregano essential oil as an antimicrobial agent and anthocyanins which extracted from purple sweet potato as a natural dye. The film showed excellent colorimetric efficiency to pH change and superior antimicrobial activity and the *E. coli* and *L. monocytogenes* inhibition levels exceeded 99.99%. With the addition of anthocyanins and essential oregano oil, the film's crystallinity index decreased. Moreover, the pH indicator developed, and the antimicrobial film activity could suggest food quality changes during storage and efficiently extend food shelf life.

Another study was conducted by Dong, He, Xiao, and Li (2020) to investigate the antimicrobial activity and sustained release kinetics of cinnamyl aldehyde and carvacrol in temperature-sensitive polyurethane (TSPU) film. For the potential use of food packaging, TSPU films combined with carvacrol and cinnamyl aldehyde have been prepared. Results showed that at relatively low addition ratio cinnamyl aldehyde and carvacrol had beneficial antimicrobial properties. The kinetic equation of first order may be used to describe its diffusion and the process of sustained release. TSPU films could greatly increase the shelf life of moon cakes in the Cantonese style by effectively inhibiting microbial growth and decreasing lipid oxidation compared with widely used food packaging for polyethylene. In a recent article by Nguyen et al. (2020), the development of physiochemical properties and antimicrobial activities of chitosan edible film with *Sonneratia caseolaris* (L.) Engl. leaf extract (SCELE) were investigated. Vietnamese banana fruit was picked to preserve food and the presence of coated bananas with and without composite films based on chitosan was evaluated. The findings suggested that the presence of SCELE could increase the antimicrobial function, water vapor barriers, while a substantial reduction in light transmittance was observed. Bananas coated with a chitosan matrix-based composite film and SCELE had a longer life than the control sample and those covered with pure chitosan film, suggesting that the incorporation of SCELE into a chitosan matrix brings potential application for active edible food packaging. The growth of tested Gram-negative bacteria was strongly inhibited, and antibacterial activity was enhanced by using chitosan film incorporated with SCELE after 12 h exposure. During 12 h exposure to *Pseudomonas aeruginosa*, the greatest antibacterial activity of chitosan film with incorporated 1% and 3% SCELE was found. It can be shown that the inhibitory efficacy of Gram-negative bacteria blend films depends on both SCELE content and exposure time. Active Polylactide Films (PLA) containing *Allium ursinum* extract L. (AU), also known as wild garlic, was successfully prepared using the electrospinning technology at 10% wt. by Radusin et al. (2019). Electrospinning of the AU-containing PLA solutions produced beaded-like morphology fibers in the range 1–2 μm, indicating that the AU extract was primarily encapsulated in certain fiber regions. Cross-sections of the film showed that the AU extract was inserted as micro-sized droplets into the PLA matrix. The thermal properties showed that the addition of the AU extract plasticized the PLA matrix and also lowered its degree of crystallinity as it interfered with the ordering of the PLA chains by impeding their folding into the crystalline lattice. Analysis of thermal stability suggested that the natural extract contributed positively to a gap in the biopolymer's thermal degradation and was thermally stable when encapsulated in the PLA film. Significant antimicrobial activity against foodborne bacteria has been achieved by electrospun PLA film containing the natural extract. The free AU extract performed very well against *E. Coli* and demonstrated a 100% reduction in viable cells compared to control. Its effectiveness against *S. aureus* was also notable, that is, a reduction of 53% after 24 h.

The study of antimicrobial, physicomechanical, and barrier properties on linseed mucilage films with addition of *Hamamelis. virginiana* extract demonstrated that increased interest in providing excellent quality and shelf-life for safe food has led to increased efforts to expand new bio-based packaging materials. The goal of this study was to develop and characterize linseed mucilage (LM)-based films at concentrations of 2.0%, 2.5%, and 3.0%, as well as to improve antimicrobial films (AFs) that contain *H. virginiana* (Hv) extract. Based on its mechanical properties, water vapor permeability and moisture sensitivity, the films with the greatest concentration of LM were selected as the best formulating. Minimum inhibitory concentrations were 1.18 mg/mL for *L. monocytogenes* and 2.37 mg/mL for *S. typhi, S. aureus, and E. coli*. Finally, by incorporating 2.37 mg/mL of Hv extract with a base of 3.0% LM, AFs were enhanced and improved antioxidant, antimicrobial and moisture activity. It also reduced water vapor permeability and tensile strength. These outcomes suggest that studied films have adequate properties for packaging material (Treviño-Garza et al., 2020).

A list of recent research conducted in this area is presented in Table 11.1.

11.2.2 Antioxidant activity aids

Antioxidant packaging is a great type of active packaging and a very promising food preservation technique to extend the shelf life of food products. As oxidation is one of the major reasons for food spoilage, insertion of antioxidants into packaging film as an antioxidant packaging has become very popular (Wu et al., 2013). The main categories of active packaging are antioxidant and antibacterial packaging which have the potential to extend the shelf life of food products and enhance the sensory properties (Bonilla & Sobral, 2016; Ma, Ren, Gu, et al., 2017). Synthetic antioxidants are commonly used for inhibiting the oxidation of food products in the food industry. Examples of common synthetic antioxidants used in food commodities are butylated hydroxyanisole (BHA) and butylated hydroxytoluene. However, some artificial additives could change the food flavors and give a different taste. Moreover, consumers today tend to not use synthetic chemicals in their products because of their concern about the harmful effects on human health (Adilah et al., 2018). The use of natural antioxidants and antimicrobials components in the protection of food is an alternative to chemical products (Bonilla & Sobral, 2016; Pastor, Sánchez-González, Chiralt, Cháfer, & González-Martínez, 2013). Owing to their high concentrations of bioactive ingredients with antioxidant and/or antimicrobial activity, edible plants, especially those rich in secondary metabolites (e.g., essential oils, polyphenols) are of increasing interest. The addition of natural extracts into

Table 11.1 Antimicrobial activity of plant extracts in food packaging.

Plant-derived	Packaging material	Packaged product	process	Antimicrobial effect	References
Grape seed extract	Pee-starch films	Pork loin	Casting	Prevent growth of *B.thermosphacta* and of undesirable pathogens in meat, enhancing quality and shelf life of food	Corrales, Han, and Tauscher (2009)
GSE or thymol	Gelidium corneum/nano-clay composite	–	Casting	Growth prevention of *Listeria monocytogenes and Escherichia coli*, prolong the shelf life	Lim et al. (2010)
Wine grape pomace water extract	Low-methoxyl pectin, sodium alginate, Ticafilm	Colorful wraps for different food	Casting	Demonstrated antibacterial activity against *Listeria innocua* and *E. coli*.	Deng and Zhao (2011)
Thyme and oregano essential oils	–	Lamb meat	–	Decrease microbial populations and lipid oxidation, extended the shelf of meat during storage	Karabagias et al. (2011)
Grapefruit seed extract	Rapeseed protein–gelatin film	Strawberry	Casting	Growth inhibition of *L. monocytogenes* and *E. coli*, prolong the shelf life of fruit	Jang, Shin, and Song (2011)
Grapefruit seed extract	Bran protein–gelatin composite film	Salmon	Casting	Reduced population of *L. monocytogenes* and *E. coli*, grapefruit seed extract (GSE) showed antimicrobial activity, impressive preserving against oxidation of lipid	Song et al. (2012)
Quillaja saponaria Mol. extracts	Milk proteins (calcium caseinate and whey protein isolate)	Strawberry	Coating	Prevent the growth of Botrytis cinerea, increased the shelf life of fruit	Zúñiga et al. (2012)
Grapefruit seed extract	Red algae	Cheese, bacon	Casting	Reduced the population of pathogenic bacteria such as *L. monocytogenes* and *E. coli*, enhanced the shelf life of food	Shin, Song, Seo, and Song (2012)
Thymus. piperella and *Thymus. moroderi* EOs	Chitosan films	Food product	Casting	Great antibacterial activity, inhibit growth of Gram-negative and Gram-positive bacteria, extended the shelf of life	Ruiz-Navajas et al. (2013)

(*Continued*)

Table 11.1 Antimicrobial activity of plant extracts in food packaging. ***Continued***

Plant-derived	Packaging material	Packaged product	process	Antimicrobial effect	References
Cinnamon EOs	Corn starch	Bread	Encapsulation	Reduced the growth of yeast and mold, decrease microbial growth, extended the shelf life	Lotfinia, Javanmard Dakheli, and Mohammadi Nafchi (2013)
Green tea extract	Agar/Agar-fish gelatin	–	Casting	Demonstrated antimicrobial activity against different microorganisms	Giménez, López de Lacey, Pérez-Santín, López-Caballero, and Montero (2013)
Grapefruit seed extract	Agar-based film	Food product	Casting	Showed antimicrobial activity against *L. monocytogenes*, *E. coli* and *Bacillus cereus*, prolonging the shelf life of food	Kanmani and Rhim (2014a)
Sweet basil hydroalcoholic extract	Pullulan film	Jonagored apples	Coating	Prevented growth of Gram-negative bacteria, showed great antimicrobial activity, antifungal properties against R. arrhizus, extended the shelf of life	Synowiec et al. (2014)
GSE	Carrageenan-based film	Food packaging	Casting	Excellent antimicrobial activity against Gram-positive and Gram-negative pathogens, enhanced the shelf life	Kanmani and Rhim (2014b)
Satureja hortensis L. water extract	Pullulan films	Pepper and apple	Casting	Growth inhibition of Gram-negative and Gram-positive bacteria, extended the shelf life of food during storage	Kraśniewska et al. (2014)
Betel leaf extract	Sago starch film	–	Casting	Improve antimicrobial activity against Gram-negative and Gram-positive bacteria except *Psuedomonas aeruginosa*	Nouri & Mohammadi Nafchi (2014)
Apple skin extract Thyme essential oil	Acaí-based film and pectin	–	Casting	Good antimicrobial activity against *L. monocytogenes*	Espitia et al. (2014)
Cinnamon essential oil	Chitosan nanoparticles	Meat	Ionic gelification	Increased *Pseudomonas spp.*, *Enterobacteriaceae* and LAB Counts, remarkably reduced the microbial growth, improve the shelf life of food during storage time	Hu et al. (2015)

Allium spp. extract	Polylactic acid	Ready-to-eat salads	Extrusion	Decrease the growth of *Fusarium spp., Alternaria spp., Peniciliumdigitatum and Zygosaccharomicesbailii, Staphylococcus aureus*	Llana-Ruiz-Cabello et al. (2015)
Oregano oil	Red pepper seed meal protein-gelatin-composite	Fatty tuna meat	Casting	Decrease the populations of *S. Typhimurium* and *L. monocytogenes*, extended the shelf life	Lee et al. (2016)
Satureja thymbra (L.) extracts	carboxymethyl-cellulose	Marine cultured gilthead seabream fillets	Casting	Lower TVC, great antimicrobial effects, enhanced the shelf life	Choulitoudi et al. (2016)
Murta leaf extract	LDPE	–	Casting	Decreased growth of *Listeria (L.) innocua*, shelf-life extension	Hauser et al. (2016)
Murta fruit extract	Methylcellulose	–	Casting	Remarkably decrease the growth of bacteria	López de Dicastillo, Bustos, Guarda, and Galotto (2016)
Cinnamon extract, guarana extract, rosemary extract and boldo-do-chile extract	Gelatin-chitosan	–	Casting	Strong inhibit growth against *E. coli* and S. *aureus*, prolonging shelf life	Bonilla and Sobral (2016)
Oregano extract	–	Sheep burger	–	TVC and LAB increased remarkably	Fernandes, Trindade, Lorenzo, Munekata, and de Melo (2016)
Moringa plant extracts	Carboxyl methylcellulose	Avocado	Coating	Great antimicrobial, preventing postharvest diseases, enhanced the shelf life of fruit	Tesfay, Magwaza, Mbili, and Mditshwa (2017)
Green tea extract	Chitosan film	Raw chicken meat	Casting	Prevent the growth of Gram-negative and Gram-positive bacteria, improved shelf life of food	Mujeeb Rahman, Abdul Mujeeb, and Muraleedharan (2017)
Grape seed extract and carvacrol	Chitosan film	Salmon meet	Casting	Lower bacterial counts observed, improvement of shelf life of food	Alves et al. (2018)
Grape seed extract	Chitosan film	Chicken breast fillets	Casting	Prevent growth of *E. coli, L. monocytogenes, S. aureus* and *Pseudomonas aeruginosa*, increase the shelf life of food	Sogut and Seydim (2018)

(Continued)

Table 11.1 Antimicrobial activity of plant extracts in food packaging. ***Continued***

Plant-derived	Packaging material	Packaged product	process	Antimicrobial effect	References
Apple peel polyphenol extract	Chitosan	–	Casting	Exhibited antimicrobial activity to control *E. coli, B. cereus, S. aureus and S. typhimurium*, extended the shelf life	Riaz et al. (2018)
Grapefruit seed extract	PLA/PBAT composite films	–	Casting	Antibacterial activity against *L. monocytogenes*, bacteriostatic activity against *E. coli*	Shankar and Rhim (2018)
Grapefruit seed extract	Poly (ε-caprolactone)/ chitosan	Salmon bread	Extrusion	Prevent the growth of *E. coli*, prolonging the shelf life of food	Wang, Yong, et al. (2019)
Supercritical CO2 hop extract	Chitosan	–	Casting	Showed antibacterial activity against foodborne pathogen *Bacillus subtilis*, enhance the shelf life	Bajić, Jalšovec, Travan, Novak, and Likozar (2019)
Rosemary extract	EVA with ZnO/ Fe-MMT	–	–	Demonstrated antibacterial activity to control *E. Coli* and *Staphylococcus Ureus*, showed significant antimicrobial activity, prolonging the shelf life	Eskandarabadi et al. (2019)
Pomegranate peel	Starch	–	Casting	Strongly showed that prevented growth of *S.aureus* and *salmonella*, increased inhibition zone	Ali et al. (2019)
Cashew nut testa extract	Cellulose	–	Interfacial self-assembly	Extend the shelf life antimicrobial activity toward the food pathogens *Escherichia* and *S. aureus*	Lee, Cui, Chai, Zhao, and Chen (2020)
Mangosteen rind powder	Chitosan	Soybean oil	Casting	Inhibit food spoilage and oxidation, exhibited great antimicrobial ability against Gram-positive bacteria than Gram-negative bacteria, prolonging the shelf life	Zhang, Liu, et al. (2020)
Chinese chives root	Chitosan	–	Casting	Preventing growth of Gram-positive and Gram-negative bacteria	Riaz et al. (2020)
Blood orange peel pectin	Fish gelatin	Cheese	Casting	Showed the antibacterial activity against four Gram-negative and Gram-positive bacteria	Jridi et al. (2020)

Red pitaya (*Hylocereuspolyrhizus*) peel extract	Starch/PVA	–	Casting	Improved antimicrobial activity against *S. aureus*, *E. coli*, *Salmonella*, *L. monocytogenes*, enhance the shelf life	Qin, Liu, Zhang, and Liu (2020)
Pectin extracted from pineapple	Sodium alginate and citric or tartaric acid	Chocolate, Indian vegetable puff	Casting	Inhibited the growth of bacteria, improve the shelf life	Singh et al. (2020)
Rheum ribes L. ethanol extract	Methylcellulose	Alternative biodegradable food packaging.	Casting	Preventing the growth of *B. cereus*, *E. coli*, *S. aureus*, *L. monocytogenes*, *Salmonella Typhimurium*, *Klebsiella pneumoniae*, and *Proteus vulgaris*	Kalkan, Otağ, and Engin (2020)
Cinnamon essential oil/ TiO2-N	Sago starch film	Fresh pistachio	Casting	Great antimicrobial activity against *E. coli*, *S. aureus* and *Salmonella typhimurium*	Arezoo, Mohammadreza, Maryam, and Abdorreza (2020)
Thyme essential oil	Orientated polypropylene or LDPE		Immersion	Better antimicrobial activity against Gram-positive bacteria	Mousavian, Mohammadi Nafchi, Nouri, and Abedinia (2020)
Mentha piperita Essential Oil	Cassava Starch	Food and non-food packaging	Solution intercalation	Antimicrobial activity against *S. aureus* and *E.coli*	Marvizadeh, Tajik, Moosavian, Oladzadabbasabadi, and Mohammadi Nafchi (2020)
Cinnamon EOs/TiO_2-NPs	Sago starch	Fresh pistachio	Casting	Decrease aflatoxins, limited the growth of *A. flavus*, extended the shelf life	Esfahani, Ehsani, Mizani, and Mohammadi Nafchi (2020)
Nigella sativa L. seeds	Sago starch	–	Casting	Higher antimicrobial properties against the Gram-negative and Gram-positive bacteria	Ekramian, Abbaspour, Roudi, Amjad, and Nafchi (2020)
Olive leaf extract	–	Gluten-free bread	Encapsulation	Enhanced antifungal properties, extended the shelf life	Moghaddam, Jalali, Nafchi, and Nouri (2020)

gelatin-based films can enhance the functional and physical properties of the film (Bonilla & Sobral, 2016). The integration of antioxidants into the packaging film is also used as oxidation is the main issue that affects the quality of the food. Antioxidants decrease the amount of chemical additives used in food because they have loaded on films could transmittance from the packaging to the food and inhibit direct contact with food. The current research based on natural active compounds rather than synthetic compounds due to consumer health concerns (Ma, Ren, Gu, et al., 2017; Maryam Adilah, Jamilah, & Nur Hanani, 2018). These natural compounds include those in green-tea extract (Siripatrawan & Harte, 2010; Wu et al., 2013), grape seed extract (Moradi et al., 2012), black soybean seed coat extract (Wang, Yong, et al., 2019; Yuan et al., 2020), α-tocopherol (Marcos et al., 2014), pomegranate peel (Ali et al., 2019; Moghadam, Salami, Mohammadian, Khodadadi, & Emam-Djomeh, 2020), thyme extract (Talón et al., 2017), curcumin (Ma, Ren, Gu, et al., 2017), banana peels extract (Zhang, Li, & Jiang, 2019) and essential oils (Ruiz-Navajas et al., 2013). There has been an increasing interest in the use of natural additives with antioxidant properties derived from plant extracts as alternatives due to reducing the use of chemical additives in the food industry. Nevertheless, the palatability of food products can be adversely affected in most cases by the direct introduction of natural compounds. The addition of these naturally occurring antioxidants into films seems to be a safe method for the gradual release of these additives into the food over its shelf life (da Rosa, Vanga, Gariepy, & Raghavan, 2020). Biodegradable films with natural antioxidant seemed to be an interesting way to preserve the quality of food and provide extra preservation for oxidative agents as compounds migrated to the food material, thereafter, promoting longer shelf life. Compared to synthetic antioxidants which could have possible toxicity, natural antioxidants such as an extract from plants were recommended. Typically, most fruits' pericarps, skins or peels are not consumed and disposed of as residue. Nevertheless, phenolic acids usually occurring in the outer sections of plants such as shells, skins, and pericarps have been identified (Nabilah, Wan Zunairah, Nor Afizah, & Nur Hanani, 2019). Researchers have recently focused on the insertion of natural bioactive compounds like phenolic compounds, α-tocopherol and essential oils into packaging materials to the produce active edible films to improve the shelf-life of food products and preserve their quality and safety without the use of synthetic additives (Moghadam et al., 2020). A variety of technological approaches may be used for active antioxidant packaging. Most of them contain either of the co-extrusion of the antioxidant together with the film-forming plastic substance or direct combining of an antioxidant agent with the plastic substances. Researchers have been outlined the promising findings of a modern active antioxidant packaging method (Camo, Lorés, Djenane, Beltrán, & Roncalés, 2011).

11.2.2.1 Polysaccharide-based film

Enhancement of active packaging chitosan-based films including mangosteen rind powder was done by Zhang, Liu, et al. (2020). Mangosteen (*Garcinia mangostana* L.) rind due to the presence of high polyphenols content and potent antioxidant and antibacterial ability has known as a traditional medicine in Southeast Asia. The goal of this study was developed chitosan (CS) active packaging film with incorporated mangosteen rind powder (MRP). Fourier transform infrared spectroscopy showed that the MRP polyphenols could interact with CS growing intermolecular hydrogen bonds. The addition of MRP greatly improved the CS film's thickness, strength and UV-visible light barrier, antioxidant, and antibacterial properties. In addition, CS-MRP film packaging effectively prevented the increase of soybean oil's peroxide value and thiobarbituric acid reactive substances during storage. The findings demonstrate that films CS-MRP could be used to improve the oxidative stability of soy oil in the food industry as active packaging. In an evaluation, the effect of guarana, cinnamon, rosemary and boldo-do-chile ethanolic extracts of gelatin: chitosan on the optical, microstructural, barrier and mechanical properties of the films were investigated as well as the antimicrobial and antioxidant activity, Bonilla and Sobral (2016) found that the microstructural and FTIR studies verified both polymers were homogeneously blended in the film matrix. Increments in the proportion of chitosan improved the elasticity of the films and produced a decrease in the permeability of water vapor, which was not substantially decreased with the incorporation of the extracts. In the TEAC test, the blends films revealed good antioxidant properties and a great inhibition of growth against *E. Coli*, and *S. aureus*, suggested that these films based on mixtures of gelatin and chitosan and ethanol extract additives Can be an option for food applications as an active packaging material. Zhang et al. (2019) was developed the antioxidant chitosan (CS)-banana peels extract (BPE) composite film. The different content of BPE (4%, 8% and 12%) was added to the CS film not only as the antioxidant but also as the cross-linking. The CS-BPE composite film displayed great antioxidant activity in food packing. The optimum concentration of CSBPE coating treatment was applied to apple fruit, and the results indicated that CS-BPE coating was more capable of enhancing apple fruit post-harvest quality than CS coating. Wang, Dong, Men, Tong, and Zhou (2013) developed chitosan active films with incorporated tea polyphenols. Fourier transform infrared spectrometry has been used to evaluate the potential interactions in the films between chitosan and tea polyphenols. The results of the tests showed that the incorporation of tea polyphenols caused interactions between tea polyphenols and chitosan and produced darker film appearances. The films showed increased water solubility after the introduction of tea polyphenols and reduced water vapor permeability. In the meantime, the incorporation of tea polyphenols has increased the total phenolic content. But with time, the chitosan film's antioxidant activity integrated tea polyphenols declined. In another study, food packaging films were enhanced based on chitosan (CS) containing anthocyanin-rich purple eggplant extract (PEE) or black eggplant extract (BEE). The amount of anthocyanin in PEE and BEE were 93.10 and 173.17 mg/g, respectively. The compositions of anthocyanin in PEE and BEE were completely different. BEE and PEE increased the thickness, blueness, UV−vis light barrier and mechanical properties of CS film. The antioxidant ability of CS film was greatly improved by adding PEE and BEE (Yong, Wang, Zhang, et al., 2019). Jamróz, Kulawik, Guzik, and Duda (2019) were developed Antioxidant and pH-sensitive chitosan-based films by incorporating various concentrations of black soybean seed coat extract (BSSCE). The addition of BSSCE could significantly alter the optical, barrier and mechanical properties of chitosan film. In particular, the chitosan-BSSCE film showed better water vapor and UV-vis light barrier properties and stronger mechanical strength than the chitosan film. Moreover, chitosan films displayed higher moisture content and transparency than chitosan-BSSCE films. Among all the films examined, the best performance was the chitosan-BSSCE film containing 15 wt.% of BSSCE on chitosan basis. Results indicated that chitosan-BSSCE films could be used as promising food packaging materials for the antioxidant and visible pH sensing. Talón et al. (2017) examined the antioxidant activity of various chitosan- and starch-based polymer matrices, incorporating a polyphenol-rich thyme extract (TE). TE had significant antioxidant activity on the films. When mixed with chitosan, enhancing the tensile behavior of films due to the polyphenols interacted with the polymer chains, acting as crosslinkers. The results suggested it could be used these antioxidant films for extending the shelf life of the products sensitive to oxidative.

Antioxidant activity and physical properties of chitosan film containing GTE were studied by Siripatrawan and Harte (2010). The aim of this study was developed prepared active film from chitosan with incorporated aqueous GTE. The results showed that addition of GTE into chitosan films enhanced water vapor barrier and mechanical properties and improved polyphenolic content and antioxidant activity of the films. Changes in the chitosan films FTIR spectra were observed when GTE was added, indicating that interactions between chitosan and GTE polyphenols occurred.

Riaz et al. (2018) developed a novel functional film with the incorporation of apple peel polyphenols (APP) into chitosan (CS). The results demonstrated that the addition of APP into CS notably enhanced the film physical properties by increasing its thickness, opacity, solubility, density, and swelling ratio while moisture content and water vapor permeability were reduced. Thermal stability was reduced in the prepared films whereas antimicrobial and antioxidant activities of the CS-based APP film were considerably increased. CS-APP film with 0.50% APP concentration performed good mechanical and antimicrobial properties that pointed out it can be enhanced as bio-composite food packaging material.

Akhtar et al. (2012) prepared HPMC films including a natural red color compound (NRC) at different concentrations. This research showed that natural plant extracts could be used with additional benefits for the functionalization of edible films. These films supply unique fruit taste, color and antioxidant ability which would greatly improve their potential applications in both food and non-food industries. Infrared spectroscopy study has verified miscibility of HPMC and NRC in composite films. Absorption bands in the FTIR spectra indicated interactions between components by hydrogen bonding. Increased peak area in this region was directly adequate to the concentration of the NRC making films more hydrophilic. During the film preparation steps NRC antioxidant capacity was stable. When NRC increased, color of edible films became darker and redder, while a rising effect of light exposure on color stability was observed. Results indicated that because of their color, plasticizing properties, good antioxidant stability and the ability to protect HPMC from photo-degradation, NRC films have good potential for food applications.

Pastor et al. (2013) carried out a study on the antioxidant properties of chitosan and methylcellulose-based films incorporating with resveratrol. New developments in edible films focus on improving their versatility by adding active compounds, such as antimicrobials or antioxidants. Resveratrol is a natural antioxidant present in a different of plant species, such as grapes which could be used to reduce or inhibit lipid oxidation in food products, postpone the development of oxidation products, preserve nutritional quality which extends the shelf life of food products. The goal of this paper was the development and characterization with the incorporation of various amounts of resveratrol in two different polymeric composite films made with chitosan and methylcellulose. By the presence of resveratrol, oxygen and water vapor permeability tend to moderately reduce and barrier properties of the film were hardly improved. Composite films demonstrate antioxidant activity, which was adequate to the concentration of resveratrol in the film. None of the films showed antimicrobial activity against *Penicillium italicum* and *Botrytis cinerea*. Hence, the result suggested these films could be used in food packing that sensitive to oxidative processes to extend their shelf life.

11.2.2.2 Gelatin-based film

Preparation, properties, and antioxidant activity of an active film from silver carp (hypophthalmichthys molitrix) skin gelatin incorporated with GTE were investigated by Wu et al. (2013).

The results of this study have shown that the incorporation of GTE into gelatin films has improved overall phenolic content, increased DPPH radical activity and reduced power. The thermal stability of gelatin-GTE films has been enhanced and increased with a rising concentration of GTE. FTIR spectra showed that the interaction of protein-polyphenol in gelatin-GTE films was involved. For gelatin-GTE films, smooth and homogeneous surfaces and compact structures were observed. As a result, the addition of GTE to gelatin film enhanced the antioxidant activity and the properties were most likely directly affected by interactions of gelatin with GTE. Zamuz et al. (2018) demonstrated that the appropriate combination of chestnut extracts in various sections (bur, leaf, and hull) can be used as a replacement for commercial antioxidants since adding chestnut extracts to bovine meat patties prevented lipid oxidation and delayed metmyoglobin formation without major sensory properties alteration. In this research through the incorporating tea polyphenol (TP) into gelatin and sodium alginate, active edible films were prepared. By increase TP concentration in the films, the tensile strength, contact angle, and cross-linking degree showed an improvement, whereas elongation at break and water vapor permeability reduce. Antioxidant capacity was enhanced by increasing the TP content in the films. The addition of TP into GSA films is an ideal choice to enhance an active edible and environmentally friendly packaging for the shelf life of food extending (Dou, Li, Zhang, Chu, & Hou, 2018).

Nabilah et al. (2019) investigated the effect of mangosteen pericarp extract (MPE) on fish gelatin (FG), corn starch (CS), and SPI films. MPE reduced the mechanical properties and increased the water vapor permeability of the films. Protein-based films have shown better mechanical properties compared to starch films. Corn starch films had good water resistance compared with protein-based films. In conclusion, MPE added in the FFS has enhanced functionality of soy protein films biodegradable film with a high antioxidant capacity may thus be a promising natural antioxidant to be used as an active film in packaging content. The results of another study by Rodriguez et al. (Rodríguez et al., 2020) demonstrated that papaya dried dehydrator edible films showed promising results in terms of their physicochemical properties, longer shelf-life under managed storage conditions and good sensory acceptance by panelists. The formulations added by Moringa provided an important nutrient for consumers considering its protein content. Enhanced papaya edible films are an efficient method for prolonging the shelf-life of oxidation-prone products, such as pears, due to their high antioxidant ability, which is why natural extracts are an appealing effect to the use of synthetic antioxidant utilized in the food industry. In addition, Moringa has had an interesting impact on the protein content of the edible films being made. Ultimately, the addition of both bioactive compounds showed an impact on the longevity of minimally processed pears 'shelf-life, with the edible film combined with ascorbic acid having a beneficial effect on the sensory acceptance of such a food matrix. In an aforementioned study Han, Yu, and Wang (2018a) on Bio-based films prepared with soybean by-products and pine (*Pinus densiflora*) bark extract, pine bark extract (PBE) has excellent antioxidant activity and is a good source of oligomeric procyanidins. Bio-based antioxidant films were prepared by adding different concentrations PBE into the SPI matrix. The structures of the films were rougher with PBE. Cross-linking interactions formed between the amino groups in the SPI matrix and the phenolic hydroxyl group in PBE. Also, PBE improved the thermal stability of the films. Antioxidant SPI films with PBE could be used in food packaging.

Table 11.2 summarizes the antioxidant activities of plant extracts in different packaging films.

Table 11.2 Antioxidant activity of plant extracts in food packaging.

Plant-derived	Packaging material	Packaged product	Process	Antioxidant compounds	Results	References
Chestnut extract	Chitosan film	Fresh pasta	Casting	Phenolic content	Total phenolic content show dependency on moisture throughout the shelf life, inhibited microbial growth during 2 months	Kõrge, Bajić, Likozar, and Novak (2020)
Pomegranate peel/ pistachio green hull extracts	–	Cooked sausages	–	Gallic acid, ellagic tannins and ellagic acid hydroxybenzoic acid, protocatechuic glucoside, acid, quercetin, naringin, and catechin	Showed great antioxidant activities reduced amounts of nitrite up to 50% in cooked sausages improve its functional properties	Aliyari, Bakhshi Kazaj, Barzegar, and Ahmadi Gavlighi (2020)
Amaranthus leaf extract	Polyvinyl alcohol and gelatin	Chicken/ fish	Casting	Phenolics: Total carotenoids, betacyanins. Flavonoids: Chlorophyll a, Betaxanthins. Tannins: Chlorophyll b, Betalamic acid	The film minimized oxidative rancidity of the products, ensuring its quality and safety, improved the antioxidant activity, prolong the shelf life	Kanatt (2020)
Sweet pepper extract	–	Canned refrigerated pork	–	Chlorogenic acid quercetin	Antioxidant capacity increased by the addition of extract. Antioxidant capacity of SPE as a canned pork additive to a reduced content of nitrite by half.	Ferysiuk, Wójciak, Materska, Chilczuk, and Pabich (2020)
Satureja thymbra extracts	Laminated film	Fried potato chips	–	Phenolic acids and flavonoids	Prooxidant activity at higher concentration delaying the deterioration of the fried product in active packaging	Choulitoudi, Velliopoulou, Tsimogiannis, and Oreopoulou (2020)
Olive leaf extract	Carrageenan films	Food product	Casting	Phenolic compounds	Exhibited good barrier properties and mechanical properties Had high antioxidant activity and has great potential for use as a functional ingredient in food packaging.	da Rosa et al. (2020)

Mangosteen (Garcinia mangostana L.) rind	Chitosan film	Soybean oil	Casting	Polyphenols	Inhibited the increase in the peroxide value and thiobarbituric acid reactive substances of soybean oil during storage enhancements in antioxidant ability	Zhang, Liu, et al. (2020)
Pine nut shell, peanut shell and jujube leaf	Chitosan film	–	Casting	5,7-Dihydroxychromone Eriodictyol Luteolin, Catechin, Epicatechin Quercetin, Apigenin Total phenolic and anthocyanin	Improved the antioxidant capacity of films reduced the homogeneity Three plant extracts had also changed the microstructure, chemical structure and thermal properties	Zhang, Lian, Shi, Meng, and Peng (2020)
Pomegranate peel	Mung bean protein film	Food product	Casting	Catechins, punicalin, pedunculagin, punicalagin, gallic acid, and ellagic acid	Significant effect on the TPC The mechanical properties, reducing power, anti-radical activity, and antibacterial attributes of mung bean protein films were improved	Moghadam et al. (2020)
Blood orange peel pectin	Fish gelatin	Cheese	Casting	Poly(1,4-galacturonic acid)	Improves the physicochemical, barrier and antioxidant properties	Jridi et al. (2020)
Moringa leaf extract	Papaya edible films	Pear	Casting	Flavonoids and phenolics contents	Lowest antioxidant activity, effect in the protein content of the developed edible films the in vitro antioxidant effect of moringa, both in DPPH assay and in contact with the pear, was rather limited	Rodríguez, Sibaja, Espitia, and Otoni (2020)

11.2.3 Biodegradable packaging aids

11.2.3.1 Polysaccharide-based films

Over the last few decades, synthetic films have grown rapidly in food production and applications, causing considerable environmental concerns because synthetic materials are resistant to degradation. Nowadays, consumers are looking to decrease environmental issues related to food packaging and health concerns, therefore, request biodegradable materials (Akhter, Masoodi, Wani, & Rather, 2019). Biodegradable and edible films due to their capabilities to prevent moisture loss, solute transport, aromas loss, water absorption in the food matrix or oxygen penetration could be a substitute for synthetic packaging materials in various applications. Hence, food scientists and engineers are seeking to develop new materials for an edible and biodegradable film focused primarily on the abundance of renewable resources. These materials are usually cheap and many are known as waste or by-products (Cazón, Velazquez, Ramírez, & Vázquez, 2017). The use of biodegradable polymers should be exploring continuously to replace the petroleum-based plastics in the food industry and to improve the shelf life of the products. The goal of biodegradable films and edible coatings is to reduce the loss of oxygen, moisture, and other gasses by promoting semipermeable barriers and increasing the shelf life of food (Sganzerla et al., 2020). Moreover, the researchers try to develop biodegradable alternatives that can be used as replacements for the existing synthetic polymers because of the environmental effect of using synthetic packaging materials as a vehicle for value-added. In addition, packaging films based on biopolymers provide multiple advantages due to their excellent biocompatibility, biodegradability, eco-friendliness and even edibility. In general, according to the source of the original polymer used, biopolymer films were classified (da Rosa et al., 2020).

Polysaccharides are feasible film-forming materials and an abundant natural resource between biopolymer materials. They have good properties as barriers, including carbon dioxide and oxygen, but poor water vapor barrier, a lot of researcher's attention has been paid to them (Alexandre et al., 2016). New opportunities to develop novel food packaging technologies may be provided by the application of polysaccharide films in food products. The film-forming ability of several polysaccharides has been studied, including cellulose, chitosan, starch, pectin, alginate, carrageenan, pullulan and kefiran. Polysaccharides such as cellulose derivatives, starches, chitosans, and gums have been known as raw material to prepare edible films and coatings that could be used for food protection as packaging material (Cazón et al., 2017). Proteins and starches have potential sources as natural polymers to formation biodegradable and edible films to replace petrochemical polymers widely used in the packaging field. The renewable sources are regarded to be promising polymers for packaging materials because they are readily available worldwide, inexpensive, biodegradability and nontoxicity. Starch-based films have a strong film-forming potential because of their potential to form a continuous matrix and their thermoplastic properties. Corn starch is one of the most used starches in food packaging applications because of its plentiful sources, renewability and low cost of this raw material (Nabilah et al., 2019). Renewable biodegradable polymers display a credible alternative to traditional oil-derived polymers in agricultural and packaging applications, to decrease environmental impact to enhance environmentally friendly sustainable, cost-effective products. Biopolymers are destroyed by the enzymatic activity of micro-organisms, such as bacteria, fungi and algae, when disposed of in bioactive environments and transformed into biomass, CO_2, CH_4, water and other natural substances. Biodegradable polymers can be synthetic or natural, depending upon their origin. Natural polymers, known as biopolymers, include microorganism-produced polysaccharides, proteins, and polyesters, while poly (vinyl alcohol) and polyesters are the most common of biodegradable synthetic polymers. Biodegradable polymers are limited in use and industrial development due to their poor chemical–physical properties, inadequate mechanical efficiency and difficult processability (Ju & Song, 2019b). A range of polysaccharides for potential application as edible packaging has been developed due to their abundant, low cost, edible, biodegradable, easy-to-handle, and good film-forming pro (Qin et al., 2019; Wyrwa & Barska, 2017). Hence, research has focused on biodegradable polymer systems. Starch is one of the most widely studied among biopolymers as it is broadly available and easily modified to produce thermoplastic polymers. However, thermoplastic starch applications are limited due to the hydrophilic nature responsible for rapid degradation through hydrolysis. In general, starch is modified by blending with synthetic polymers, such as polyesters or vinyl alcohol copolymers, to vanquish this experimental drawback (Cerruti et al., 2011). Native starch depending on the source of starch is extremely variable in its function and structure. The commercially used starches are obtained from many plant origins such as potato, wheat, cassava, corn and rice (Qin et al., 2019). Carrageenan is a polysaccharide natural product derived from seaweed (Wyrwa & Barska, 2017).

As an additive to a starch-based polymer (Mater-Bi), a polyphenol-containing extract from winery bio-waste (EP) was used by Cerruti et al. (2011). It was observed that the processing, mechanical, thermal, and biodegradation properties were effectively modulated by EP. Also decrease in melt viscosity has shown that EP could enhance the polymer processing productivity. Larger values of elongation at break were noticed because of the additive's plasticizing behavior. Eventually, doped Mater-Bi's bio-disintegration rate decreased, indicating that EP acted as an antimicrobial agent by interacting with the polymer film's bio-digestion. Kanmani and Rhim (2014a) developed antimicrobial packaging films based on natural biopolymers as an alternative for synthetic packaging films. The GSE has been incorporated into agar as a natural antimicrobial agent in various concentrations to prepare antimicrobial packaging film by casting method. With the addition of GSE the UV barrier, color, water solubility, moisture content and water vapor permeability increased, while a reduction in tensile strength, elastic modulus surface and hydrophobicity of the films was observed. The incorporation of GSE changed the film's microstructure but did not affect the agar-based films ' crystallinity and thermal stability. These findings indicate that agar/GSE films could be used to preserve food protection and increase the shelf-life of the packaged product in active food packaging systems. Comparative studies were done by Norajit, Kim, and Ryu (2010). In this research, the physical and antioxidant properties of biodegradable alginate film containing green, red, and extruded green ginseng extracts was studied. The highest moisture content of the ginseng extract was incorporated in film samples, but no differences in the moisture content of all alginate film samples were observed. The adding of ginseng extract to alginate film reduced elastic modulus and tensile strength but increased the percent of elongation at break. Such tests for free-radical scavenging operation were accompanied by film tests containing red and white ginseng extracts, respectively. These findings have shown that the extruded white ginseng extract has a strong potential to be integrated into alginate to create an antioxidant biodegradable film or coating for different food applications. Wyrwa and Barska (2017) reported that the addition of plant oils greatly improved the thickness of the film. Nevertheless, the moisture content, solubility and tensile strength of the film decreased dramatically as plant oils were incorporated. The addition of plant oils also led to a plasticizing effect, which significantly increased the elongation values at break. In conclusion, the plant oils used in this research greatly improved kappa-carrageenan films properties, thereby demonstrating the ability of these products to be used as films and coatings for food packaging. Other study aimed with to develop novel functional films based on chitosan (CS) containing Chinese chive root extract (CRE) by casting method. SEM showed that a higher concentration of the extracts triggered agglomerate formation within the films. The WVP has been reduced due to the good barrier property of CS-based CRE film. Additionally, overall color attributes have been improved from transparent to opaque. The films made by incorporating CRE into CS showed strong antioxidant and antimicrobial activity suggesting that it could be produced for the food industry as a bio-composite food packaging material (Riaz et al., 2020). Rubilar et al. (2013) demonstrated that mixtures of GSE and carvacrol can be added to a matrix of chitosan films. They successfully prepared transparent biodegradable films and the application of these natural agents affected the mechanical, barrier and color properties of the chitosan films. The addition of compounds such as GSE and carvacrol can also be used to protect foods from degradation caused by UV light. These films may also offer alternatives to synthetic materials, potentially leading to improved food safety and extended shelf-life. Sganzerla et al. (2020) studied the biodegradable packaging produced with pinhão starch and citric pectin used the agroindustrial waste of Acca sellowiana by-product (feijoa peel flour, FPF). Regarding the morphological, physicochemical, antioxidant and antimicrobial properties of the packaging, positive effects were achieved. The results obtained show that, for all the parameters tested, FPF addition had a positive effect on the packaging characteristics. The packaging that was developed retained apple consistency during storage, after 5 days of storage with constant weight. Based on their findings, the bioactive packaging can be considered as a potential alternative to food packaging. Da Rosa et al. (2020) analyzed the bioactive compounds from olive leaf extract and to produce biodegradable carrageenan films with antioxidant properties by adding varying olive leaf extract concentrations. The extract of MAE solvent-free olive leaf has a high antioxidant activity and has a high potential for use as a functional ingredient in food packaging. Given the addition of extract to biofilm resulting in a slight increase in stretching capacity, reduced tensile strength and increased water vapor permeability, the carrageenan-based biodegradable films containing olive leaf exhibited strong barrier properties and mechanical properties. With an increase in concentration of olive leaf extract, the overall phenolic compounds and antioxidant content of films dramatically increased. The addition of natural antioxidants appears to be a possible technique for applying additives to food-suitable packaging materials. In a study investigating the optical and antioxidant and mechanical properties of funoran films containing various concentrations of yellow onion peel extract (YOPE) (0.3%, 0.5%, and 1.0%), Ju and Song (2019b) reported that after YOPE's addition, the total phenolic and flavonoid content of the funoran films increased. These results indicate that the funoran-based film combined with YOPE, prepared from underused red algae and discarded onion peel, can be used as a biodegradable packaging material with antioxidant benefit. In a study conducted by Ju, Baek, Kim, and Song (2019), it was shown that the develop and characterize the properties of Khorasan wheat starch (KWS) films with addition of moringa leaf extract (MLE). The isolated KWS from Khorasan wheat was mixed with Different amounts of MLE and the film properties were investigated. With increasing MLE content elongation at break increased and tensile strength of the KWS films reduced. KWS films including MLE revealed acceptable antioxidant and ultraviolet light-blocking properties. Moreover, KWS films containing 1.0% MLE were biodegradable within 30 days. The result suggested that Developed KWS film containing MLE may be used as a biodegradable packaging material with antioxidant activity. Three plant extracts (pine nutshell, peanut shell and jujube leaf) enhanced the antioxidant capacity of chitosan-based biodegradable films. Compared to control films, the DPPH radical scavenging activity of chitosan-jujube leaf films 3.8-times increased. The chitosan film with incorporation of pine nutshell had the greatest oxygen permeability and water vapor, while chitosan film with adding peanut shell demonstrated the highest increase in CO_2 permeability and the highest thermal stability. However, three plants extract decreased homogeneity and induced chitosan film porous structure. The research presented a target for preparing high antioxidant and gas-permeability polysaccharide-based active films (Zhang, Lian, et al., 2020). In an investigation into a teff starch (TFS) biodegradable film was enhanced with incorporation of camu-camu extract (CCE), Ju and Song (2019a) found the TFS films with CCE revealed great radical scavenging activities. With increasing CCE content, the elongation at break increased, whereas the tensile strength of TFS films decreased. The TFS films containing CCE completely blocked the ultraviolet light in the 200–360 nm range. In addition, the surface of the TFS films became rough and the contact angle of the TFS films increased from 36 angle to 69 angle as the amount of CCE incorporated increased. These results indicate that TFS films containing CCE could be used as an antioxidant packaging material in food industry. A significant analysis and discussion on develop polylactic acid biodegradable packaging materials (PLA) and extracted starch from cassava tubers was presented by Kaushalya, Dhanushka, Samarasekara, and Weragoda (2019). Biodegradable PLA and starch-based blends were prepared through incorporation starch as the main additive. During the soil burial test, weight loss in starch-containing samples increased slowly with time. Experimental findings also revealed that product biodegradability increased, and mechanical properties such as tensile strength and elongation also decreased with adding starch content. These results showed that the incorporation of starch to PLA can be a good method for enhancing the biodegradability of PLA—starch blends.

11.2.3.2 Protein-based films

Food packaging is an essential mechanism in food processing and plays a major role in preserving the food from possible contaminations, protecting the quality of the food, and extending its shelf life during preparation, transport, and storage (Han, Shin, Park, & Kim, 2015; Yuan et al., 2020). If the packaging material can not completely prevent moisture from entering properly, there would be a reduction in shelf-life due to the degradation and spoilage of the product and ultimately result in soggy foods. Besides, microbial contamination and lipid oxidation may happen after food packaging, and these have become the main problems in the food industry. Therefore, the development of food packaging materials has been growing up significance with the quick progress of the global food industry (Han et al., 2015). The use of non-edible and non-biodegradable synthetic products for food packaging, such as polyamide and polyethylene, causes significant environmental issues and imposes several health risks (Jridi et al., 2020; Tulamandi et al., 2016). In addition, Synthetic films contained harmful molecules, which, when added to the surface, would migrate to the food product. Therefore, scientist efforts to search for new packaging from biodegradable, natural and sustainable biopolymers have been increased. Natural biopolymers are known as a potential source for the development of new food packaging materials and eco-friendly (Hoque, Benjakul, & Prodpran, 2011; Jridi et al., 2020). In last few years, food-packaging industries have indicated an interest in edible films. According to environmental and consumers health concerns, conventional plastic packaging is replaced by edible and biodegradable biopolymers films. Edible films also enhance the quality of food by providing barriers to moisture, gas, and oxygen they have been used during storage to preserve the quality of the food (Alexandre et al., 2016; Jang et al., 2011). Food characteristics can be provided by edible coatings based on these natural materials, performing as an excellent barrier to water vapor and oxygen and enhancing their physicochemical, microbial, and sensory properties.

Protein film-forming ingredients are various and can be easily obtained from sources of animals and fish, such as collagen, whey protein, and gelatin (Jridi et al., 2020). The physicochemical properties of proteins depend entirely on the arrangement and relative amount of amino acid substitutes alongside the chain of the polymer (Hassan, Chatha, Hussain, Zia, & Akhtar, 2018). Materials based on protein are typically biodegradable and derived from the sources of renewables. Protein-based films have appropriate optical, mechanical and oxygen barrier characteristics while because of its hydrophilic nature have a high sensitivity to moisture and low water vapor barrier properties (Zink, Wyrobnik, Prinz, & Schmid, 2016). Thus due to its biodegradability, good gas barrier property and oil resistance capacity, protein-based films have gained more attention these last years (Insaward, Duangmal, & Mahawanich, 2015).

Moreover, gelatin has a linear structure and minimal monomer composition, contributing to excellent film-forming and is biodegradable (Abedinia et al., 2020; Oladzadabbasabadi, Ebadi, Mohammadi Nafchi, Karim, & Kiahosseini, 2017). Gelatin is a highly processable substance and sensitive to moisture.

Some authors have driven the further development of bioplastic materials for food packaging by the addition of a natural antioxidant agent into films based on soy protein Han et al. (2015). The films were prepared by SPI with incorporation catechin (CT) and/or carboxymethylcellulose (CMC). The original SPI film turned opaque after being blended with CMC or CT, whereas the original SPI film showed good optical transparency. The SPI film with CMC exhibited improved tensile strength and water solubility and reduced percentage of elongation and water vapor permeability as compared with pure SPI film. The CT-incorporated SPI or SPI/CMC films also revealed a synergistic, free radical scavenging effect. The results indicated that the bioplastic soy protein-based film produced in this study could be used in the food industry as a potent antioxidant packaging material.

Rapeseed protein–gelatin edible film containing antimicrobial GSE has been developed to manufacture a packaging film for "Maehyang" strawberries. The incorporation of GSE to the protein—gelatin (RG) rapeseed film inhibited pathogenic bacteria growth. GSE-RG film-packaged strawberries provide better sensory scores than the control. These findings indicate the GSE- RG film can be used for packaging strawberries and extending the shelf life (Jang et al., 2011). Previous research by Jridi et al. (2020) showed that the efficacy of the incorporation of gray triggerfish gelatin and blood orange peel pectin resulting in enhanced composite film properties as demonstrated by thermal, mechanical, and structural analysis. In addition, the blend film demonstrated strong antioxidant and antibacterial properties compared to the gelatin-based film. The composite films have double positive benefits according to their fascinating biological activities, either by improving the microbial consistency of cheese during cold storage or by offering consumer health benefits. Recent findings suggest the further use of gelatin pectin films as active packaging material in the food industry. In another study, the mixed of gelatin and defatted soy protein to the papaya films had enhanced the barrier, mechanical, optical properties as well as structural properties. The optical properties were similarly comparable with the packaging materials based on polymers. Higher seal strength and tear strength suggested the use of edible films for functional packaging. The best indices for correlation between polymeric materials are the thermal properties of papaya edible films. It is reported that there were highly important edible film properties in the combination of 8 g papaya puree, 3 g gelatin and 4 g defatted soy protein composite films (Tulamandi et al., 2016).

The effect of incorporating blueberry-extract into a soybean-protein-isolate edible film on the consistency of packaged lard was compared with individual incorporations of vitamin E or butylated hydroxyl anisole (BHA) during storage at 36C and relative humidity at 40% over 5 weeks by lard. The incorporation of blueberry extract into soybean-protein-isolate film demonstrated higher tensile strength and lower oxygen and water vapor permeability than individual incorporations of vitamin E or BHA. At the other side, the soybean-protein-isolate film's antioxidant potential integrated with the blueberry extract was greater than that integrated with vitamin E, and comparable to that incorporated with BHA. Therefore, blueberry-extract incorporations in the soybean-protein-isolate film not only strengthened mechanical and barrier properties but also delayed packaged lard's oxidation and hydrolysis. These films have potential as a packaging material that protects the quality of stored lard (Zhang et al., 2010)

A recent study by Yuan et al. (2020) has found that the combination of oolong tea, corn silk, and black soybean seed coat to shrimp shell, protein-based films not only could change the physicochemical properties of the final film but also improve its antioxidant activity. The resulting films with increased antioxidant of complete marine origin would be promising for application in several types of foods.

Hoque et al. (2011) concluded that partially hydrolyzed gelatin incorporation with herbal extracts containing clove, cinnamon, and star anise extracts showed enhanced TS and decreased WVP. Nevertheless, to some degree, those extracts may affect the color of the resulting films. Star anise extract became the most effective to enhance the gelatin film's mechanical and water barrier properties. Oxidation extracts demonstrated greater efficiency than the non-oxidized equivalent in increasing the strength of the films. However, the molecular weight or length of the gelatin chain has been implicated in the cross-linking of phenolic compounds in the extracts as well as the microstructure of the resulting film. Herbal extracts may be used as natural protein cross linkers capable of modifying the properties of gelatin film or other proteins. Thus the incorporation of various herbal extracts directly affected the film properties with and without hydrolysis of cuttlefish skin gelatin.

In a recent study, collagen hydrolysate (CH) films were developed by addition of thyme essential oil (TO). Incorporating of TO led to increases in the thickness, elongation at break (%), and light barrier performance of CH–TO films as well as there was a considerable reduction in film solubility and tensile strength of the CH films. Moreover, the increase of TO showed higher lightness and yellowness while lower redness amount compared to CH film observed hence they can be used for active packaging purposes (Ocak, 2020).

Also, Mohammadi, Mirabzadeh, Shahvalizadeh, and Hamishehkar (2020) enhanced a biodegradable whey protein isolate-based film incorporated with cinnamon essential oil and chitosan nanofiber, and the antibacterial activity of films was improved by incorporating cinnamon oil.

11.2.4 Active packaging aids

The demand for food products has risen over recent years and can long sustain their quality, which leads to a decline in economic losses and health issues caused by food spoilage. While the use of chemicals increases food shelf life, it is not considered nutritious or safe and is avoided by consumers (Eskandarabadi et al., 2019). Over the last few decades, synthetic film development and use in food packaging have increased rapidly, leading to significant environmental issues as synthetic plastics are not degradation (Cazón et al., 2017). Currently, the growing demand for healthier and safer food has led to new preservation methods being investigated (Gómez-Estaca, López-de-Dicastillo, Hernández-Muñoz, Catalá, & Gavara, 2014).

As a response, a refined and enhanced protective feature of packaging has led to the evolution of modern packaging innovations such as active packaging (Yildirim et al., 2018). The principle of incorporation in packaging system of certain components that release or absorb substances from or into the packaged food or the surrounding environment to extend shelf life and preserve the consistency, health and sensory characteristics of the foodstuff (Realini & Marcos, 2014).

Natural sources such as GTE (Mujeeb Rahman et al., 2017; Peng et al., 2013; Siripatrawan & Harte, 2010; Wambura, Yang, & Mwakatage, 2011; Wu et al., 2013), olive leaf extract (Amaro-Blanco, Delgado-Adámez, Martín, & Ramírez, 2018; da Rosa et al., 2020), mango kernel extract (Maryam Adilah et al., 2018), rosemary extract (Eskandarabadi et al., 2019), mangosteen pericarp extract (Nabilah et al., 2019; Zhang, Liu, et al., 2020), murta fruit and leaf extract (Hauser et al., 2016; López de Dicastillo et al., 2016; Silva-Weiss, Bifani, Ihl, Sobral, & Gómez-Guillén, 2013), blueberry extract (Gutiérrez & Alvarez, 2018) moringa leaf extract (Ju et al., 2019) camucamu extract (Ju & Song, 2019a) onion peel extract (Ju & Song, 2019b), Rheum ribes L. extract (Kalkan et al., 2020) Chinese hawthorn fruit extract (Kan et al., 2019), GSE (Kanmani & Rhim, 2014a, 2014b; Shankar & Rhim, 2018; Song et al., 2012; Tan et al., 2015; Wang, Lim, et al., 2019), plant extracts (Bonilla & Sobral, 2016; Kraśniewska et al., 2014; Mir, Dar, Wani, & Shah, 2018), ginger essential oil (Alexandre et al., 2016), hemp and sage oils (Mihaly Cozmuta et al., 2015), oregano essential oil (Chen et al., 2020; Lee et al., 2016), and plant oils (Wyrwa & Barska, 2017) are some examples of plant extracts that have been used in the development of food packaging.

Fish gelatin films prepared with mango peels extract (MPE) with three different concentrations of 1%–5% by casting method caused excellent free radical scavenging activity by Adilah et al. (2018). Films incorporated with MPE exhibited a reduction of water vapor permeability and fewer films solubility. Also, the more rigid and less flexible film formation observed with a high level of MPE. This study showed that the mango peel extract incorporated into gelatin-based films could be used as a potential material for active packaging.

Active packaging with olive leaf extract was not effective in preserving the iberian sliced dry-cured shoulder, either alone or in combination with HPP. The more lipid-soluble active principle could probably be more effective for Iberian dry- products due to their high lipid content. The high stability of this product would probably be somewhat responsible for the quality characteristics of the shoulders of pigs reared in Montanera (Amaro-Blanco et al., 2018).

A comparative study on natural plant extracts as active components in chitosan-based films was done by Bajić, Ročnik, et al. (2019) were done. Extracts obtained from oak three, hop plant and algae were added separately to the films for evaluated and mutually compared. Of the blended films, the film containing the oak extract had fewer amount of water and total soluble matter in addition to the highest phenolic content. This ultimately led to the creation of material with desirable mechanical properties in the lowest pliability and the highest strength and stiffness relative to the other two blended films. It can be claimed that the OE film overcomes the performance of HE and AE films. Therefore, it has a high potential to be used as active food packaging material made from natural-based materials.

Effect of active packaging containing Satureja thymbra extracts on the oxidative stability of fried potato chips evaluated by Choulitoudi et al. (2020) Satureja thymbra extracts that have rich in phenolic acids and flavonoids, were gained with ethyl acetate and ethanol by successive extractions and analyzed as natural antioxidants to extend the shelf- of fried potato chips. The extracts were more effective when spraying them on the fried product surface or adding to the frying oil when coated on a laminated film used as active packaging for the chips. Partial migration was observed of the natural polyphenols into the product.

The incorporation of GTE into the film-forming solution produced active biodegradable films based on agar and agar-gelatin. Agar-gelatin films were less resistant than agar films and more deformable. The use of GTE in both agar and agar gelatin films reduced the tensile strength and elongation at break. Water vapor permeability and water resistance were not impaired either by the substitution of agar with gelatin or by the addition of GTE. But the water solubility in films containing GTE improved significantly. Total phenolic compounds, catechins and flavonols not released into the water, which was reported due to the presence of gelatin in the Green Tea Matrix Agar film. As a result, the antioxidant power released by the films was lower for gelatin-containing films. Nevertheless, the presence of gelatin did not affect the antimicrobial activity of the films (Giménez et al., 2013).

Characterization of SPI films containing licorice residue extract was made by Han, Yu, and Wang (2018b). Through adding liquorice residue extract (LRE) into SPI films were made antioxidant films. They have investigated the effects of different concentrations of LRE on the physical properties, microstructure and antioxidant activity of SPI films. Evaluation of FTIR suggested the formation of hydrogen bonds between active protein matrix groups and phenolic hydroxyl LRE groups. Adding LRE enhanced the film's mechanical, water-, oxygen-, and light barrier as well as antioxidant properties. The water affinity of the films was still strong, due to the hydrophilic nature of SPI films. TPC release test showed that, due to the rapid swelling of the hydrophilic matrix in 10% ethanol, the release rate of TPC from active films into 95% ethanol was slower than in 10% ethanol. SPI films containing LRE have a high potential as an active packaging material to preserving fatty foods.

A new active packaging film was made based on murta leaf extract which was inserted into a layer of methylcellulose that was based on polyethylene (LDPE) film that was low density. They were observed slight changes in the color, optical and mechanical properties; although, thermal and water vapor transmission properties did not change through the active coating on the LDPE film. Eventually, a sensory analysis revealed that the active coating did not affect the odor and flavor properties of a fatty food packed within the active substances. It indicates that this active packaging film could be used to expand packaged food's shelf-life (Hauser et al., 2016).

The effect of the polyphenolic extract of N. sativa seeds was evaluated on the functional property of edible chitosan film by Kadam, Shah, Palamthodi, and Lele (2018). The changes in bonding between CH–CH and CH-water and the presentation of CH–NSE interactions led to considerable changes in the crystallinity, WVP, mechanical properties and thermal characteristics of chitosan films. This suggests that it could be used in active packing based on the requirements of a particular packaging.

Kan et al. (2019) developed active packaging by polyphenols that was extracted from the fruits of Chinese hawthorn and further added into chitosan-gelatin blend films. Investigated the microstructure, physical, mechanical, barrier and antioxidant properties of the films were revealed that procyanidin B2, chlorogenic acid and epicatechin were the main polyphenols in the hawthorn fruits extract. When the extract was incorporated, the inner microstructure of chitosan-gelatin blend films became more compact. The addition of extract significantly increased the tensile strength, thickness, and elongation at break of chitosan-gelatin blend films. However, the water vapor permeability, moisture content and light transmittance of chitosan-gelatin blend films were remarkably decreased by the incorporation of the extract. Also, chitosan-gelatin blend films containing the extract have ability the potent of free radical scavenging. In the consequences, they suggested the extract of Chinese hawthorn fruit could be used to enhance the barrier, mechanical and antioxidant properties of chitosan-gelatin blend films as a natural antioxidant.

Active films were made from polyvinyl alcohol (PVA) and chitosan (Ch) incorporating pomegranate peel extract (PE) and aqueous mint extract (ME) by Kanatt, Rao, Chawla, and Sharma (2012). Incorporation of extract into the films increased protection against UV light. The tensile strength of the films was enhanced by the addition of ME/PE. The highest tensile strength was observed in the Ch-PVA films incorporated with PE. The results demonstrated that the Ch-PVA film incorporating ME/PE could be used for the improvement of active packaging.

In another study, carrageenan-based films were developed by the addition of GSE at different concentrations, and their physical and mechanical properties were investigated by Kanmani and Rhim (2014b). Polyphenolic compounds in the GSE led to the carrageenan/GSE composite films appeared yellowish tint. SEM analysis showed on the cross-section of the films, rough surface with sponge-like structures. Based on the FT-IR results, it can be concluded that the GSE is in good compatibility with the carrageenan. The amorphous structure of polymer films has remained constant by the incorporation of GSE. While the incorporation of GSE increased water vapor permeability, moisture content and surface hydrophilicity of the films. As the GSE content increased, the tensile strength and elastic modulus decreased. The thermal stability of the film did not change with the addition of GSE. This suggests that carrageenan-based composite films with GSE could be used in active food packing application.

Active films were made of polylactic acid (PLA) containing different concentrations of an Allium spp extract to developed packaging of ready-to-eat salads (Llana-Ruiz-Cabello et al., 2015). At the incorporation of the active material, the mechanical and optical properties of PLA films did not exhibit significant changes. Also, no remarkable antioxidant activity was recorded, although significant antimicrobial activity was observed, mostly in films containing 5% and 6.5% of the Allium spp. extract. Relating to aerobic bacteria, the film with the highest active agent concentration (6.5%) was effectual for up to 5 days of storage, as well as 7 days for molds. Thus, this film could be used as active packing, especially for ready-to-eat salads.

Active films based on methylcellulose (MC) and murta fruit extract (MU) were prepared by the casting method of Lopez de Dicastillo et al. (López de Dicastillo et al., 2016). The addition of GA and MU had a slight effect on thermal properties. The incorporation of GA notably reduced swelling index and enhanced mechanical properties. Antioxidant and antimicrobial activity reached optimum efficiency when GA was applied at the lowest amount.

Kanatt and Chawla (2018) demonstrated that mango peel has bioactive compounds, and that Langra peel had the strongest antioxidant and antibacterial activities among the four varieties tested. An active composite film was developed from PVA cyclodextrin-gelatin with addition LMPE. During the film formation process, the bioactive properties of LMPE were retained. UV-blocking and mechanical properties of the film enhanced with the incorporation of LMPE. When kept at chilled temperatures, the shelf life of chicken meat packed in these films was increased by 10 days. Although 80% acetone and 70% ethanol extract of LMP were effective as an active ingredient in film, aqueous ethanol extract is preferred for food packaging films. The film produced has the potential to act as an active film for food packaging and to reduce the burdens of food packaging waste dependent on petroleum.

11.2.5 Intelligent packaging aids

The packaging is a critical part of modern trade, which assigns the protection of the quality of food products and is one of the most important means of advertising. In fact, it performs the main task by safeguarding the packaged product against environmental factors, affecting its consistency and health, and promoting transport, storage, and dosing. Besides, new packaging technologies are expected to play an increasingly important role by providing various and innovative prolong the shelf-life solutions. Active and intelligent packaging is a modern idea, provided for further enhanced food product safety and control (Balbinot-Alfaro et al., 2019). Mainly, intelligent packaging technology is mighty of providing functions such as detecting, recording, monitoring, communicating, and applying scientific logic to promote decision-making, improve protection, enhance quality, as well as provide information and warning about potential issues (Alexandre et al., 2016; Balbinot-Alfaro et al., 2019; Ezati & Rhim, 2020). While the share of so-called advanced packaging is projected to represent approximately 5% of the packaging market's total value but there are signs that its sales will develop rapidly in the following years. The sum of patent applications and patents awarded represents the competition in these solutions (Balbinot-Alfaro et al., 2019). In the past few years, smart (active and intelligent) packaging systems based on biopolymers and natural extracts have recently attracted growing interest within the food industry (Yong, Wang, Bai, et al., 2019). This type of packaging system is not only intended to ensure food protection, quality and prolong shelf life but also to reduce the negative effects on the environment and improve health safety. Moreover, intelligent packaging is one type of smart material packaging system, which can provide consumers with information on the current food conditions within the packaging. Indeed, intelligent packaging serves as a communicator between the packaged product and the customer (Jamróz et al., 2019). Several kinds of intelligent packaging systems exist which are depending on the use of sensitive dyes, for example, time-temperature sensors (Pavelková, 2013; Pereira, de Arruda, & Stefani, 2015), pH indicators (Peralta, Bitencourt-Cervi, Maciel, Yoshida, & Carvalho, 2019) and freshness sensors (Kuswandi et al., 2011).

Traditionally, food packaging functions include four categories: containment, protection, convenience, and communication (Jamróz et al., 2019). As today's society has become more complex, they are not adequate for our society's quality standards. Novel concepts for intelligent packaging have been developed respond to the rising concerns of consumers about food safety and quality. Intelligent packaging is described as a food packaging system capable of monitoring and informing consumers in real-time about the food conditions. As pH changes are a significant factor in informing spoilage in many food items, numerous attempts have been made to enhancement visual pH indicators, as one type of intelligent food packaging system, due to several advantages include small size, high sensitivity, and lower cost (Choi, Lee, Lacroix, & Han, 2017). The pH indicators are applied for monitoring the freshness of stored food via changes in atmospheric pH that occur during food spoilage. It is good for the consumer, who gets clear information without opening the package about the freshness of the product (Alexandre et al., 2016; Jamróz et al., 2019). A pH change can be observed when food process deterioration occurs. Thus, allows pH changes to be used as indicators of food quality as they indicate the product status. In intelligent packaging, colorimetric pH indicators can be combined in packing and monitored by visual changes in color. Natural pigments responsible for blue, purple, violet, or red coloration in fruit and vegetables may be used as natural indicators for the changes of pH which known as anthocyanin (Andretta, Luchese, Tessaro, & Spada, 2019).

A colorimetric pH indicator film was enhanced using agar, potato starch, and natural dyes extracted from purple sweet potato, to form pH indicator films. Potato starch and agar as a biodegradable material were used to immobilize anthocyanins which extracted from purple sweet potatoes, Ipomoea batatas. Hence, the developed pH indicator films have great potential to assure food safety and detection of food spoilage as a diagnostic tool. Also, it has potential in the intelligent food packaging application (Choi et al., 2017). The colorimetric pH indicator film application was investigated with immobilized black chokeberry pomace extract in chitosan by Halász and Csóka (2018). During this analysis, various quantities of pomace extract from black chokeberry were incorporated with chitosan to prepare colorimetric pH indicator films. The pomace extract contained sufficient active ingredients to greatly reduce the solubility and swelling of chitosan. The indicator films retained their integrity even at acid pH due to the interactions between the polymer chains and the extract's phenolic components. Increased resistance to water was observed with higher extract content, which also resulted in reduced dye migration. The immobilized dye in chitosan films reacted well to the change in pH and showed a high difference in color from pH 1 to pH 10. Recently, a novel colorimetric pH indicator film was developed using agar and natural dye extracted from *Arnebiae uchroma* root by casting/solvent evaporation method. Enhanced colorimetric indicator films have revealed high potential as "real-time" intelligent packing for the convenient, anti-destructive and visual monitoring of fish spoilage due to their non-toxicity and visible color response (Huang et al., 2019). In other study, the fish spoilage test evaluated by microbiological analysis showed that between 6th and 10th days of storage the fish was spoiled. The film's color changes were not sufficient to effectively notify the trained sensory panelists about the spoilage of the stored Atlantic mackerel (Jamróz et al., 2019). Jancikova, Jamróz, Kulawik, Tkaczewska, and Dordevic (2019) studied the effect of rosemary extract from dry leaves (DRE) and fresh leaves (FRE) on intelligent and active properties of furcellaran (FUR), gelatin hydrolysate (GELH)-based film. Water content, tensile strength, and thickness were increased with the addition of rosemary extracts into FUR/GELH films. The incorporation of rosemary extracts into the FUR/GELH matrix improved the UV barrier properties of the tested films. The antioxidant activity (DPPH and FRAP) has not changed with the addition of FRE while notably increased with the addition of DRE. Color changes in varying pH have been noted; moreover, the fish spoilage test has shown that these films are not acceptable as an intelligent film to control the freshness for this sort of food product. Li et al. (2019) enhanced a novel pH-indicating based on chitosan and surface-deacetylated chitin nanofibers (CN) with incorporation of purple potato extractions (PPE). The Fourier transform infrared spectrophotometry (FT-IR) and differential scanning calorimetry results showed that PPE was effectively joined to the chitosan film. In contrast, higher CN content did not further enhance mechanical properties and due to aggregation in the films, CN would distribute unequally. Besides the ability to indicate pH, CS-CN-PPE has demonstrated exceptional antioxidant activity, and this presents a further advantage to the packaging of oxidized substances. The existing literature on the effects of purple onion peel extract (POPE) on the rheological properties of the *Artemisia sphaerocephala* Krasch, gum ASKG film-forming solutions was investigated. The rheological findings showed that the solutions-formed films were non-Newtonian fluids. Results of FTIR revealed that hydrogen bonds formed between POPE and ASKG compromise with rheological findings. The results of the TGA showed that the POPE reduced the stability of the films whereas The SEM observation showed a homogeneous cross-section of the films developed. Adding POPE reduced the film's TS, EB, WVP and light transmission rate. In buffer solutions the film color changed from red (pH 350 3.0) to brown (pH 11.0). The results indicated that ASKG films which contain POPE display great potential as intelligent packaging materials (Liang, Sun, Cao, Li, & Wang, 2018). A recent study by Liu et al. (2019) involved characterization fish gelatin-based film and haskap berries extract as active and intelligent packaging. In this research, to produce active and intelligent packaging films, polyphenols were obtained from the haskap berries and incorporated with FG. Results showed that the major polyphenols in the haskap berry extract (HBE) were anthocyanins and phenolic acids. FG film's crystallinity has been improved by incorporating HBE. In addition, the addition of HBE notably increased some factors such as water vapor, tensile strength, total color difference value and UV−vis light barrier properties and antioxidant ability of FG film. The results demonstrate that FG-HBE films can be used in the food industry as novel antioxidant and intelligent packaging. Ma, Liang, Cao, and Wang (2018) published this study aimed to prepare an intelligent film based on PVA, chitosan nanoparticles (CHNPs) with incorporating of mulberry extracts. Throughout the fish spoilage cycle, this film also changed from red to green, showing that it can detect changes in fish quality. The results suggested that films have great potential application as an intelligent film to detect food spoilage. Owing to their biodegradability and accessibility from reproducible materials, bio-based films have become preferred. The *Vitis amurensis* husk is a white-wine processing by-product. By integrating *V. amurensis* husk extracts into the tara gum/cellulose matrix, a new bio-based intelligent colorimetric film was made. This research shows that an intelligent colorimetric film can be used as a visual indicator for assuring food safety (Ma, Ren, Gu, et al., 2017). Intelligent packaging technology was improved based on film (gelatin, starch, and chitosan) as a natural polymeric, including aqueous hibiscus extract (HAE) bioactive compounds by Peralta et al. (2019). Under various pH conditions, anthocyanin, a compound present on HAE, changes the color. In the area of the intelligent packaging technology, the application of HAE as a natural pH indicator with noticeable color variation integrated with renewable materials to allow quick, economical, and easy. In another study, researchers enhanced and characterized a time-temperature indicator (TTI) of PVA/chitosan-based polymer film interfered with anthocyanins to indirectly specify changes in food quality by monitoring changes in the pH of packaged food products when subject to inappropriate storage temperatures. The TTI was made from *Brassica oleracea var.capitata* (Red Cabbage) extract as anthocyanins, chitosan and PVA. The developed TTI offers attractive features for application in intelligent food packaging due to its physicochemical specifications. The application of the TTI presented here is accompanied by an activation test on pasteurized milk, with obvious changes in the coloring of the film, which is necessary to show consumers that the product has undergone changes in its chemical composition (Pereira et al., 2015). Qin et al. (2020) improved intelligent and active packaging. The starch/polyvinyl alcohol film that contains 1.00 wt.% of the betalains-rich red pitaya *(Hylocereus polyrhizus)* PE has become more sensitive to ammonia than other films. Once used to monitor shrimp freshness, the film containing 1.00 wt.% of the extract shows noticeable color variations during the shrimp spoiling process due to the compacted volatile nitrogen compounds. The findings revealed that betacyanins were the main components of the extract, which showed significant changes in color under alkaline conditions. These findings indicated the film that contains 1.00 wt.% of the extract has the potential to utilized as active and intelligent packaging. Yong, Liu, et al. (2019) prepared pH-sensing films based on chitosan and incorporation of anthocyanin-rich, purple-fleshed sweet potato extract (PSPE). Based on the mass spectroscopic analysis, eight main compounds in PSPE were known as anthocyanins. PSPE could change its colors with increasing pH due to the plenty of anthocyanin content. PSPE incorporation could dramatically increase chitosan film thickness, water solubility, UV–vis light barrier property and thermal stability, Whereas the moisture content, elongation at break and crystalline character of chitosan film could decrease. Chitosan-PSPE films can also be used as antioxidant and intelligent pH-sensing films to prolong the shelf-life and track the consistency of food products. In the same study, Yong, Liu, et al. (2019) developed pH-sensitive based on chitosan by adding black and purple rice extracts. With the incorporation of PRE or BRE into the CS matrix, active and intelligent packaging films were successfully developed. The addition of a low content (1 wt.%) of PRE or BRE greatly improved the water barrier property and TS of CS film due to the hydrogen bonds formed between CS and the extract. Besides, the incorporation of PRE or BRE greatly improved CS film's light barrier and antioxidant ability due to the polyphenol in the extract. The reason for the pH-sensitive property of CS-PRE and CS-BRE films to the plenty anthocyanins in the extract has been reported. CS-BRE films showed stronger water vapor permeability, antioxidant properties and UV–vis light barrier than CS-PRE films at the same extract incorporation levels. CS-PRE films were ideally suited for monitoring pork spoilage due to moderate anthocyanin content. They suggested that CS-PRE films could be used for packaged food to monitor the freshness of product as an intelligent packaging. The production of cassava starch films without and with the incorporation of blueberry residues was carried out successfully by thermocompression. The colorimetric analysis demonstrated the potential use of blueberry residue anthocyanins as a color change indicator due to a modification in the pH value of the food product by altering the film color. Therefore, the film with blueberry could be used as intelligent packaging due to the phenolic compounds (Andretta et al., 2019).

11.3 Conclusion and further remarks

In summary, recent research has shown that plant extracts are awesome candidates to be used as packaging aids. The active components in some plant extracts have antimicrobial or/and antioxidant activities to be used as active packaging agents to improve the shelf life of foods. Some other plants are rich in anthocyanin components that are sensitive to pH changing and can detect food spoilage, especially in animal-based food products as an intelligent packaging agent. Meanwhile, the plant extracts are sensitive to heat and light. Food packaging materials mostly produce in high temperature processing. So, future research is essential in the area of stabilizing the active components from common plant sources or finding alternative plants to extract more stable active agents.

References

Abedinia, A., Mohammadi Nafchi, A., Sharifi, M., Ghalambor, P., Oladzadabbasabadi, N., Ariffin, F., & Huda, N. (2020). Poultry gelatin: Characteristics, developments, challenges, and future outlooks as a sustainable alternative for mammalian gelatin. *Trends in Food Science & Technology*, *104*, 14–26. Available from https://doi.org/10.1016/j.tifs.2020.08.001.

Adilah, A. N., Jamilah, B., Noranizan, M. A., & Hanani, Z. A. N. (2018). Utilization of mango peel extracts on the biodegradable films for active packaging. *Food Packaging and Shelf Life*, *16*, 1–7. Available from https://doi.org/10.1016/j.fpsl.2018.01.006.

Akhtar, M. J., Jacquot, M., Jasniewski, J., Jacquot, C., Imran, M., Jamshidian, M., ... Desobry, S. (2012). Antioxidant capacity and light-aging study of HPMC films functionalized with natural plant extract. *Carbohydrate Polymers*, *89*(4), 1150–1158. Available from https://doi.org/10.1016/j.carbpol.2012.03.088.

Akhter, R., Masoodi, F. A., Wani, T. A., & Rather, S. A. (2019). Functional characterization of biopolymer based composite film: Incorporation of natural essential oils and antimicrobial agents. *International Journal of Biological Macromolecules*, *137*, 1245–1255. Available from https://doi.org/10.1016/j.ijbiomac.2019.06.214.

Alexandre, E. M. C., Lourenço, R. V., Bittante, A. M. Q. B., Moraes, I. C. F., & Sobral, P. J. d A. (2016). Gelatin-based films reinforced with montmorillonite and activated with nanoemulsion of ginger essential oil for food packaging applications. *Food Packaging and Shelf Life*, *10*, 87–96. Available from https://doi.org/10.1016/j.fpsl.2016.10.004.

Ali, A., Chen, Y., Liu, H., Yu, L., Baloch, Z., Khalid, S., ... Chen, L. (2019). Starch-based antimicrobial films functionalized by pomegranate peel. *International Journal of Biological Macromolecules*, *129*, 1120–1126. Available from https://doi.org/10.1016/j.ijbiomac.2018.09.068.

Aliyari, P., Bakhshi Kazaj, F., Barzegar, M., & Ahmadi Gavlighi, H. (2020). Production of functional sausage using pomegranate peel and pistachio green hull extracts as natural preservatives. *Journal of Agricultural Science and Technology*, *22*(1), 159–172.

Alves, V. L. C. D., Rico, B. P. M., Cruz, R. M. S., Vicente, A. A., Khmelinskii, I., & Vieira, M. C. (2018). Preparation and characterization of a chitosan film with grape seed extract-carvacrol microcapsules and its effect on the shelf-life of refrigerated Salmon (Salmo salar). *LWT*, *89*, 525–534. Available from https://doi.org/10.1016/j.lwt.2017.11.013.

Amaro-Blanco, G., Delgado-Adámez, J., Martín, M. J., & Ramírez, R. (2018). Active packaging using an olive leaf extract and high pressure processing for the preservation of sliced dry-cured shoulders from Iberian pigs. *Innovative Food Science & Emerging Technologies*, *45*, 1–9. Available from https://doi.org/10.1016/j.ifset.2017.09.017.

Andretta, R., Luchese, C. L., Tessaro, I. C., & Spada, J. C. (2019). Development and characterization of pH-indicator films based on cassava starch and blueberry residue by thermocompression. *Food Hydrocolloids*, *93*, 317–324. Available from https://doi.org/10.1016/j.foodhyd.2019.02.019.

Arezoo, E., Mohammadreza, E., Maryam, M., & Abdorreza, M. N. (2020). The synergistic effects of cinnamon essential oil and nano TiO_2 on antimicrobial and functional properties of sago starch films. *International Journal of Biological Macromolecules*, *157*, 743–751. Available from https://doi.org/10.1016/j.ijbiomac.2019.11.244.

Bajić, M., Jalšovec, H., Travan, A., Novak, U., & Likozar, B. (2019). Chitosan-based films with incorporated supercritical CO_2 hop extract: Structural, physicochemical, and antibacterial properties. *Carbohydrate Polymers*, *219*, 261–268. Available from https://doi.org/10.1016/j.carbpol.2019.05.003.

Bajić, M., Ročnik, T., Oberlintner, A., Scognamiglio, F., Novak, U., & Likozar, B. (2019). Natural plant extracts as active components in chitosan-based films: A comparative study. *Food Packaging and Shelf Life*, *21*, 100365. Available from https://doi.org/10.1016/j.fpsl.2019.100365.

Balbinot-Alfaro, E., Craveiro, D. V., Lima, K. O., Costa, H. L. G., Lopes, D. R., & Prentice, C. (2019). Intelligent packaging with pH indicator potential. *Food Engineering Reviews*, *11*(4), 235–244. Available from https://doi.org/10.1007/s12393-019-09198-9.

Bhavaniramya, S., Vishnupriya, S., Al-Aboody, M. S., Vijayakumar, R., & Baskaran, D. (2019). Role of essential oils in food safety: Antimicrobial and antioxidant applications. *Grain & Oil Science and Technology*, *2* (2), 49–55. Available from https://doi.org/10.1016/j.gaost.2019.03.001.

Bonilla, J., & Sobral, P. J. A. (2016). Investigation of the physicochemical, antimicrobial and antioxidant properties of gelatin-chitosan edible film mixed with plant ethanolic extracts. *Food Bioscience*, *16*, 17–25. Available from https://doi.org/10.1016/j.fbio.2016.07.003.

Bouarab Chibane, L., Degraeve, P., Ferhout, H., Bouajila, J., & Oulahal, N. (2019). Plant antimicrobial polyphenols as potential natural food preservatives. *Journal of the Science of Food and Agriculture*, *99*(4), 1457–1474.

Camo, J., Lorés, A., Djenane, D., Beltrán, J. A., & Roncalés, P. (2011). Display life of beef packaged with an antioxidant active film as a function of the concentration of oregano extract. *Meat Science*, *88*(1), 174–178. Available from https://doi.org/10.1016/j.meatsci.2010.12.019.

Cazón, P., Velazquez, G., Ramírez, J. A., & Vázquez, M. (2017). Polysaccharide-based films and coatings for food packaging: A review. *Food Hydrocolloids*, *68*, 136–148. Available from https://doi.org/10.1016/j.foodhyd.2016.09.009.

Cejudo Bastante, C., Cran, M. J., Casas Cardoso, L., Mantell Serrano, C., Martínez de la Ossa, E. J., & Bigger, S. W. (2019). Effect of supercritical CO_2 and olive leaf extract on the structural, thermal and mechanical properties of an impregnated food packaging film. *The Journal of Supercritical Fluids*, *145*, 181–191. Available from https://doi.org/10.1016/j.supflu.2018.12.009.

Cerruti, P., Santagata, G., Gomez d'Ayala, G., Ambrogi, V., Carfagna, C., Malinconico, M., & Persico, P. (2011). Effect of a natural polyphenolic extract on the properties of a biodegradable starch-based polymer. *Polymer Degradation and Stability*, *96*(5), 839–846. Available from https://doi.org/10.1016/j.polymdegradstab.2011.02.003.

Chen, S., Wu, M., Lu, P., Gao, L., Yan, S., & Wang, S. (2020). Development of pH indicator and antimicrobial cellulose nanofibre packaging film based on purple sweet potato anthocyanin and oregano essential oil. *International Journal of Biological Macromolecules*, *149*, 271–280.

Choi, I., Lee, J. Y., Lacroix, M., & Han, J. (2017). Intelligent pH indicator film composed of agar/potato starch and anthocyanin extracts from purple sweet potato. *Food Chemistry*, *218*, 122–128. Available from https://doi.org/10.1016/j.foodchem.2016.09.050.

Choulitoudi, E., Bravou, K., Bimpilas, A., Tsironi, T., Tsimogiannis, D., Taoukis, P., & Oreopoulou, V. (2016). Antimicrobial and antioxidant activity of Satureja thymbra in gilthead seabream fillets edible coating. *Food and Bioproducts Processing*, *100*, 570–577. Available from https://doi.org/10.1016/j.fbp.2016.06.013.

Choulitoudi, E., Velliopoulou, A., Tsimogiannis, D., & Oreopoulou, V. (2020). Effect of active packaging with Satureja thymbra extracts on the oxidative stability of fried potato chips. *Food Packaging and Shelf Life*, *23*, 100455. Available from https://doi.org/10.1016/j.fpsl.2019.100455.

Corrales, M., Han, J. H., & Tauscher, B. (2009). Antimicrobial properties of grape seed extracts and their effectiveness after incorporation into pea starch films. *International Journal of Food Science & Technology*, *44*(2), 425–433. Available from https://doi.org/10.1111/j.1365-2621.2008.01790.x.

da Rosa, G. S., Vanga, S. K., Gariepy, Y., & Raghavan, V. (2020). Development of biodegradable films with improved antioxidant properties based on the addition of carrageenan containing olive leaf extract for food packaging applications. *Journal of Polymers and the Environment*, *28*(1), 123–130. Available from https://doi.org/10.1007/s10924-019-01589-7.

Deng, Q., & Zhao, Y. (2011). Physicochemical, nutritional, and antimicrobial properties of wine grape (cv. Merlot) pomace extract-based films. *Journal of Food Science*, *76*(3), E309–E317. Available from https://doi.org/10.1111/j.1750-3841.2011.02090.x.

Dong, H., He, J., Xiao, K., & Li, C. (2020). Temperature-sensitive polyurethane (TSPU) film incorporated with carvacrol and cinnamyl aldehyde: Antimicrobial activity, sustained release kinetics and potential use as food packaging for Cantonese-style moon cake. *International Journal of Food Science & Technology*, *55*(1), 293–302. Available from https://doi.org/10.1111/ijfs.14276.

Dou, L., Li, B., Zhang, K., Chu, X., & Hou, H. (2018). Physical properties and antioxidant activity of gelatin-sodium alginate edible films with tea polyphenols. *International Journal of Biological Macromolecules*, *118*, 1377–1383. Available from https://doi.org/10.1016/j.ijbiomac.2018.06.121.

Ekramian, S., Abbaspour, H., Roudi, B., Amjad, L., & Nafchi, A. M. (2020). Influence of Nigella sativa L. extract on physico-mechanical and antimicrobial properties of sago starch film. *Journal of Polymers and the Environment*. Available from https://doi.org/10.1007/s10924-020-01864-y.

Esfahani, A., Ehsani, M. R., Mizani, M., & Mohammadi Nafchi, A. (2020). Application of bio-nanocomposite films based on nano-TiO_2 and cinnamon essential oil to improve the physiochemical, sensory, and microbial properties of fresh pistachio. *Journal of Nuts*, *11*(3), 195–212. Available from https://doi.org/10.22034/JON.2020.1903741.1091.

Eskandarabadi, S. M., Mahmoudian, M., Farah, K. R., Abdali, A., Nozad, E., & Enayati, M. (2019). Active intelligent packaging film based on ethylene vinyl acetate nanocomposite containing extracted anthocyanin, rosemary extract and ZnO/Fe-MMT nanoparticles. *Food Packaging and Shelf Life*, *22*, 100389. Available from https://doi.org/10.1016/j.fpsl.2019.100389.

Espitia, P. J. P., Avena-Bustillos, R. J., Du, W.-X., Chiou, B.-S., Williams, T. G., Wood, D., ... Soares, N. F. F. (2014). Physical and antibacterial properties of açaí edible films formulated with thyme essential oil and apple skin polyphenols. *Journal of Food Science*, *79*(5), M903–M910. Available from https://doi.org/10.1111/1750-3841.12432.

Ezati, P., & Rhim, J.-W. (2020). pH-responsive chitosan-based film incorporated with alizarin for intelligent packaging applications. *Food Hydrocolloids*, *102*, 105629. Available from https://doi.org/10.1016/j.foodhyd.2019.105629.

Fernandes, R. P. P., Trindade, M. A., Lorenzo, J. M., Munekata, P. E. S., & de Melo, M. P. (2016). Effects of oregano extract on oxidative, microbiological and sensory stability of sheep burgers packed in modified atmosphere. *Food Control*, *63*, 65–75. Available from https://doi.org/10.1016/j.foodcont.2015.11.027.

Ferysiuk, K., Wójciak, K. M., Materska, M., Chilczuk, B., & Pabich, M. (2020). Modification of lipid oxidation and antioxidant capacity in canned refrigerated pork with a nitrite content reduced by half and addition of sweet pepper extract. *LWT*, *118*, 108738. Available from https://doi.org/10.1016/j.lwt.2019.108738.

Giménez, B., López de Lacey, A., Pérez-Santín, E., López-Caballero, M. E., & Montero, P. (2013). Release of active compounds from agar and agar–gelatin films with green tea extract. *Food Hydrocolloids*, *30*(1), 264–271. Available from https://doi.org/10.1016/j.foodhyd.2012.05.014.

Gómez-Estaca, J., López-de-Dicastillo, C., Hernández-Muñoz, P., Catalá, R., & Gavara, R. (2014). Advances in antioxidant active food packaging. *Trends in Food Science & Technology*, *35*(1), 42–51. Available from https://doi.org/10.1016/j.tifs.2013.10.008.

Guillard, V., Gaucel, S., Fornaciari, C., Angellier-Coussy, H., Buche, P., & Gontard, N. (2018). The next generation of sustainable food packaging to preserve our environment in a circular economy context. *Frontiers in Nutrition*, *5*, 121.

Gutiérrez, T. J., & Alvarez, V. A. (2018). Bionanocomposite films developed from corn starch and natural and modified nano-clays with or without added blueberry extract. *Food Hydrocolloids*, *77*, 407–420. Available from https://doi.org/10.1016/j.foodhyd.2017.10.017.

Halász, K., & Csóka, L. (2018). Black chokeberry (Aronia melanocarpa) pomace extract immobilized in chitosan for colorimetric pH indicator film application. *Food Packaging and Shelf Life*, *16*, 185–193. Available from https://doi.org/10.1016/j.fpsl.2018.03.002.

Han, J., Ruiz-Garcia, L., Qian, J.-P., & Yang, X.-T. (2018). Food packaging: A comprehensive review and future trends. *Comprehensive Reviews in Food Science and Food Safety*, *17*(4), 860–877. Available from https://doi.org/10.1111/1541-4337.12343.

Han, J., Shin, S.-H., Park, K.-M., & Kim, K. M. (2015). Characterization of physical, mechanical, and antioxidant properties of soy protein-based bioplastic films containing carboxymethylcellulose and catechin. *Food Science and Biotechnology*, *24*(3), 939–945. Available from https://doi.org/10.1007/s10068-015-0121-0.

Han, Y., Yu, M., & Wang, L. (2018a). Bio-based films prepared with soybean by-products and pine (Pinus densiflora) bark extract. *Journal of Cleaner Production*, *187*, 1–8. Available from https://doi.org/10.1016/j.jclepro.2018.03.115.

Han, Y., Yu, M., & Wang, L. (2018b). Preparation and characterization of antioxidant soy protein isolate films incorporating licorice residue extract. *Food Hydrocolloids*, *75*, 13–21. Available from https://doi.org/10.1016/j.foodhyd.2017.09.020.

Hassan, B., Chatha, S. A. S., Hussain, A. I., Zia, K. M., & Akhtar, N. (2018). Recent advances on polysaccharides, lipids and protein based edible films and coatings: A review. *International Journal of Biological Macromolecules*, *109*, 1095–1107. Available from https://doi.org/10.1016/j.ijbiomac.2017.11.097.

Hauser, C., Peñaloza, A., Guarda, A., Galotto, M. J., Bruna, J. E., & Rodríguez, F. J. (2016). Development of an active packaging film based on a methylcellulose coating containing murta (ugni molinae turcz) leaf extract. *Food and Bioprocess Technology*, *9*(2), 298–307. Available from https://doi.org/10.1007/s11947-015-1623-8.

Hoque, M. S., Benjakul, S., & Prodpran, T. (2011). Properties of film from cuttlefish (Sepia pharaonis) skin gelatin incorporated with cinnamon, clove and star anise extracts. *Food Hydrocolloids*, *25*(5), 1085–1097. Available from https://doi.org/10.1016/j.foodhyd.2010.10.005.

Hu, J., Wang, X., Xiao, Z., & Bi, W. (2015). Effect of chitosan nanoparticles loaded with cinnamon essential oil on the quality of chilled pork. *LWT - Food Science and Technology*, *63*(1), 519–526. Available from https://doi.org/10.1016/j.lwt.2015.03.049.

Huang, S., Xiong, Y., Zou, Y., Dong, Q., Ding, F., Liu, X., & Li, H. (2019). A novel colorimetric indicator based on agar incorporated with Arnebia euchroma root extracts for monitoring fish freshness. *Food Hydrocolloids*, *90*, 198–205. Available from https://doi.org/10.1016/j.foodhyd.2018.12.009.

Insaward, A., Duangmal, K., & Mahawanich, T. (2015). Mechanical, optical, and barrier properties of soy protein film as affected by phenolic acid addition. *Journal of Agricultural and Food Chemistry*, *63*(43), 9421–9426.

Jamróz, E., Kulawik, P., Guzik, P., & Duda, I. (2019). The verification of intelligent properties of furcellaran films with plant extracts on the stored fresh Atlantic mackerel during storage at 2 °C. *Food Hydrocolloids*, *97*, 105211. Available from https://doi.org/10.1016/j.foodhyd.2019.105211.

Jancikova, S., Jamróz, E., Kulawik, P., Tkaczewska, J., & Dordevic, D. (2019). Furcellaran/gelatin hydrolysate/rosemary extract composite films as active and intelligent packaging materials. *International Journal of Biological Macromolecules*, *131*, 19–28. Available from https://doi.org/10.1016/j.ijbiomac.2019.03.050.

Jang, S.-A., Shin, Y.-J., & Song, K. B. (2011). Effect of rapeseed protein–gelatin film containing grapefruit seed extract on 'Maehyang' strawberry quality. *International Journal of Food Science & Technology*, *46*(3), 620–625. Available from https://doi.org/10.1111/j.1365-2621.2010.02530.x.

Jridi, M., Abdelhedi, O., Salem, A., Kechaou, H., Nasri, M., & Menchari, Y. (2020). Physicochemical, antioxidant and antibacterial properties of fish gelatin-based edible films enriched with orange peel pectin: Wrapping application. *Food Hydrocolloids*, *103*, 105688. Available from https://doi.org/10.1016/j.foodhyd.2020.105688.

Ju, A., Baek, S.-K., Kim, S., & Song, K. B. (2019). Development of an antioxidative packaging film based on khorasan wheat starch containing moringa leaf extract. *Food Science and Biotechnology*, *28*(4), 1057–1063. Available from https://doi.org/10.1007/s10068-018-00546-9.

Ju, A., & Song, K. B. (2019a). Development of teff starch films containing camu-camu (Myrciaria dubia Mc. Vaugh) extract as an antioxidant packaging material. *Industrial Crops and Products*, *141*, 111737. Available from https://doi.org/10.1016/j.indcrop.2019.111737.

Ju, A., & Song, K. B. (2019b). Incorporation of yellow onion peel extract into the funoran-based biodegradable films as an antioxidant packaging material. *International Journal of Food Science & Technology, n/a(n/a)*. Available from https://doi.org/10.1111/ijfs.14436.

Kadam, D., Shah, N., Palamthodi, S., & Lele, S. S. (2018). An investigation on the effect of polyphenolic extracts of Nigella sativa seedcake on physicochemical properties of chitosan-based films. *Carbohydrate Polymers*, *192*, 347–355. Available from https://doi.org/10.1016/j.carbpol.2018.03.052.

Kalkan, S., Otağ, M. R., & Engin, M. S. (2020). Physicochemical and bioactive properties of edible methylcellulose films containing Rheum ribes L. extract. *Food Chemistry*, *307*, 125524. Available from https://doi.org/10.1016/j.foodchem.2019.125524.

Kan, J., Liu, J., Yong, H., Liu, Y., Qin, Y., & Liu, J. (2019). Development of active packaging based on chitosan-gelatin blend films functionalized with Chinese hawthorn (Crataegus pinnatifida) fruit extract. *International Journal of Biological Macromolecules*, *140*, 384–392. Available from https://doi.org/10.1016/j.ijbiomac.2019.08.155.

Kanatt, S. R. (2020). Development of active/intelligent food packaging film containing Amaranthus leaf extract for shelf life extension of chicken/fish during chilled storage. *Food Packaging and Shelf Life*, *24*, 100506. Available from https://doi.org/10.1016/j.fpsl.2020.100506.

Kanatt, S. R., & Chawla, S. P. (2018). Shelf life extension of chicken packed in active film developed with mango peel extract. *Journal of Food Safety*, *38*(1), e12385. Available from https://doi.org/10.1111/jfs.12385.

Kanatt, S. R., Rao, M. S., Chawla, S. P., & Sharma, A. (2012). Active chitosan–polyvinyl alcohol films with natural extracts. *Food Hydrocolloids*, *29*(2), 290–297. Available from https://doi.org/10.1016/j.foodhyd.2012.03.005.

Kanmani, P., & Rhim, J.-W. (2014a). Antimicrobial and physical-mechanical properties of agar-based films incorporated with grapefruit seed extract. *Carbohydrate Polymers*, *102*, 708–716. Available from https://doi.org/10.1016/j.carbpol.2013.10.099.

Kanmani, P., & Rhim, J.-W. (2014b). Development and characterization of carrageenan/grapefruit seed extract composite films for active packaging. *International Journal of Biological Macromolecules*, *68*, 258–266. Available from https://doi.org/10.1016/j.ijbiomac.2014.05.011.

Karabagias, I., Badeka, A., & Kontominas, M. G. (2011). Shelf life extension of lamb meat using thyme or oregano essential oils and modified atmosphere packaging. *Meat Science*, *88*(1), 109–116. Available from https://doi.org/10.1016/j.meatsci.2010.12.010.

Kargozari, M., & Hamedi, H. (2019). Incorporation of essential oils (EOs) and nanoparticles (NPs) into active packaging systems in meat and meat products: A review. *Food & Health*, *2*(1), 16–30, http://fh.srbiau.ac.ir/article_13598_46f800791e3022d71e3c01c95d6c6258.pdf.

Kaushalya, R. A. N. C., Dhanushka, M. K. D. T., Samarasekara, A. M. P. B., & Weragoda, V. S. C. (2019, July 3–5). *Development of biodegradable packaging materials using polylactic acid (PLA) and locally extracted starch*. Paper presented at the 2019 Moratuwa Engineering Research Conference (MERCon).

Kõrge, K., Bajić, M., Likozar, B., & Novak, U. (2020). Active chitosan-chestnut extract films used for packaging and storage of fresh pasta. *International Journal of Food Science & Technology*.

Kraśniewska, K., Gniewosz, M., Synowiec, A., Przybył, J. L., Bączek, K., & Węglarz, Z. (2014). The use of pullulan coating enriched with plant extracts from Satureja hortensis L. to maintain pepper and apple quality and safety. *Postharvest Biology and Technology*, *90*, 63–72. Available from https://doi.org/10.1016/j.postharvbio.2013.12.010.

Kumar, S., Boro, J. C., Ray, D., Mukherjee, A., & Dutta, J. (2019). Bionanocomposite films of agar incorporated with ZnO nanoparticles as an active packaging material for shelf life extension of green grape. *Heliyon*, *5*(6), e01867. Available from https://doi.org/10.1016/j.heliyon.2019.e01867.

Kuswandi, B., Wicaksono, Y., Jayus., Abdullah, A., Heng, L. Y., & Ahmad, M. (2011). Smart packaging: Sensors for monitoring of food quality and safety. *Sensing and Instrumentation for Food Quality and Safety*, *5*(3), 137–146. Available from https://doi.org/10.1007/s11694-011-9120-x.

Lee, J., Yang, H.-J., Lee, K.-Y., & Song, K. B. (2016). Physical properties and application of a red pepper seed meal protein composite film containing oregano oil. *Food Hydrocolloids*, *55*, 136–143. Available from https://doi.org/10.1016/j.foodhyd.2015.11.013.

Lee, J. J. L., Cui, X., Chai, K. F., Zhao, G., & Chen, W. N. (2020). Interfacial assembly of a cashew nut (Anacardium occidentale) testa extract onto a cellulose-based film from sugarcane bagasse to produce an active packaging film with pH-triggered release mechanism. *Food and Bioprocess Technology*, *13*(3), 501–510. Available from https://doi.org/10.1007/s11947-020-02414-z.

Li, Y., Ying, Y., Zhou, Y., Ge, Y., Yuan, C., Wu, C., & Hu, Y. (2019). A pH-indicating intelligent packaging composed of chitosan-purple potato extractions strength by surface-deacetylated chitin nanofibers. *International Journal of Biological Macromolecules*, *127*, 376–384. Available from https://doi.org/10.1016/j.ijbiomac.2019.01.060.

Liang, T., Sun, G., Cao, L., Li, J., & Wang, L. (2018). Rheological behavior of film-forming solutions and film properties from Artemisia sphaerocephala Krasch. gum and purple onion peel extract. *Food Hydrocolloids*, *82*, 124–134. Available from https://doi.org/10.1016/j.foodhyd.2018.03.055.

Lim, G.-O., Jang, S.-A., & Song, K. B. (2010). Physical and antimicrobial properties of Gelidium corneum/nano-clay composite film containing grapefruit seed extract or thymol. *Journal of Food Engineering*, *98*(4), 415–420. Available from https://doi.org/10.1016/j.jfoodeng.2010.01.021.

Liu, J., Yong, H., Liu, Y., Qin, Y., Kan, J., & Liu, J. (2019). Preparation and characterization of active and intelligent films based on fish gelatin and haskap berries (Lonicera caerulea L.) extract. *Food Packaging and Shelf Life*, *22*, 100417. Available from https://doi.org/10.1016/j.fpsl.2019.100417.

Llana-Ruiz-Cabello, M., Pichardo, S., Baños, A., Núñez, C., Bermúdez, J. M., Guillamón, E., ... Cameán, A. M. (2015). Characterisation and evaluation of PLA films containing an extract of Allium spp. to be used in the packaging of ready-to-eat salads under controlled atmospheres. *LWT - Food Science and Technology*, *64*(2), 1354–1361. Available from https://doi.org/10.1016/j.lwt.2015.07.057.

Lloyd, K., Mirosa, M., & Birch, J. (2019). Active and intelligent packaging. In L. Melton, F. Shahidi, & P. Varelis (Eds.), *Encyclopedia of food chemistry* (pp. 177–182). Oxford: Academic Press.

López de Dicastillo, C., Bustos, F., Guarda, A., & Galotto, M. J. (2016). Cross-linked methyl cellulose films with murta fruit extract for antioxidant and antimicrobial active food packaging. *Food Hydrocolloids*, *60*, 335–344. Available from https://doi.org/10.1016/j.foodhyd.2016.03.020.

López-de-Dicastillo, C., Gómez-Estaca, J., Catalá, R., Gavara, R., & Hernández-Muñoz, P. (2012). Active antioxidant packaging films: Development and effect on lipid stability of brined sardines. *Food Chemistry*, *131*(4), 1376–1384. Available from https://doi.org/10.1016/j.foodchem.2011.10.002.

Lotfinia, S., Javanmard Dakheli, M., & Mohammadi Nafchi, A. (2013). Application of starch foams containing plant essential oils to prevent mold growth and improve shelf life of packaged bread. *Journal of Chemical Health Risks*, *3*(4). Available from https://doi.org/10.22034/JCHR.2018.544043.

Ma, Q., Liang, T., Cao, L., & Wang, L. (2018). Intelligent poly (vinyl alcohol)-chitosan nanoparticles-mulberry extracts films capable of monitoring pH variations. *International Journal of Biological Macromolecules*, *108*, 576–584. Available from https://doi.org/10.1016/j.ijbiomac.2017.12.049.

Ma, Q., Ren, Y., Gu, Z., & Wang, L. (2017). Developing an intelligent film containing Vitis amurensis husk extracts: The effects of pH value of the film-forming solution. *Journal of Cleaner Production*, *166*, 851–859. Available from https://doi.org/10.1016/j.jclepro.2017.08.099.

Ma, Q., Ren, Y., & Wang, L. (2017). Investigation of antioxidant activity and release kinetics of curcumin from tara gum/ polyvinyl alcohol active film. *Food Hydrocolloids, 70*, 286–292. Available from https://doi.org/10.1016/j.foodhyd.2017.04.018.

Marcos, B., Sárraga, C., Castellari, M., Kappen, F., Schennink, G., & Arnau, J. (2014). Development of biodegradable films with antioxidant properties based on polyesters containing α-tocopherol and olive leaf extract for food packaging applications. *Food Packaging and Shelf Life, 1*(2), 140–150. Available from https://doi.org/10.1016/j.fpsl.2014.04.002.

Marvizadeh, M. M., Oladzadabbasabadi, N., Mohammadi Nafchi, A., & Jokar, M. (2017). Preparation and characterization of bionanocomposite film based on tapioca starch/bovine gelatin/nanorod zinc oxide. *International Journal of Biological Macromolecules, 99*, 1–7. Available from https://doi.org/10.1016/j.ijbiomac.2017.02.067.

Marvizadeh, M. M., Tajik, A., Moosavian, V., Oladzadabbasabadi, N., & Mohammadi Nafchi, A. (2020). Fabrication of Cassava starch/Mentha piperita Essential oil biodegradable film with enhanced antibacterial properties. *Journal of Chemical Health Risks*. Available from https://doi.org/10.22034/jchr.2020.1900584.1135.

Maryam Adilah, Z. A., Jamilah, B., & Nur Hanani, Z. A. (2018). Functional and antioxidant properties of protein-based films incorporated with mango kernel extract for active packaging. *Food Hydrocolloids, 74*, 207–218. Available from https://doi.org/10.1016/j.foodhyd.2017.08.017.

Mihaly Cozmuta, A., Turila, A., Apjok, R., Ciocian, A., Mihaly Cozmuta, L., Peter, A., ... Benković, T. (2015). Preparation and characterization of improved gelatin films incorporating hemp and sage oils. *Food Hydrocolloids, 49*, 144–155. Available from https://doi.org/10.1016/j.foodhyd.2015.03.022.

Mir, S. A., Dar, B. N., Wani, A. A., & Shah, M. A. (2018). Effect of plant extracts on the techno-functional properties of biodegradable packaging films. *Trends in Food Science & Technology, 80*, 141–154. Available from https://doi.org/10.1016/j.tifs.2018.08.004.

Mir, S. A., Shah, M. A., Dar, B. N., Wani, A. A., Ganai, S. A., & Nishad, J. (2017). Supercritical impregnation of active components into polymers for food packaging applications. *Food and Bioprocess Technology, 10* (9), 1749–1754.

Moghadam, M., Salami, M., Mohammadian, M., Khodadadi, M., & Emam-Djomeh, Z. (2020). Development of antioxidant edible films based on mung bean protein enriched with pomegranate peel. *Food Hydrocolloids Development of Antioxidant Edible Films Based on Mung Bean Protein Enriched with Pomegranate Peel, 104*, 105735. Available from https://doi.org/10.1016/j.foodhyd.2020.105735.

Moghaddam, M. F. T., Jalali, H., Nafchi, A. M., & Nouri, L. (2020). Evaluating the effects of lactic acid bacteria and olive leaf extract on the quality of gluten-free bread. *Gene Reports, 21*, 100771. Available from https://doi.org/10.1016/j.genrep.2020.100771.

Mohammadi, M., Mirabzadeh, S., Shahvalizadeh, R., & Hamishehkar, H. (2020). Development of novel active packaging films based on whey protein isolate incorporated with chitosan nanofiber and nano-formulated cinnamon oil. *International Journal of Biological Macromolecules, 149*, 11–20. Available from https://doi.org/10.1016/j.ijbiomac.2020.01.083.

Moradi, M., Tajik, H., Razavi Rohani, S. M., Oromiehie, A. R., Malekinejad, H., Aliakbarlu, J., & Hadian, M. (2012). Characterization of antioxidant chitosan film incorporated with Zataria multiflora Boiss essential oil and grape seed extract. *LWT - Food Science and Technology, 46*(2), 477–484. Available from https://doi.org/10.1016/j.lwt.2011.11.020.

Mousavian, D., Mohammadi Nafchi, A., Nouri, L., & Abedinia, A. (2020). Physicomechanical properties, release kinetics, and antimicrobial activity of activated low-density polyethylene and orientated polypropylene films by Thyme essential oil active component. *Journal of Food Measurement and Characterization*. Available from https://doi.org/10.1007/s11694-020-00690-z.

Mujeeb Rahman, P., Abdul Mujeeb, V. M., & Muraleedharan, K. (2017). Chitosan–green tea extract powder composite pouches for extending the shelf life of raw meat. *Polymer Bulletin*, *74*(8), 3399–3419. Available from https://doi.org/10.1007/s00289-016-1901-2.

Nabilah, B., Wan Zunairah, W. I., Nor Afizah, M., & Nur Hanani, Z. A. (2019). Effect of mangosteen (Garcinia mangostana L.) pericarp extract on the different carriers for antioxidant active packaging films. *Journal of Packaging Technology and Research*, *3*(2), 117–126. Available from https://doi.org/10.1007/s41783-019-00058-9.

Nguyen, T. T., Dao, U. T. T., Bui, Q. P. T., Bach, G. L., Thuc, C. H., & Thuc, H. H. (2020). Enhanced antimicrobial activities and physiochemical properties of edible film based on chitosan incorporated with Sonneratia caseolaris (L.) Engl. leaf extract. *Progress in Organic Coatings*, *140*, 105487.

Norajit, K., Kim, K. M., & Ryu, G. H. (2010). Comparative studies on the characterization and antioxidant properties of biodegradable alginate films containing ginseng extract. *Journal of Food Engineering*, *98*(3), 377–384. Available from https://doi.org/10.1016/j.jfoodeng.2010.01.015.

Nouri, L., & Mohammadi Nafchi, A. (2014). Antibacterial, mechanical, and barrier properties of sago starch film incorporated with betel leaves extract. *International Journal of Biological Macromolecules*, *66*, 254–259. Available from https://doi.org/10.1016/j.ijbiomac.2014.02.044.

Ocak, B. (2020). Properties and characterization of thyme essential oil incorporated collagen hydrolysate films extracted from hide fleshing wastes for active packaging. *Environmental Science and Pollution Research*. Available from https://doi.org/10.1007/s11356-020-09259-1.

Oladzadabbasabadi, N., Ebadi, S., Mohammadi Nafchi, A., Karim, A. A., & Kiahosseini, S. R. (2017). Functional properties of dually modified sago starch/κ-carrageenan films: An alternative to gelatin in pharmaceutical capsules. *Carbohydrate Polymers*, *160*, 43–51. Available from https://doi.org/10.1016/j.carbpol.2016.12.042.

Oladzadabbasabadi, N., Karazhiyan, H., & Keyhani, V. (2017). Addition of the Chubak extract and egg white on biophysical properties of grape juice during evaporation process. *Journal of Food Process Engineering*, *40*(5), e12538. Available from https://doi.org/10.1111/jfpe.12538.

Pastor, C., Sánchez-González, L., Chiralt, A., Cháfer, M., & González-Martínez, C. (2013). Physical and antioxidant properties of chitosan and methylcellulose based films containing resveratrol. *Food Hydrocolloids*, *30*(1), 272–280. Available from https://doi.org/10.1016/j.foodhyd.2012.05.026.

Pavelková, A. (2013). Time temperature indicators as devices intelligent packaging. *Acta Universitatis Agriculturae et Silviculturae Mendelianae Brunensis*, *61*(1), 245–251.

Peng, Y., Wu, Y., & Li, Y. (2013). Development of tea extracts and chitosan composite films for active packaging materials. *International Journal of Biological Macromolecules*, *59*, 282–289. Available from https://doi.org/10.1016/j.ijbiomac.2013.04.019.

Peralta, J., Bitencourt-Cervi, C. M., Maciel, V. B. V., Yoshida, C. M. P., & Carvalho, R. A. (2019). Aqueous hibiscus extract as a potential natural pH indicator incorporated in natural polymeric films. *Food Packaging and Shelf Life*, *19*, 47–55. Available from https://doi.org/10.1016/j.fpsl.2018.11.017.

Pereira de Abreu, D., Cruz, J. M., & Paseiro Losada, P. (2012). Active and intelligent packaging for the food industry. *Food Reviews International*, *28*(2), 146–187. Available from https://doi.org/10.1080/87559129.2011.595022.

Pereira, V. A., de Arruda, I. N. Q., & Stefani, R. (2015). Active chitosan/PVA films with anthocyanins from Brassica oleraceae (Red Cabbage) as time–temperature indicators for application in intelligent food packaging. *Food Hydrocolloids*, *43*, 180–188. Available from https://doi.org/10.1016/j.foodhyd.2014.05.014.

Qin, Y., Liu, Y., Yong, H., Liu, J., Zhang, X., & Liu, J. (2019). Preparation and characterization of active and intelligent packaging films based on cassava starch and anthocyanins from Lycium ruthenicum Murr. *International Journal of Biological Macromolecules*, *134*, 80–90. Available from https://doi.org/10.1016/j.ijbiomac.2019.05.029.

Qin, Y., Liu, Y., Zhang, X., & Liu, J. (2020). Development of active and intelligent packaging by incorporating betalains from red pitaya (Hylocereus polyrhizus) peel into starch/polyvinyl alcohol films. *Food Hydrocolloids*, *100*, 105410. Available from https://doi.org/10.1016/j.foodhyd.2019.105410.

Radusin, T., Torres-Giner, S., Stupar, A., Ristic, I., Miletic, A., Novakovic, A., & Lagaron, J. M. (2019). Preparation, characterization and antimicrobial properties of electrospun polylactide films containing Allium ursinum L. extract. *Food Packaging and Shelf Life*, *21*, 100357.

Rakholiya, K. D., Kaneria, M. J., & Chanda, S. V. (2013). Chapter 11—Medicinal plants as alternative sources of therapeutics against multidrug-resistant pathogenic microorganisms based on their antimicrobial potential and synergistic properties. In M. K. Rai, & K. V. Kon (Eds.), *Fighting multidrug resistance with herbal extracts, essential oils and their components* (pp. 165–179). San Diego: Academic Press.

Realini, C. E., & Marcos, B. (2014). Active and intelligent packaging systems for a modern society. *Meat Science*, *98*(3), 404–419. Available from https://doi.org/10.1016/j.meatsci.2014.06.031.

Riaz, A., Lagnika, C., Luo, H., Dai, Z., Nie, M., Hashim, M. M., ... Li, D. (2020). Chitosan-based biodegradable active food packaging film containing Chinese chive (Allium tuberosum) root extract for food application. *International Journal of Biological Macromolecules*, *150*, 595–604. Available from https://doi.org/10.1016/j.ijbiomac.2020.02.078.

Riaz, A., Lei, S., Akhtar, H. M. S., Wan, P., Chen, D., Jabbar, S., ... Zeng, X. (2018). Preparation and characterization of chitosan-based antimicrobial active food packaging film incorporated with apple peel polyphenols. *International Journal of Biological Macromolecules*, *114*, 547–555. Available from https://doi.org/10.1016/j.ijbiomac.2018.03.126.

Rodríguez, G. M., Sibaja, J. C., Espitia, P. J. P., & Otoni, C. G. (2020). Antioxidant active packaging based on papaya edible films incorporated with Moringa oleifera and ascorbic acid for food preservation. *Food Hydrocolloids*, *103*, 105630. Available from https://doi.org/10.1016/j.foodhyd.2019.105630.

Rubilar, J. F., Cruz, R. M. S., Silva, H. D., Vicente, A. A., Khmelinskii, I., & Vieira, M. C. (2013). Physico-mechanical properties of chitosan films with carvacrol and grape seed extract. *Journal of Food Engineering*, *115*(4), 466–474. Available from https://doi.org/10.1016/j.jfoodeng.2012.07.009.

Ruiz-Navajas, Y., V.-M., M., Sendra, E., Perez-Alvarez, J. A., & Fernández-López, J. (2013). In vitro antibacterial and antioxidant properties of chitosan edible films incorporated with Thymus moroderi or Thymus piperella essential oils. *Food Control*, *30*(2), 386–392. Available from https://doi.org/10.1016/j.foodcont.2012.07.052.

Sani, M. A., Ehsani, A., & Hashemi, M. (2017). Whey protein isolate/cellulose nanofibre/TiO_2 nanoparticle/rosemary essential oil nanocomposite film: Its effect on microbial and sensory quality of lamb meat and growth of common foodborne pathogenic bacteria during refrigeration. *International Journal of Food Microbiology*, *251*, 8–14.

Santos, N. L., Braga, R. C., Bastos, M. S. R., Cunha, P. L. R., Mendes, F. R. S., Galvão, A. M. M. T., ... Passos, A. A. C. (2019). Preparation and characterization of Xyloglucan films extracted from Tamarindus indica seeds for packaging cut-up 'Sunrise Solo' papaya. *International Journal of Biological Macromolecules*, *132*, 1163–1175. Available from https://doi.org/10.1016/j.ijbiomac.2019.04.044.

Sganzerla, W. G., Rosa, G. B., Ferreira, A. L. A., da Rosa, C. G., Beling, P. C., Xavier, L. O., ... de Lima Veeck, A. P. (2020). Bioactive food packaging based on starch, citric pectin and functionalized with Acca sellowiana waste by-product: Characterization and application in the postharvest conservation of apple. *International Journal of Biological Macromolecules*, *147*, 295–303. Available from https://doi.org/10.1016/j.ijbiomac.2020.01.074.

Shankar, S., & Rhim, J.-W. (2018). Preparation of antibacterial poly(lactide)/poly(butylene adipate-co-terephthalate) composite films incorporated with grapefruit seed extract. *International Journal of Biological Macromolecules*, *120*, 846–852. Available from https://doi.org/10.1016/j.ijbiomac.2018.09.004.

Shin, Y. J., Song, H. Y., Seo, Y. B., & Song, K. B. (2012). Preparation of red algae film containing grapefruit seed extract and application for the packaging of cheese and bacon. *Food Science and Biotechnology*, *21* (1), 225–231. Available from https://doi.org/10.1007/s10068-012-0029-x.

Silberbauer, A., & Schmid, M. (2017). Packaging concepts for ready-to-eat food: Recent progress. *Journal of Packaging Technology and Research*, *1*(3), 113–126. Available from https://doi.org/10.1007/s41783-017-0019-9.

Silva-Weiss, A., Bifani, V., Ihl, M., Sobral, P. J. A., & Gómez-Guillén, M. C. (2013). Structural properties of films and rheology of film-forming solutions based on chitosan and chitosan-starch blend enriched with murta leaf extract. *Food Hydrocolloids*, *31*(2), 458–466. Available from https://doi.org/10.1016/j.foodhyd.2012.11.028.

Singh, P., Baisthakur, P., & Yemul, O. S. (2020). Synthesis, characterization and application of crosslinked alginate as green packaging material. *Heliyon*, *6*(1), e03026. Available from https://doi.org/10.1016/j.heliyon.2019.e03026.

Siripatrawan, U., & Harte, B. R. (2010). Physical properties and antioxidant activity of an active film from chitosan incorporated with green tea extract. *Food Hydrocolloids*, *24*(8), 770–775. Available from https://doi.org/10.1016/j.foodhyd.2010.04.003.

Sogut, E., & Seydim, A. C. (2018). The effects of Chitosan and grape seed extract-based edible films on the quality of vacuum packaged chicken breast fillets. *Food Packaging and Shelf Life*, *18*, 13–20. Available from https://doi.org/10.1016/j.fpsl.2018.07.006.

Song, H. Y., Shin, Y. J., & Song, K. B. (2012). Preparation of a barley bran protein–gelatin composite film containing grapefruit seed extract and its application in salmon packaging. *Journal of Food Engineering*, *113*(4), 541–547. Available from https://doi.org/10.1016/j.jfoodeng.2012.07.010.

Synowiec, A., Gniewosz, M., Kraśniewska, K., Przybył, J. L., Bączek, K., & Węglarz, Z. (2014). Antimicrobial and antioxidant properties of pullulan film containing sweet basil extract and an evaluation of coating effectiveness in the prolongation of the shelf life of apples stored in refrigeration conditions. *Innovative Food Science & Emerging Technologies*, *23*, 171–181. Available from https://doi.org/10.1016/j.ifset.2014.03.006.

Talón, E., Trifkovic, K. T., Nedovic, V. A., Bugarski, B. M., Vargas, M., Chiralt, A., & González-Martínez, C. (2017). Antioxidant edible films based on chitosan and starch containing polyphenols from thyme extracts. *Carbohydrate Polymers*, *157*, 1153–1161. Available from https://doi.org/10.1016/j.carbpol.2016.10.080.

Tan, Y. M., Lim, S. H., Tay, B. Y., Lee, M. W., & Thian, E. S. (2015). Functional chitosan-based grapefruit seed extract composite films for applications in food packaging technology. *Materials Research Bulletin*, *69*, 142–146. Available from https://doi.org/10.1016/j.materresbull.2014.11.041.

Tesfay, S. Z., Magwaza, L. S., Mbili, N., & Mditshwa, A. (2017). Carboxyl methylcellulose (CMC) containing moringa plant extracts as new postharvest organic edible coating for Avocado (Persea americana Mill.) fruit. *Scientia Horticulturae*, *226*, 201–207. Available from https://doi.org/10.1016/j.scienta.2017.08.047.

Treviño-Garza, M., Yañez-Echeverría, S., García, S., Mora-Zúñiga, A., & Arévalo-Niño, K. (2020). Physico-mechanical, barrier and antimicrobial properties of linseed mucilague films incorporated with H. virginiana extract. *Revista Mexicana de Ingeniería Química*, *19*(2), 983–996.

Tulamandi, S., Rangarajan, V., Rizvi, S. S. H., Singhal, R. S., Chattopadhyay, S. K., & Saha, N. C. (2016). A biodegradable and edible packaging film based on papaya puree, gelatin, and defatted soy protein. *Food Packaging and Shelf Life*, *10*, 60–71. Available from https://doi.org/10.1016/j.fpsl.2016.10.007.

Wambura, P., Yang, W., & Mwakatage, N. R. (2011). Effects of sonication and edible coating containing rosemary and tea extracts on reduction of peanut lipid oxidative rancidity. *Food and Bioprocess Technology*, *4* (1), 107–115. Available from https://doi.org/10.1007/s11947-008-0150-2.

Wang, K., Lim, P. N., Tong, S. Y., & Thian, E. S. (2019). Development of grapefruit seed extract-loaded poly (ε-caprolactone)/chitosan films for antimicrobial food packaging. *Food Packaging and Shelf Life*, *22*, 100396. Available from https://doi.org/10.1016/j.fpsl.2019.100396.

Wang, L., Dong, Y., Men, H., Tong, J., & Zhou, J. (2013). Preparation and characterization of active films based on chitosan incorporated tea polyphenols. *Food Hydrocolloids*, *32*(1), 35–41. Available from https://doi.org/10.1016/j.foodhyd.2012.11.034.

Wang, S., Marcone, M., Barbut, S., & Lim, L.-T. (2012a). The impact of anthocyanin-rich red raspberry extract (ARRE) on the properties of edible soy protein isolate (SPI) films. *Journal of Food Science*, *77*(4), C497–C505. Available from https://doi.org/10.1111/j.1750-3841.2012.02655.x.

Wang, S., Marcone, M. F., Barbut, S., & Lim, L.-T. (2012b). Fortification of dietary biopolymers-based packaging material with bioactive plant extracts. *Food Research International*, *49*(1), 80–91. Available from https://doi.org/10.1016/j.foodres.2012.07.023.

Wang, X., Yong, H., Gao, L., Li, L., Jin, M., & Liu, J. (2019). Preparation and characterization of antioxidant and pH-sensitive films based on chitosan and black soybean seed coat extract. *Food Hydrocolloids*, *89*, 56–66. Available from https://doi.org/10.1016/j.foodhyd.2018.10.019.

Wu, J., Chen, S., Ge, S., Miao, J., Li, J., & Zhang, Q. (2013). Preparation, properties and antioxidant activity of an active film from silver carp (Hypophthalmichthys molitrix) skin gelatin incorporated with green tea extract. *Food Hydrocolloids*, *32*(1), 42–51. Available from https://doi.org/10.1016/j.foodhyd.2012.11.029.

Wyrwa, J., & Barska, A. (2017). Innovations in the food packaging market: Active packaging. *European Food Research and Technology*, *243*(10), 1681–1692. Available from https://doi.org/10.1007/s00217-017-2878-2.

Yilar, M., Kadioglu, I., & Telci, I. (2018). Chemical composition and antifungal activity of Salvia Officinalis (L.), S. Cryptantha (Montbret et aucher ex Benth.), S. Tomentosa (MILL.) plant essential oils and extracts. *Fresen. Environ. Bull*, *27*, 1695–1706.

Yildirim, S., Röcker, B., Pettersen, M. K., Nilsen-Nygaard, J., Ayhan, Z., Rutkaite, R., ... Çoma, V. (2018). Active packaging applications for food. *Comprehensive Reviews in Food Science and Food Safety*, *17*(1), 165–199. Available from https://doi.org/10.1111/1541-4337.12322.

Yong, H., Liu, J., Qin, Y., Bai, R., Zhang, X., & Liu, J. (2019). Antioxidant and pH-sensitive films developed by incorporating purple and black rice extracts into chitosan matrix. *International Journal of Biological Macromolecules*, *137*, 307–316. Available from https://doi.org/10.1016/j.ijbiomac.2019.07.009.

Yong, H., Wang, X., Bai, R., Miao, Z., Zhang, X., & Liu, J. (2019). Development of antioxidant and intelligent pH-sensing packaging films by incorporating purple-fleshed sweet potato extract into chitosan matrix. *Food Hydrocolloids*, *90*, 216–224. Available from https://doi.org/10.1016/j.foodhyd.2018.12.015.

Yong, H., Wang, X., Zhang, X., Liu, Y., Qin, Y., & Liu, J. (2019). Effects of anthocyanin-rich purple and black eggplant extracts on the physical, antioxidant and pH-sensitive properties of chitosan film. *Food Hydrocolloids*, *94*, 93–104. Available from https://doi.org/10.1016/j.foodhyd.2019.03.012.

Yousefi, H., Su, H.-M., Imani, S. M., Alkhaldi, K., M. Filipe, C. D., & Didar, T. F. (2019). Intelligent food packaging: A review of smart sensing technologies for monitoring food quality. *ACS sensors*, *4*(4), 808–821.

Yuan, G., Jia, Y., Pan, Y., Li, W., Wang, C., Xu, L., ... Chen, H. (2020). Preparation and characterization of shrimp shell waste protein-based films modified with oolong tea, corn silk and black soybean seed coat extracts. *Polymer Testing*, *81*, 106235. Available from https://doi.org/10.1016/j.polymertesting.2019.106235.

Zamuz, S., López-Pedrouso, M., Barba, F. J., Lorenzo, J. M., Domínguez, H., & Franco, D. (2018). Application of hull, bur and leaf chestnut extracts on the shelf-life of beef patties stored under MAP: Evaluation of their impact on physicochemical properties, lipid oxidation, antioxidant, and antimicrobial potential. *Food Research International*, *112*, 263–273. Available from https://doi.org/10.1016/j.foodres.2018.06.053.

Zhang, C., Guo, K., Ma, Y., Ma, D., Li, X., & Zhao, X. (2010). Original article: Incorporations of blueberry extracts into soybean-protein-isolate film preserve qualities of packaged lard. *International Journal of Food Science & Technology*, *45*(9), 1801–1806. Available from https://doi.org/10.1111/j.1365-2621.2010.02331.x.

Zhang, W., Li, X., & Jiang, W. (2019). Development of antioxidant chitosan film with banana peels extract and its application as coating in maintaining the storage quality of apple. *International Journal of Biological Macromolecules*. Available from https://doi.org/10.1016/j.ijbiomac.2019.10.275.

Zhang, X., Lian, H., Shi, J., Meng, W., & Peng, Y. (2020). Plant extracts such as pine nut shell, peanut shell and jujube leaf improved the antioxidant ability and gas permeability of chitosan films. *International Journal of Biological Macromolecules*, *148*, 1242–1250. Available from https://doi.org/10.1016/j.ijbiomac.2019.11.108.

Zhang, X., Liu, J., Yong, H., Qin, Y., Liu, J., & Jin, C. (2020). Development of antioxidant and antimicrobial packaging films based on chitosan and mangosteen (Garcinia mangostana L.) rind powder. *International Journal of Biological Macromolecules*, *145*, 1129–1139. Available from https://doi.org/10.1016/j.ijbiomac.2019.10.038.

Zink, J., Wyrobnik, T., Prinz, T., & Schmid, M. (2016). Physical, chemical and biochemical modifications of protein-based films and coatings: An extensive review. *International Journal of Molecular Sciences*, *17*(9), 1376. Available from https://doi.org/10.3390/ijms17091376.

Zúñiga, G. E., Junqueira-Gonçalves, M. P., Pizarro, M., Contreras, R., Tapia, A., & Silva, S. (2012). Effect of ionizing energy on extracts of Quillaja saponaria to be used as an antimicrobial agent on irradiated edible coating for fresh strawberries. *Radiation Physics and Chemistry*, *81*(1), 64–69. Available from https://doi.org/10.1016/j.radphyschem.2011.08.008.

CHAPTER

Health benefits of plant extracts 12

Toiba Majeed and Naseer Ahmad Bhat
Department of Food Science & Technology, University of Kashmir, Srinagar, India

12.1 Introduction

Around 250,000–500,000 plants have been documented throughout the world to date, out of which relatively 1%–10% have been biologically screened by humans (Dekebo, 2019). WHO reports that 80% of individuals in developing nations use traditional plant- derived medicine for their health care (Van Wyk & Prinsloo, 2018). Although developed countries are highly advanced in medicinal science, over one-fourth of leading medicines are originated directly or indirectly from plants (Srivastava, Singh, Devi, & Chaturvedi, 2014). Of all life-saving medicines for clinical use worldwide are extracted from natural produce, almost 25% is contributed by higher plants (Dekebo, 2019). New plants are always been in research, and their extracts are studied for various health benefits in the search to find new plants with a different action spectrum. Globalization has increasingly renewed the interest of plant extracts in functional foods with a balanced nutritional profile, healthy ingredients, and disease prevention properties. Thus researchers have focused on unfolding and redesigning traditionally known products by replacing some preservatives and flavor components with ingredients having positive physiological effects including natural plant extracts with rich bioactivity (Franzen & Bolini, 2019).

The word "extract" is derived from "*extractus*" a Latin word meaning "things extracted from another," a process that aims to extract certain components present in plant matrix (Franzen & Bolini, 2019). Extracts of plants are complex in nature, and different parts of plants also vary in composition. Usually plants contain various types of polar and non-polar chemical components. These may be organic acids, alkaloids, glycosides, resins, oils, carbohydrates, amino acids, proteins, enzymes, tannins, plant pigments, waxes, and other inorganic components in traces (Regnier, Combrinck, & Du Plooy, 2012). These bioactive compounds are not easily accessible. Therefore, for the extraction of these phytochemical compounds and for the elimination of other inactive fractions, which showcase no or low relevant benefits, plants are subjected to several works. This in turn will lead to the achievement of phytochemicals, in a concerted form, which could be readily incorporated in other produces than whole plant inclusion (Veiga, Costa, Silva, & Pintado, 2018). With the advance of numerous studies such as extraction, isolation, techniques of structural and compositional determinations and pharmacological effects, separated and purified extracts has initiated to be taken earnestly. There are more than a million types of extracts derived from plants, which have been validated for various beneficial purposes like for food stuffs (antioxidant, stabilizer, texturizer, antifungal, substitute

Plant Extracts: ***Applications in the Food Industry***. **DOI: https://doi.org/10.1016/B978-0-12-822475-5.00013-2**

for food colorants, etc.), processing aids (chemical replacers, bio preservation), pharmaceutical properties (preventive and curative) such as hepatoprotection, antitumoral, antimutagenic, antimicrobial, antidiabetic, antiallergic, vaso-dilatory, antiinflammatory, protection against oral diseases etc. (Mir, Shah, Ganai, Ahmad, & Gani, 2019; Veiga, et al., 2018). These plant bioactives also present antioxidant actions particularly against low-density lipoproteins (LDLs) and nucleic acid oxidative changes (Kiokias, Proestos, & Oreopoulou, 2018).

The past decade has seen an upsurge in life expectation as well as the public concerns with quality of life. Several researchers scientifically proved that proper nutrition is one of the major factors behind the well-being of an individual. Proper nutrition plays a prominent role on one's health, with a stable diet being ultimate for the maintenance of homeostasis without causing any health risk and hence the suitable functioning of human body. The exploitation of plant extracts is gaining more popularity in the food industry because of their low cost, functional properties, and renewable source of biologically relevant compounds (Mourtzinos et al., 2018). This chapter is aimed to highlight several bioactive compounds and the prevailing evidence about various probable health benefits of consuming plant extracts and extract-based substances supported by in vivo and epidemiological studies.

12.2 Plant polyphenolic composition

Plants are richest sources of secondary metabolites like phenolics, carotenoids, anthocyanins, xanthophylls, or other constituents that play a photo-protective role and several other functions vital for plant metabolism and survival such as defense against pathogenic diseases. Moreover, they are also responsible for pigmentation and other organoleptic attributes of plants like flavor and color (Cheynier, 2012; Lattanzio, 2013). About 8000 distinctive phenolic species have been described in nature so far. These substances may be present in leaf, stem, fruit, flower, or roots of the plants and can be divided into several groups according to their chemical structures ranging from comparatively simple to highly polymerized compounds. Chemically, phenolics could be defined as the molecules containing at least one phenolic unit, and the molecules comprised of more than one of these subunits are usually categorized as polyphenols (Veiga et al., 2018).

The various other phenolic compounds present in the plant materials are flavanols like kaempferol, quercetin, myricetin, gallocatechin, catechin gallate, catechin, gallocatechin gallate, epicatechin, epicatechin gallate, epigallocatechin, and epigallocatechin gallate (Da Silva et al., 2013). The plants are also richest sources of phenolic acids like gallic, *p*-coumaric, caffeic and quinic acids, which provide one of the most broadly exploited metabolic pathways in plant research (Deotale, Dutta, Moses, & Anandharamakrishnan, 2019). In fruits like berry, apple, and pears, caffeic acid (CA) together with p-coumaric acid was present in higher amounts (75%–100%) of total hydroxycinnamic acids (Da Silva et al., 2013). Also, CA is reported to be present in *Eucalyptus globulus* bark and was reported as the major phenolic compound in coffee and its oil (Deotale et al., 2019). The 3,4,5-trihydroxybenzoic acid also known as Gallic Acid (GA) is the major tea phenolic acid, also present in higher concentrations in berries and chestnuts (Pandurangan, Mohebali, Norhaizan, & Looi, 2015). Recently, Souza et al. (2020) isolated GA from extracts of black tea at 0.8 mg/kg concentration. It is also found among a number of land plants, like *Cynomorium coccineum* (a parasitic plant), aquatic plants, and the some blue-green algal species (Liu, Carver, Calabrese, & Pukala, 2014). Rosmarinic acid (RA) is present in some plants of family Lamiaceae like *Perilla* spp., *Rosmarinus offcinalis*, *Origanum* spp., and *Salvia offciniali*s as the main phenolic component (Oreopoulou et al., 2018). The various other researchers also reported it in several herbs like oregano, thyme and rosemary in 0.05 and 26 g/kg concentrations (Yashin, Yashin, Xia, & Nemzer, 2017). Additionally, Tsimogiannis et al. (2017) reported 19.5 (g/kg) of RA in pink savory leaves. Carnosic acid, a phenolic diterpene of labdane-type is present in a number of plants of family Lamiaceae, like rosemary, and common salvia (Loussouarn et al., 2017). Carnosic acid is lipophilic compound and is mainly found in the dry sage leaves at a concentration of 1.5%–2.5% (Raes, Doolaege, Deman, Vossen, & De Smet, 2015).

The phenolic acid (ferulic acid) arises from the metabolism of tyrosine and phenylalanine and is present in both free and conjugated forms in seeds and leaves. FA is found in the coffee, peanut, apple, artichoke, and orange seeds. Flaxseed has been reported as the richest FA glucoside source (4.1–0.2 g/kg) occurring naturally (Bagchi, Moriyama, & Swaroop, 2016). According to Mojica, Meyer, Berhow, and de Mejía (2015), an average 0.8 g/kg concentration, FA has been isolated from black beans. In addition, FA is also present in Brassica species and tomatoes.

p-Coumaric acid is present in huge number of plants species, and the richest sources of p-coumaric acids are fungi, peanut, navy beans, tomatoes, carrot, basil, and garlic (Trisha, 2018). The p-Coumaric acid substance is also abundant in most of the fruits and cereals especially in pears and berries (Bento-Silva et al., 2020). A dihydroxybenzoic acid derivative, Vanillic Acid is obtained from various fruits, olives, and cereals like wheat, and also found in wine, beer, and cider products (Siriamornpun & Kaewseejan, 2017).

Carotenoids are natural colored substances present in birds, plants and fish meat, algae, fungi, and insect's cuticle. Based on their function, carotenoids may be grouped under two categories: xanthophylls including lutein, zeaxanthin and carotenes like lycopene, α-carotene and β-carotene (Kumar et al., 2014). These carotenoids prevent various diseases and help the human body in several positive ways. The Astaxanthin has strong antioxidant, antiinflammatory, anticancerous properties and promotes cardiac health (Fasano et al., 2014). Lutein prevents cataract, muscle age-related degeneration and circulatory diseases and has antioxidative and anticancer functions (Manayi et al., 2016). β-Carotene checks night blindness, acts as antioxidant and protects against liver fibrosis (Virtamo et al., 2014).

Anthocyanins exist in numerous plant parts, like stem, fruit, flower, leaf, and root. In higher plants, commonly anthocyanidins comprises of cyanidin, petunidin, pelargonidin, malvidin, peonidin and delphinidin. These anthocyanidins are distributed in edible plant parts in the concentration of 50%, cyanidin, 12% pelargonidin, 12% peonidin, 12% delphinidin, 7% petunidin and 7% malvidin. Among these most abundant glycosides comprises of delphinidin, cyanidin and pelargonidin, which represents 80% of leaf, 69% of fruits and 50% of flower pigments, respectively. The furthermost prevalent anthocyanin in fruits is cyanidin-3-glucoside (Pascual-Teresa & Sanchez-Ballesta, 2008).

12.3 Health benefits of plant extracts

The plant extracts obtained from various plant parts impart health benefits through their numerous active principles and bio complexes prepared by different methods and processes (Table 12.1).

Table 12.1 Health benefits of plant extracts.

Part of plant	Species/name of plant	Health beneficial properties	Reference
Fruit	Mulberry (*Morus alba*)	Hypolipidemic/cardio-protective; antidiabetic; antitumor; antiobesity; antioxidant; hepatoprotective and protection against brain damage	Zhang, Ma, Luo, and Li (2018, 2016)
	Berberis vulgaris	Antidiabetic; anticancer; antiacne; cardio-protective and antihypertensive.	Rahimi-Madiseh, Lorigoini, Zamani-gharaghoshi, and Rafieian-Kopaei (2017)
	Dragon fruit (*Hylocereus undatus*)	Antioxidant; anticancerous; antihypertensive; antidiabetic and decrease arterial stiffness and lipid peroxidation.	Divakaran, Lakkakula, Thakur, Kumawat, and Srivastava (2019)
	Raspberry (*Rubus idaeus*)	Antiinflammatory; chemopreventive; antidiabetic; antioxidative; anticancer; antiproliferative; immunoregulatory and improves lipid metabolism.	Kowalska, Olejnik, Zielińska-Wasielica, and Olkowicz (2019), Stagos (2019)
	Persimmon (*Diospyros kaki*)	Anticancer; antioxidant; antiinflammatory; gastro-protective; antidiarrheal, hypolipidemic and antidiabetic.	Guler et al. (2021), Dhawefi et al. (2021)
	Kiwi (*Actinidia arugata*)	Gastro and hepatoprotective; cardiovascular protective; antiinflammatory; antidiabetic; antioxidant, anticancerous; antiangiotensin converting enzyme activity; attenuates DNA damage and antiplatelet aggregation activity.	D'Eliseo et al. (2019), Hussein et al. (2015)
Seed	Fennel (*Foeniculum vulgare*)	Immune enhancement; antioxidant; antimutagenic; antioxidant; antiinfertility; hepatoprotective; antiinflammatory, hypolipidemic; regulates hormonal & biochemical changes associated with PCOS syndrome; antiosteoarthritic and protection against nephrotoxicity.	Bayrami et al. (2020), Samadi-Noshahr, Hadjzadeh, Moradi-Marjaneh, and Khajavi-Rad (2021)
	Moringa (*Moringa oleifera*)	Antiproliferative against breast and colorectal cancer cells; antioxidant; antiinflammatory; pro-apoptotic activity on cancer cells; immunomodulatory; healing power; antidiabetic.	Potestà et al. (2019), Xu, Chen, and Guo (2019)
	Fenugreek (*Trigonella foenum-graecum*)	Antiinflammatory; antioxidative; antifibrotic; PCOS reduction; antidiabetic; antiobesity; prevents pancreatic damage and have phytoestrogenic effect.	Thomas et al. (2020), Rahmani et al. (2018)
	Pomegranate (*Punica granatum*)	Antitumor; antioxidant; antiradical; antiatherogenic; prevents dental plaque and gingival inflammation; hypolipidemic; effects on insulin resistance; neuroprotective; antiosteoporosis and impressive on neurodegenerative diseases.	Doostan et al. (2017), Thitipramote et al. (2019)
	Parkia biglobosa	Antianemic; antimutagenic; antioxidant; antiinflammatory; antiulcer; antihypertensive; anticancer/antitumor and hypoglycemic.	Saleh et al. (2021)

Flower	Dandelion *(Taxaxacum officinale)*	Hepatoprotective; antidiabetic; antifibroblastic; mito-inhibitory or stimulative (cell division); antioxidant, antiplatelet aggregation and anticoagulant.	Li, Yang, Yang, and Zu (2018), Grauso, Emrick, de Falco, Lanzotti, and Bonanomi (2019)
	Lavender *(Lavandula augustifolia mill)*	Antioxidant; sedative & hypnotic; antiproliferative; neuroprotective (Alzheimer's disease); antiinflammatory; antianxiety and antidiabetic.	Dhasthakeer, Kavitha, and Vishnupriya (2020), Nurzyńska-Wierdak, and Zawiślak (2016).
	Rose *(Rosa rubiginosa)*	Cardiovascular prevention; antioxodative; antiinflammatory; anticonvulsant; hypoglycemic; analgesic; preventive against neurological atrophy; bronchodilatory; antiHIV; antiaging, antilipase and ophthalmic effects.	Yang and Shin (2017), Liu, Tang, Zhao, Xin, and Aisa (2017)
	Hibiscus rosa	Gastro-protective; cardio-protective; anticancer; antioxidative; antidiabetic; antifertility; antihyperlipidemic; wound healing activity; hair growth promoter and immune response.	Missoum (2018).
	Passiflora	Antihypertensive; antioxidative; antianxiety; antiinflammatory; anticancerous; prevents psychological disorders and relieves menopausal transition.	Taïwe and Kuete (2017), Hameed, Cotos, and Hadi (2017)
Leaf	*Aloe vera*	Immunomodulatory; antiinflammatory; hepatoprotective; antioxidant; anticancer; purgative; wound and cell proliferation; antihyperlipidemic; antiulcer.	Salehi et al. (2018)
	Ginseng *(Panax ginseng)*	Hypoglycemic; antiaging; prevents oxidative damage; antiobesity; cytotoxic against cancer cells; protection against testicular damage; cardio-protective and neuroprotective.	Shin, Lee, Son, Park, and Jung (2017), Lee, Lee, Jang, Kwon, and Nam (2017)
	Peppermint *(Mentha piperita)*	Cytotoxic and anticarcinogenic; antihepatotoxic; antibreast carcinoma; antioxidant; antidiabetic; nephron-protective; antiinflammatory.	Konda et al. (2020), Li, Al-Misned, El-Serehy, and Yang (2021)
	Artichoke *(Cynara scolymus)*	Cholesterol lowering; hypoglycemic; hepatoprotective; antioxidant; pre/probiotic; antiatherosclerotic; immune-modulatory; cholerectic; cardio-protective; anticancer; antiinflammatory; gastro-protective.	Salem et al. (2015)
	Olive *(Olea europaea)*	Anticancer; antiinflammatory; hypotensive; hypolipidemic; hypoglycemic; antiproliferative (breast cancer cell line); antioxidative; prevention against hepato toxicity; immune-modulatory and antiatherosclerotic.	Lins, Pugine, Scatolini, and de Melo (2018), Qabaha, Al-Rimawi, Qasem, and Naser (2018)
Stem/ bark	Cinnamon *(Cinnamomum verum)*	Antidiabetic; antioxidant; anticholinergic; hypolipidemic; antiinflammatory; antiallergic; anticancer and hepatoprotective.	Vijayakumar et al. (2020), Gulcin et al. (2019).
	Berberis aristata	Anticancer; anti spasmodic and antidiarrheal; anticonvulsant; hypoglycemic; hypolipidemic; protection against mitochondrial damage and tyrosinase inhibitor.	Rahimi-Madiseh et al. (2017), Chander, Aswal, Dobhal, and Uniyal (2017)
	Ginger *(Zingiber officinale)*	Antiinflammatory; antioxidant; antihypercholesteraemic; hypotensive; antiulcer and antiemetic; cardio-protective; hypoglycemic; vasodialatory; antirheumatic.	Al-Awwadi (2017), Wu et al. (2018)
	Pine bark *(Pinus radiata)*	Antioxidant/radical scavenging; antiinflammatory; anticancer; immune enhancement; reduction in side effects of radio and chemotherapy; cardiovascular protection; neuroprotective and antiatherosclerotic.	Li, Feng, Zhang, and Cui (2015).

(Continued)

Table 12.1 Health benefits of plant extracts. ***Continued***

Part of plant	Species/name of plant	Health beneficial properties	Reference
Root/ tuber	Ginseng *(Panax ginseng)*	Antiinflammatory; antiParkinson activity; antiarthritic; cardio-protective; antidiabetic; anti stress; antiAlzheimer activity; antiischemic and antihypoxic activity.	Dar, Hamid & Ahmad (2015)
	Turmeric *(Curcuma longa)*	Antiarthritic; hypoglycemic; antioxidant; antiinflammatory; immune stimulation; anticancer; hepatoprotective; cardiovascular protection chemo protective; neuroprotective; antiallergic.	Mohammed et al. (2017), Daily, Yang, and Park (2016)
	Beet root *(Beta vulgaris)*	Lipid lowering; antihypertensive; antiinflammatory; antioxidant; cytotoxic against cancer cell lines (prostate and breast); osteoarthritis prevention; anticancer; antiobesity; antidiabetic; neuroprotective.	Hadipour, Taleghani, Tayarani-Najaran, and Tayarani-Najaran (2020), Kapadia et al. (2011)
	Berberis integerrima	Cardiovascular protection; antihypertensive; anticonvulsant; antidiabetic/ hypoglycemic; improves renal dysfunction; hypolipidemic; antioxidant; antiarthritic and antihepatopathic.	Rahimi-Madiseh et al. (2017), Aryaeian, et al. (2020)

12.3.1 Fruit extracts

Fruits are primary sources of nutritional compounds in human diet. Fruit extracts from different plants have numerous health benefits that include antioxidant, antidiabetic, antiinflammatory, and antitumoral activity among several other activities. The main biologically active compounds present in fruit extract mainly include vitamins, polyphenols, carotenoids, polysaccharides, dietary fiber, alkaloids, essential oils etc. Their natural origin makes them an excellent substitute for synthetic compounds present in fruit extracts that have carcinogenic and toxicological effects. These biologically active compounds are proven to be more beneficial when present together.

Early studies on kiwi (*Actinidia deliciosa*) fruit extracts (hexane, acetone and methanol) have revealed its potential in cancer prevention (Motohashi et al., 2002). Cardio-protective properties of kiwi fruit extract (water and 70% ethanol) were also investigated in in vitro models by analyzing its hypotensive, antihypercholesterolemic and antioxidative activities (Jung et al., 2005). In vitro studies on lyophilized aqueous extracts of kiwi fruit also demonstrate its antioxidant activity by monitoring its radical scavenging efficiency (Bursal & Gülçin, 2011).

Mulberry (*Morus* spp.) fruit extract offers high amounts of polyphenols like anthocyanins, phenolic acids, flavanol derivatives, chlorogenic acid, and quercetin glycoside. Ability of mulberry fruit extract (MFE) in prevention of liver fibrosis is of vital importance. Mulberry water extracts reduced lipid peroxidation and hampered pro-inflammatory gene expression displaying their protective and therapeutic effects against Carbon tetrachloride (CCl_4)-induced fibrosis and liver damage (Hsu et al., 2012). In a study, protective effect of polyphenolic-rich MFE has been investigated against Ethyl Carbamate (EC)-generated cytotoxicity and oxidative stress. The mechanism for protection of human liver Hep G2 cells from EC generated cytotoxicity included scavenging of excess cellular reactive oxygen species (ROS). Also, pretreatment of MFE significantly reduced the EC induced mitochondrial membrane potential collapse, mitochondrial lipid peroxidation plus intracellular glutathione (GSH) depletion. It further prevented GSH depletion and returned the function of mitochondrial membrane (Wei, Yuting, Tao, & Vemana, 2017).

An investigation was carried out by Dahham, Agha, Tabana, and Majid (2015) on antiangiogenic and anticancerous activities of banana peel extracts. Results revealed that highest antiangiogenic activity (85.32% inhibition at concentration of 100 μg/mL) was exhibited by banana extract and it also effectively reduced the growth of colon cancer cell line.

Berberis plant species are largely consumed as fresh, dehydrated, or used in the production of juices (Farhadi Chitgar, Aalami, Maghsoudlou, & Milani, 2017). Among these species, *Berberis vulgaris* is widespread due to their nutritional and phytochemical significance as these are abundant sources of vitamins, minerals, alkaloids, flavonoids, anthocyanins, and antioxidants (obtained from all plant parts), which can be extensively used in a collection of pharmaceutical and nutraceutical stuffs. Most important alkaloids in the plant are berberin, bermamine, oxycontin, palmatine, bervulcine, columbamine, and coptisine (Sarraf, Beig-babaei, & Naji-Tabasi, 2019). The crude extract of these phytochemicals demonstrate a number of biological activities including antiinflammatory, antioxidant, antipyretic, hypotensive, antinociceptive anticholinergic and antiseptic Zarei, A., Ashtiyani, S. C., Taheri, S. & Ramezani, M. (2015). Phytochemical analysis of *B. vulgaris* extracts detected several alkaloids like tertandrine, berbamine, and chondocurine which are wee-known for their antioxidant and antiinflammatory activities (Salehi et al., 2019). The possible antioxidant property of *B. vulgaris* is through inhibiting NF-κB, 1,1-diphenyl-2-picrylhydrazyl (DPPH) scavenging and inhibiting lipoxygenase activity owing to its phenolic and flavonoid constituents (Eddouks, Maghrani, Lemhadri, Ouahidi, & Jouad, 2002). The berberine rich fruit extracts of *Berberis* species demonstrate significant antitumor properties, with reported efficiencies in mitigating various eye ailments (Srivastava, Srivastava, Misra, Pandey, & Rawat, 2015). Furthermore, reported plasma glucose lowering and hyper-lipidemic properties of *Berberis* may be due to barberine-induced adjustment in adipokine secretion which in turn results in improvement in insulin sensitivity (Zhu, Bian, & Gao, 2016). Efficiency of *Berberis* extracts in maintaining cardiovascular health confers to its ability of improving hypertension, cardiac arrhythmias, cardiomyopathy and ischemic heart disease (Imenshahidi & Hosseinzadeh, 2016).

Dragon fruit (*Hylocereus polyrhizus*) is a store house of variety of biochemical constituents like alkaloids, saponins, flavonoids, steroids, tannins, and terpenoids that have excellent health-promoting effects. Dried dragon fruit extracts were evaluated for antioxidant capacity using total phenol essay, which was found to be equivalent to that of GA (Rebecca, Boyce, & Chandran, 2010). Also, ethanolic extract of dragon fruit flesh was determined for antiinflammatory effects in animal model (mice) induced by tetramethyl benzidine. The results showed a significant reduction in the expression of pro-inflammatory molecules and degradation of protein NF-κB levels (Macias-Ceja et al., 2016).

Apple fruit extracts and its rich phenolics confer potent antioxidant activities against free radicals generated through lipid oxidation. Also, results on the treatment of apple extracts on colon cancer cells showed a dose-dependent inhibition of cell propagation with a maximum inhibition (43%) at a concentration of 50 mg/mL. Similar results with highest cell inhibition (57%) were seen in Hep G2 liver melanoma cells at a dosage of 50 mg/mL (Eberhardt, Lee, & Liu, 2000). *Carica papaya* extract has been seen to have wound healing properties in animal models. Diabetic wounds that are difficult to manage due to their slow, non-healing nature continue for weeks in spite of adequate and proper care. Topical application of *C. papaya* extract in streptozotocin-generated diabetic rats showed that wound size reduced within 5days of treatment (Nayak, Pereira, & Maharaj, 2007).

12.3.2 Leaf extracts

Although fruits are rich in numerous bioactive compounds, leaves also have been reported to possess significant amount of beneficial compounds. Several studies have documented that leaves contain higher amount of phenolic compounds as compared to fruits (Veiga et al., 2018). The possible reason for the difference could be the photosynthetic function of leaves and subsequent production of oxygen resulting in the generation of ROS that are injurious to tissues. Thus taking into consideration the known antoxidant properties of phenolics, these may be present in higher concentration in leaves and protect its tissues to against the stress caused by solar radiations. The aqueous extracts of passion fruit leaves were found to possess significantly high antioxidant properties including leaves of other berries like blackberry, strawberry and raspberry than the fruit themselves (Da Silva et al., 2013). Tea (*Camellia sinensis*) leaf extracts have been investigated for their preventive effect against *cardiovascular diseases*. The strong antioxidant activity of tea polyphenols (catechins) decrease free radical damage to cells and reduce oxidation of LDL cholesterol which may inhibit the development of atherosclerotic plaques, thus preventing heart diseases (Lorenzo & Munekata, 2016). Much focus has also been put on its *anticancerous properties* due to the presence of tea polyphenols (epigallocatechin gallate, epigallocatechin and epicatechin gallate). In animal model,

Zhang, Duan, Owusu, Wu, and Xin (2015) have documented that green tea supplementation in rats resulted in induction of apoptosis and cell cycle arrest inhepatoma cells (AH109A cell line) and murine melanoma cells (B16 cell line) Also, *antiinflammatory properties* of an antioxidant rich polyphenolic extract from green tea were reported in rats. The study showed a significant reduction in the incidence of arthritis in the mice fed with green tea polyphenols which is attributed to the marked decrease in the expression of inflammatory mediators including cyclooxygenase 2, interferon (IFN) and tumor necrosis factor (TNF) (Cooper, Morré, & Morré, 2005).

Artichoke (*Cynara scolymus*) leaf extracts have been investigated for antioxidant, bile-enhancing, hepatoprotective, lipid lowering, and choleretic effects which also corresponded with its historical use. Numerous in vitro studies have documented that the antioxidant activity of Artichoke leaf extracts is mainly due to radical scavenging and metal ion chelating effects of its components like chlorogenic acid, cynarin, and flavonoids (Pérez-García, Adzet, & Cañigueral, 2000). *Hepatoprotective effects* of artichoke leaf extracts may be due to its ability to remove harmful toxins and digest fats by increasing the bile production of the liver. The *antiatherosclerosis action* of Artichoke leaf extracts was supposed to be the outcome of two mechanisms of action: an antioxidant effect of its bioactive compound cynarin that reduced LDL oxidation and inhibition of cholesterol synthesis. Studies have also shown many powerful phenol-type antioxidants like GA, rutin and quercetin found in Artichoke leaf extracts that could play a crucial in the prevention and management of various type of *cancers* (leukemia, prostate and breast cancer) by inducing apoptosis and reduction in cancer cell proliferation. Researchers confirm that Artichoke leaf extracts help to maintain *cardiovascular health* by increasing the cholesterol breakdown to bile salts thereby enhancing its elimination by increasing the production of bile and also prevents the internal production of cholesterol in the liver, thus preventing atherosclerotic deposits (Salem et al., 2015).

Papaya (*C. papaya L.*) leaf extract has been documented in literature for its antiinflammatory, antitumor, and antidiabetic effects due to the presence of many complex bioactive compounds including esterified phenolics, flavonols, organic acids, carpaine alkaloids and other constituents. Papaya leaf extracts have been used from ancient time as a remedy to treat many disorders, such as cancer and other infectious diseases. Immunomodulatory, antiinflammatory and antiarthritic activity of papaya leaf extracts has also been investigated which may be due to the presence biologically active compounds like carpaine and nicotinic acid. Besides, antibacterial, antiviral, antihelminthic, gastro-protective, cardio-protective and antioxidative activities of papaya leaf extracts has also been documented (Tatyasaheb, Snehal, Anuprita, & Shreedevi, 2014).

Artemisia absinthium have various biological activities like antiinflammatory, antidiabetic, anticancer, antitumor, antihelminthic, antipyretic, antioxidant, hepatoprotective, neuroprotective, bile stimulant, antiarthritic, antifertility, menopause, premenstrual syndrome, dysmenorrhea, analgesic, and antidote to insect poison (Koul, Taak, Kumar, Khatri, & Sanyal, 2017; Nigam et al., 2019 Goud & Swamy, 2015). Goud and Swamy (2015) investigated *antidiabetic effect* of methanolic and ethanolic leaf extract of *A. absinthium* in both normal and diabetic animals. It lead to a significant reduction in blood glucose level in a dose-dependent manner which may be due to the presence of active components such as α- and β-thujone, thujyl alcohol, azulenes, bisaboline, cadinine, sabinene and pinene (Dabe & Kefale, 2017). Li, Zheng, et al. (2015) also studied antidiabetic effect of powdered leaf extracts in humans and they concluded that plant possesses good hypoglycemic effect in a dose-dependent manner through insulinotropic (to increase insulin secretion) action. These herbs are also believed to be involved in the repairment and regeneration of pancreatic β-cells (Bhat et al., 2019). Also, animal model (rats) based study on aereal parts of *A. absinthium* extracts (methanol, hexane, ethanol and CCl4) revealed its antiulcer property. This was followed by depletion in secretion of gastric juice and peptic activity with the elevation in mucin levels (Shafi, Khan & Ghauri (2004).

Aloe vera plants have been extensively known and used by mankind by centuries in folklore for therapeutic purposes due to their health-promoting properties (Surjushe, Vasani & Saple, 2008). *Aloe vera* leaf extracts promotes a variety of antiinflammatory responses in the body by reduction of leukocyte adhesion and pro-inflammatory cytokine production (Duansak, Somboonwong, & Patumraj, 2003). *Aloe vera* gel extracts also exhibit hepatoprotective effects by inhibiting ethanol-induced fatty liver by suppressing mRNA expression of lipogenic genes in liver (Radha & Laxmipriya, 2015). *Aloe vera* contains many physiologically active substances which prove beneficial for its *antidiabetic activity*. Several in vitro and in vivo studies conducted on water soluble leaf fractions of *Aloe vera* demonstrated its glucose lowering activities. Polysaccharides play a major role in antidiabetic activities by preventing β-cells from oxidative damage (Das et al., 2011), increase insulin levels and hence show hypoglycaemic effects. The main anthraquinone of *Aloe* namely Aloin has been proposed to be a potential therapeutic option against cancer by having anti-proliferation effect on some cancer cell types like lung, squamous, glioma and neuroectodermal cancer cells by inhibiting both N-acetyl transferase activity and gene expression (Masaldan & Iyer, 2014).

12.3.3 Stem and bark extracts

Stem and bark extracts possess several potential health benefits due to their phytochemical content including phenols, flavonoids, tannins, saponins, alkaloids, glycosides, steroids, anthocyanins, and resins. *Ficus racemosa* stem bark extracts have antioxidant, hypoglycemic, hepatoprotective, antiinflammatory, antibacterial/antifungal, gastro-protective, analgesic, antidiarrheal, hypotensive, antipyretic, and wound healing activities, which may be contributed by its phytochemical compounds like steroids, alkaloids, tannins, quercetin, gluanol acetate, stigmasterol, β-sitosterol, and β-sitosterol-D-glucoside (Ahmed & Urooj, 2010). Hypoglycemic capacity of different solvent extracts of *F. racemosa* stem bark powder displayed a noticeable long term effect on reducing glucose levels of blood (up to 80%) in diabetic rats (alloxan-induced). This glucose lowering capacity of bark extracts was analogous to that of compound glibenclamide, a standard antidiabetic agent (Vasudevan, Sophia, Balakrishanan, & Manoharan, 2007). Also, in diabetic rats (alloxan-induced), aqeous and ethanolic extracts of *F. racemosa* bark showed a noteworthy radical quenching activity. This in turn ominously improved the radical quenching status by reducing Thiobarbituric acid reactive substances and enhancing GSH levels and other antioxidant defense system (Vasudevan et al., 2007). Methanol extract of its bark on oral administration along with carbon tetrachloride at 250 and 500 mg/kg body weight showed liver protection as is clearly evident by serum transaminase reversal elevations (Channabasavaraj, Badami, & Bhojraj, 2008). Studies conducted on in vitro antiinflammatory potential of bark extracts (ethanolic) of *F. racemosa* presented a significant hindrance in COX-1 enzyme activity to an extent of 89%, 71%, and 41%, respectively (Li, Myers, Leach, Lin, & Leach, 2003). Gastro-protective effect of bark extracts (ethanolic) of *F. racemosa* may be attributed to its antiulcerogenic activity (Malairajan, Gopalakrishnan, Narasimhan, & Kavimani, 2007).

Artocarpus chaplasa stem bark extract was used traditionally for the treatment of various ailments and skin diseases in north eastern India. Antioxidant activity of stem extracts of *A. chaplasa* was evaluated by using superoxide radical quenching assay and was found to have a significant radical scavenging activity for both superoxide and DPPH. This may be attributed to the phenolic contents of the plant showing a positive relationship between total polyphenol constituents and DPPH free radical quenching action (Siriwardhana & Shahidi, 2002). Methanolic stem extracts of *A. chaplasa* may be used as a potential supplement for treating noninsulin dependent diabetes mellitus by inhibiting α-glucosidase activity (Bhattacharjee, Singha, Banik, Dinda, & Maiti, 2012).

Several bioactive components from asparagus stem like polyphenols, dietary fiber, saponins, and anthocyanins have achieved growing courtesy in recent times due to their anticancer, antitumor, antioxidative, immune-modulatory, hypotensive, and hypoglycemic effects studied through in vitro and in vivo studies. Anticancer activities of Asparagus extracts have been indicated by many studies. In early 1996, some studies reported antitumor potential of Asparagus saponins extracted from shoots in a dose-dependent manner (Shao et al., 1996). antitumor action of ethanolic asparagus extract saponins extracted from matured stalk was evaluated by using prostate cancer (PC-3) cell lines. Results illustrated these extracts have cytotoxic impact against these cells at 1.5 mg/mL concentrations. Stems of Asparagus displayed a major dose-dependent cytotoxic capacity against 3 tumor cell lines comprising colon, breast and pancreatic cancers. This cytotoxic action is possibly achieved by inhibiting the invasive capability of human breast cancer cell MDAMB-231 and by regulation of small G protein action to hinder motility of tumor cells (Jaramillo et al., 2016). Also, in animal model (Wistar rats) it was observed that extracts of asparagus stem showed positive action against Bisphenol (toxic substance) by boosting antioxidant capacity (Poormoosavi, Najafzadehvarzi, Behmanesh, & Amirgholami, 2018). Asparagus polysaccharides reportedly showed a significant improvement in the phagocytic activity of peritoneal macrophages of normal mice by enhancing cellular and humoral immunity. Also, these polysaccharide extracts improve the conversion rate of lymphocytes and encourage the development of hemolysin and hemolytic plaques (Zhang, 2003). It is also supposed that polysaccharides of asparagus accelerate release of NO in macrophages by expression of iNOS gene activation. Zhao, Xie, & Yan, 2012 reported the hypoglycemic effects of old matured stem extracts of asparagus which indicated extraordinary results in model rat. Hypoglycemic activity of AEO may be due to its ability to alleviate the oxidative damage by improvement of antioxidant shielding enzymes activity which in turn stimulate insulin secretion and adjust blood glucose metabolism. Hypolipidemic potential of Asparagus were also been studied. It has been reported that old stem extract of asparagus could remarkably inhibit the elevated serum cholesterol in hyperglycemic rats and had a regulatory influence on the lipid metabolism syndrome. The polysaccharides from Asparagus stem extracts also reduce the levels of LDL cholesterol and the manifestation of atherosclerosis to uphold blood lipid metabolism (Guo et al., 2019; Guo, Wang, & Liu, 2020; Zhao, Xie, & Yan, 2012).

Information on the ethano-botanical and pharmacological use of plants from genus *Parkia* was regained owing to its phytochemistry and it was found that their extracts possess anticancer, antimicrobial, antihypertensive, antiinflammatory, antiulcer, antidiabetic, antioxidant, hepatoprotective and antidiarrheal activities (Saleh et al., 2021). The stem bark of *Parkia biglobosa* comprises of phenols, sugars, saponin, flavonoid, tannin, terpenoid, steroid, alkaloid and other glycoside components. In vitro study on anticancer activity of methanol extracts of *Parkia* species has been studied on various human cancer cell lines. The results show that methanolic extract of barks of *Parkia filicoidea and P. biglobosa* exhibit varying degrees of antiproliferative activities on prostate cancer (T-549 and BT-20), acute T cell leukemia (PC-3) and colon cancer (SW-480) cells at concentration ranging between 20–200 μg/mL (Fadeyi, Fadeyi, Adejumo, Okoro, & Myles, 2013). Also, the antitumor ability of the extracts of some *Parkia* species such as *Parkia biglandulosa and Parkia speciosa* could be a characteristic of their antiangiogenic activities (Shete, Mundada, & Dhande, 2017). Aqueous extracts of *P. biglobosa* stem also demonstrates sound antihypertensive effect in adrenaline-induced hypertensive female rabbits. The hypotensive potential of *P. biglobosa* could be attributed to its key phytochemical compounds like phenols and flavonoids. These compounds promote vaso-relaxation by hampering angiotensin converting enzyme (ACE), and by regulating nitric oxide availability and decreasing oxidative stress that ultimately leads to blood pressure lowering effects (Takagaki & Nanjo, 2015; Yi et al., 2016). antidiarrheal activities of aqueous bark extracts of *P. biglobosa* was investigated in mice. The study revealed that the aqueous stem extract possess antidiarrheal activities which may be linked to their direct inhibitory effect on the propulsive movement of smooth muscles of gastrointestinal (GI) tract and antimicrobial effect on the diarrhea causing pathogenic organisms (Tijani et al., 2009).

Rhubarb is a perennial plant having laxative tendency and is used to treat constipation problems. Study conducted by Fallah Hossini, FakhrZade, Larijani, & Sheikh Samani, 2005; revealed a significant correlation between rhubarb extract use and blood glucose, total cholesterol and LDL declination in diabetic patients (type II). Another investigation was conducted on the efficacy of extracts of rhubarb stem on the HbA1C and blood glucose levels in patients with type II diabetes. The results revealed a significant reduction in HbA1C after rhubarb use. Also, fasting blood sugar level showed a significant decrease after consuming oral capsules of rhubarb stem extracts for 3 months in type II diabetic patients. The blood glucose reducing effect of rhubarb stem extract was attributed to the presence of tannins which stimulates pancreatic beta cells causing a reduction in blood glucose level (Shad & Haghighi, 2018).

12.3.4 Seed extract

Seed extracts have proven numerous health benefits. There are several studies that have shown the protective effect of seed extracts on human health. Extracts from seeds such as grape seed, nigella seeds, pumpkin seeds, fenugreek seeds, sunflower seeds, chia seeds, etc. have been studied extensively for their phytochemical properties. The phytochemicals present in seed extracts have antioxidant, antidiabetic, anticancerous effects. These beneficial effects of seed extract have led to increased use of these as food substitutes. The amount of phytochemicals present in seeds varies with the kind of seed and other conditions under which the seed has developed.

Grape seeds contain proanthocyanidins, which are polyhydroxyflavan polymers. The conjugated and colonic metabolites of proanthocyanidins present in grape seed promote beneficial health properties. Numerous in vitro and in vivo studies have shown that the proanthocyanidin have pharmacological properties. The effects include antioxidant, antineurodegenerative, antiobesity, anticancer, antidiabetic, antiosteoarthritis, and cardio-protective capabilities. In a study, effect of procyanidin on Wister female rats was carried out. In this study the rats were treated for 30 days with 25 mg/kg grape seed extracts. The results showed an enhanced homeostatic model valuation insulin resistance index attended by primers down regulation: The 4(Glut4) Glucose transporter, 1(Irs1) Insulin receptor, and (PPARg2) peroxisome proliferator-activated receptor gamma isoform-2 in mesenteric white adipose tissue. Therefore this study suggested that seed procyanidin of grape have positive effect homeostasis glucose (Montagut et al., 2010). The procyanidin antiinflammatory effects of grape seeds were also studied by Terra et al. (2009).

Cumin, a widely used spice mainly used for its unique flavor in several cultural cuisines, contains major compounds that possess antioxidant and antispoilage capabilities. Cumin seed extracts are used as natural antioxidants in foods. The antiradical and antioxidant activities of cumin seed extract decreases the occurrence of several health issues (Al-Juhaimi & Ghafoor, 2013).

Nigella sativa (black seed), an annual herb, has many pharmacological assets due to its active compounds. Most health-promoting effects black seeds are attributed to thymoquinone components present in it. A study on *Nigella sativa* seed extract for its antihypertensive effects was carried out by Dehkordi and Kamkhah (2008). The results showed that consumption of *Nigella* extracts for 2 weeks at 200 or 400 mg/day decreased systolic as well as diastolic blood pressure in patients.

Avocado (*Persea americana*) seed, which is an agro-industrial residue, contains ample amount of extractable polyphenolic compounds which confer various food and health benefits due to their antioxidant capacity. Methanolic extracts of avocado seeds exhibited anticancer and antiinflammatory activities against colon and liver cancer cell line in a dose-dependent manner (Alkhalaf, Alansari, Ibrahim, & ELhalwagy, 2019). Antioxidant activity of avocado seed extracts were assessed by its DPPH radical scavenging action with highest activity observed in methanolic seed extracts than aqueous extracts (Bahru, Tadele, & Ajebe, 2019). antidiabetic effects of aqueous seed extracts of avocado were also confirmed on alloxan-induced diabetic mice models (Alhassan et al., 2012). In vitro research findings have also determined significant reduction in heart rate and blood pressure in rat models by the treatment of aqueous avocado seed extract (Anaka, Ozolua, & Okpo, 2009).

The papaya seeds contain benzyl isothiocyanate, which is sulfur-containing compound having various positive effects. These substances play an important role in plant defense systems also (El Moussaoui et al., 2008). Papaya seed extracts have therapeutic such as carminative, antifertility effects in males. These isothiocyanate compounds present in papaya seeds have been seen to prevent various cancers like breast, lung, colon pancreas and prostate in humans. Isothiocyanate inhibits and prevents formation and development of cancer cells through numerous mechanisms (Barbra & Minton, 2008). Also, methanol seed extracts of papaya was confirmed for its antiinflammatory and antinociceptic action in a dose-dependent manner on rat models (Anaga & Onehi, 2010).

The seeds of *Moringa oleifera* have antioxidant property and decrease the oxidative damage. These seeds possess hepatoprotective, antiinflammatory and antifibrotic activities against carbon tetrachloride induced damage and fibrosis of liver (Hamza, 2010). Rats were intoxicated with CCl_4 and simultaneously were given these seed extracts 1 g/kg body weight. After 8 weeks of this treatment the results showed lower AST and ALT serum levels and elevated albumin levels, demonstrating better liver synthesis compared to control rats. Lower levels of globulin, diminishes myeloperoxidase action and also lower hepatic inflammatory cell infiltration and hence reduces inflammation.

12.3.5 Flower extracts

Flower extracts possess many health benefits due to the presence of various phytochemicals like polysaccharides, phenolics, essential oils, alkaloids, tannins, saponins, sterols, carotenoids, and prebiotics (inulin) (Zheng et al., 2021; Takahashi, Rezende, Moura, Dominguete, & Sande, 2020). Saffron (*Crocus sativus* L.) flower extracts are abundant source of biologically active compounds such as safranal, picrocrocin, and crocin, which have been explored for several health benefits and pharmacological properties. Flower extracts of saffron has been studied for *antitumor* and anticancer properties. Crocin is suggested to be the major antitumor ingredient in saffron. Also, safranal can act against HeLa cell line growth, proliferation of MCF-7 cell lines and also suppresses some toxic biochemical markers. *antitumoral potential* of crocetin is due to its capability to inhibit nucleic acid synthesis, blocking growth factor signaling pathways, causing apoptosis and improving antioxidative system (Gutheil, Reed, Ray, Anant, & Dhar, 2012). Saffron also reported *antidiabetic* response due to its active constituents (crocin, safranal and crocetin) which exhibit insulin sensitizing effects without interfering serum glutamic-pyruvic transaminase and creatinine levels. Also, saffron stimulates 5′-AMP-triggered protein kinase which encourages glucose uptake in skeletal muscles which in turn improves insulin sensitivity, thus preventing excess glucose accumulation in blood serum (Razak, Anwar Hamzah, Yee, Kadir, & Nayan, 2017). Methanolic and water−methanol extracts of saffron stigma has been evaluated for their antioxidant, antiinflammatory and cardio-protective properties by Poma, Fontecchio, Carlucci, and Chichiricco (2012). Reduction in serum triglycerides, LDLs, very LDLs and total cholesterol levels, blocking apoptosis signaling pathways and restraining myocardial cell mortality by saffron extracts could be possible events in reduction of cardiovascular diseases. In addition, saffron extracts has been reported to improve neurodegenerative diseases like Alzheimer's disease by decreasing acetylcholinesterase activity defecating the accumulation of amyloid β and improving cerebral antioxidant markers in human brain (Razak et al., 2017).

Aqueous and ethanolic extracts obtained from *Chrysanthemum morifolium* display antioxidant activities, which are characteristics of its phytochemical complexes such as phenolic acids, flavonoids and terpenes. The potential antioxidant property of *C. morifolium* is due to its ability to boost the antioxidant enzyme activity like superoxide dismutase, catalase and glutathione peroxidase and scavenging free radicals (ROS) and peroxidation of melonaldehyde (Li et al., 2019; Yang, Yang, Feng, Jiang, & Zhang, 2019). A polysaccharide (JHBOS2) obtained from aqueous extracts of *C. morifolium* flowers exhibited antiangiogenic activity at (150 μg/mL) by the possible inhibition of tube formation in HMEC-1 cells (human mammary epithelial cells) (Zheng, Dong, Du). Researchers also showed antiinflammatory effects of flower extracts (hot water and methanol) of *C. morifolium* against lipopolysaccharide-initiated human leukemia monocytic THP-1 cells at low levels of 1, 10 and 100 (μg/mL). Interestingly, it was noticed that hot aqueous fraction of *C. morifolium* restrain lipopolysaccharide stimulated emergence of pro-inflammatory mediators (Interlukin-6, Interlukin-1β and Cyclooxygenase 2). Significantly higher concentrations of phenolic acids and flavonoids in flower extracts of *C. morifolium* are the basis for its antiinflammatory potential (Zheng, Lu, & Xu, 2021). In a study, strong neuroprotective activity of *C. morifolium* was discussed in different cell and animal models. Phenolic glycoside [2,6-Dimethoxy-4-hydroxymethyl-phenol 1-D-(6-O-Caffeoyl)-β-D glucopyranoside] and ligans isolated from flower extracts of *C. morifolium* were tested to estimate antineurotoxicity in H_2O_2-provoked SH-SY5y cells. The results showed that these compounds significantly improved cell viability at a concentration of 10 μM after treatment with H_2O_2. *C. morifolium* flower extracts also pose antidiabetic activity against obese diabetic KK-AY mice in a dose-dependent sequence. Also, the extract was able to improve insulin resistance as a result of elevation in adiponectin levels after its administration (Yamamoto et al.,

2015). Moreover, flavonoids obtained from the methanolic flower extracts of *C. morifolium* showed *antiosteoporotic* properties due to the inhibition of osteoclast growth by repressing the manifestation of tartrate-resistant acid phosphate, an enzyme responsible for the production of ROS which damage bone structure (Zheng et al., 2021). *M. oleifera* flower is an abundant store house of bioactive compounds like ethyl oleate, quinic acid and cis-9-hexadecenal which displayed promising antioxidant, anticancer and antiinflammatory properties. In vitro antioxidant potential of *M. oleifera* ethanolic flower extract was tested by DPPH free radical scavenging activity and was related with standard ascorbic acid. Results revealed a satisfactory free radical scavenging activity of flower extract. Thus consumption of *M. oleifera* flower extracts can be advantageous in preventing oxidative stress and related complications (Alhakmani, Kumar, & Khan, 2013). In another study, a detailed examination of *M. oleifera* flower extracts (ethanolic) on its antiinflammatory potential was carried out which determines successful inhibition of NO and pro-inflammatory interluekins like IL-6, IL-1β, TNF-α, and PGE 2. Simultaneously, these flower extracts aid in the formation of antiinflammatory IL-10 and IƘB-α expressions which improves inflammatory damage via NF-κB pathway in microphages. Hepatoprotective property of *M. oleifera* flower extract tested in mice models is basically the extension of its antiinflammatory and antioxidative capability (Kalappurayil & Joseph, 2017).

Hibiscus (*Hibiscus rosa sinensis)* flower extracts possess antioxidant, antiinflammatory, antidiabetic, anticancer, and antifertility properties. Besides these, hibiscus flower extracts also exhibit cardio-protective, neuroprotective, gastro-protective, and hepatoprotective characteristics (Missoum, 2018). Antioxidant potential of *H. rosa* flower extracts was determined by using DPPH essay with BHT used as a control. This study compared DPPH radical scavenging activity of methanolic and ethanolic flower extracts with BHT. The results revealed higher radical scavenging action of methanolic extracts than ethanolic extracts due to its higher concentration of flavonoid and phenolic substances. Anticancer activity of hibiscus flower (acetone) extracts was executed on viability of HeLa cell lines and the results show only 12.96% cell viability (Durga, Kumar, Hameed, Dheeba, & Saravanan, 2018). Another study was conducted on antidiabetic property of *H. rosa* flower extracts on pregnant Wister rats and albino rabbits. Ethanolic flower extracts was observed to exhibit best antidiabetic activity against alloxan initiated diabetes within female pregnant rats and reduced plasma glucose levels in albino rabbits in a progressive manner (Pethe, Yelwatkar, Gujar, Varma, & Manchalwar, 2017). Effect of aqueous flower extracts on gastro-protective activity against aspirin, pylorus ligation and ethanol-induced ulceritis on albino Wister rats was studied. The results demonstrated best gastroprotectivity of hibiscus flower extracts against these models owing to the radical scavenging activity of flower tannins and flavonoids (Kumar et al., 2014).

12.3.6 Roots and tuber extracts

Roots and tuber crops have been extensively studied for their possible health benefits and as a source of functional constituents. Since ancient times, roots and tuber crops have been a part of different foods and are being used in the modern diet to add variety in addition to providing numerous desired nutritional and health benefits including, antidiabetic, antiobesity, antioxidative, and immune-modulatory activities. The health benefits of roots and tuber crops have been reported to be due to the presence of bioactive compounds including phenolic compounds, bioactive proteins, saponins, phytic acids, glycoalkaloids, carotenoids, ascorbic acid and hydroxycoumarins

(Chandrasekara, 2018). In a study the extracts from tuber *Solanum jamesii* were found to have antiproliferative effect on intestinal cancer cells. The authors reported that the compounds present in the extracts have significantly high biological activity when interacting synergistically among themselves which was the basis for their antiproliferative activity (Chandrasekara, 2016). A comparative analysis was carried to determine the antiproliferative activity of anthocyanins in root tubers (var. Bhu Krishna) and leaves against colon, cervical and breast cancer cells. This was related to the ability of anthocyanins to induce apoptosis in the cancer cells and thus producing significant antitumor effect against these cells (Vishnu et al., 2019). The methanolic extracts of the peel and peel bandage of the sweet potatoes has been evaluated for their wound healing properties by excision and incision wound models on Wistar rats. The authors documented that hydroxyproline content increased significantly in the test group in comparison to that of wounded control group. The increase in the content of hydroxyproline results in enhanced collagen synthesis which ultimately improves wound healing. Furthermore, the malondialdehyde content decreased in test groups when compared to wounded control group which depicts the role of sweet potato peels in inhibiting lipid oxidation (Panda, Sonkamble, Sanjeev Panda, & Kundnani, 2012).

Yacon tubers are a rich source of phenolic compounds and fructo-oligosaccharides and have the potential ability to treat diabetes, kidney problems, and reduce the risk of colon cancer (Cocato et al., 2019). In addition, the yacon tuber extracts increase the population of health-promoting bacteria and decrease the numbers of pathogenic bacteria in the gut and thus show significantly high prebiotic effect and gut modulating properties. The prebiotic effect of yacon extract has been studied using guinea pig model and found that yacon extract increases beneficial bacteria including lactobacillus and promotes the formation of short chain fatty acids (Cocato et al., 2019). In a similar study, the inulin extracted from Jerasalem artichoke was documented to have significantly high prebiotic effect than a commercial prebiotic with inulin and was found to be associated with the reduction in colorectal cancer and intestinal pH (Barszcz, Taciak, & Skomiał, 2016). The antioxidant properties of ethanolic and aqueous extracts of yam peel on tert-butyl hydroperoxide (t-BHP) induced oxidative stress in mice liver cells have been studied. The ethanolic extracts of yam peel showed a pronounced protective effect on t-BHP treated cells as compared to aqueous extracts. Moreover, the catalase enzyme activity was enhanced by ethanolic extracts while aqueous extracts reduced it. The phytochemicals present in yam improve the activities of endogenous antioxidant enzymes. The levels of γ-glutamyl transpeptidase (GGT), triacylglycerol, and LDL were decreased in serum of rats in which carbon tetrachloride was used for inducing hepatic fibrosis (Chandrasekara, 2016; Hsu, Yeh, & Wei, 2011). Chinese yam flour extract was evaluated for its effects on the GI tract of rats and was found to enhance digestive capacity and convert some intestinal flora to helpful bacteria (Jeong et al., 2006).

12.4 Conclusion

Plant extracts have been used traditionally as medicine, flavor, tonic, and preservatives in foods from centuries. These extracts contribute to the biological activity through its phytochemical nature and thus people prefer their consumption than medicines. The plant extracts are richest sources of essential components having antimicrobial, antioxidant, antiproliferative, and laxative effects. Additionally, extracts rich in potential bioactive components might bestow some extra benefits when incorporated to the food products, contributing improvements of the overall strength and well-being. Globalization has greatly paved the way to renew the interest in functional foods, and nowadays majority of people prefer to take healthy diet enriched with different plant extracts as a nutraceutical remedy. This flexibility in plant remedies is essentially due to the fact that these plant extracts are highly effective in curing diseases due to their diverse functional properties and low cost. Although numerous studies have proved the health-promoting potential of these leaves, fruit, stem and root-derived extracts, additional human studies are still compulsory to ascertain their true efficacy.

References

Ahmed, F., & Urooj, A. (2010). Traditional uses, medicinal properties, and phytopharmacology of *Ficus racemosa*: A review. *Pharmaceutical Biology*, *48*(6), 672–681. Available from https://doi.org/10.3109/13880200903241861.

Al-Awwadi, N. A. J. (2017). Potential health benefits and scientific review of ginger. *Journal of Pharmacognosy and Phytotherapy*, *9*(7), 111–116.

Alhakmani, F., Kumar, S., & Khan, S. A. (2013). Estimation of total phenolic content, in–vitro antioxidant and anti–inflammatory activity of flowers of Moringa oleifera. *Asian Pacific Journal of Tropical Biomedicine*, *3*(8), 623–627.

Alhassan, A. J., Sule, M. S., Atiku, M. K., Wudil, A. M., Abubakar, H., & Mohammed, S. A. (2012). Effects of aqueous avocado pear (Persea americana) seed extract on alloxan induced diabetes rats. *Greener Journal of Medical Sciences*, *2*(1), 005–011.

Al-Juhaimi, F., & Ghafoor, K. (2013). Extraction optimization and in vitro antioxidant properties of phenolic compounds from cumin (*Cuminum cyminum* L.) seed. *International Food Research Journal*.

Alkhalaf, M. I., Alansari, W. S., Ibrahim, E. A., & ELhalwagy, M. E. (2019). Anti-oxidant, anti-inflammatory and anti-cancer activities of avocado (Persea americana) fruit and seed extract. *Journal of King Saud University-Science*, *31*(4), 1358–1362.

Anaga, A. O., & Onehi, E. V. (2010). Antinociceptive and anti-inflammatory effects of the methanol seed extract of Carica papaya in mice and rats. *African Journal of Pharmacy and Pharmacology*, *4*(4), 140–144.

Anaka, O. N., Ozolua, R. I., & Okpo, S. O. (2009). Effect of the aqueous seed extract of Persea americana Mill (Lauraceae) on the blood pressure of Sprague-Dawley rats. *African Journal of Pharmacy and Pharmacology*, *3*(10), 485–490.

Aryaeian, N., Sedehi, S. K., Khorshidi, M., Zarezadeh, M., Hosseini, A., & Shahram, F. (2020). Effects of hydroalcoholic extract of Berberis Integerrima on the anthropometric indices and metabolic profile in active rheumatoid arthritis patients. *Complementary Therapies in Medicine*, *50*, 102331.

Bagchi, D., Moriyama, H., & Swaroop, A. (Eds.), (2016). *Green coffee bean extract in human health*. CRC Press.

Bahru, T. B., Tadele, Z. H., & Ajebe, E. G. (2019). A review on avocado seed: Functionality, composition, antioxidant and antimicrobial properties. *Chemical Science International Journal*, *27*(2), 1–10.

Barbra Minton, L. (2008). Papaya is tasty way to fight cancer and poor digestion. *International Journal of Oncology*, *2008*.

Barszcz, M., Taciak, M., & Skomiał, J. (2016). The effects of inulin, dried Jerusalem artichoke tuber and a multispecies probiotic preparation on microbiota ecology and immune status of the large intestine in young pigs. *Archives of Animal Nutrition*, *70*, 278–292. Available from https://doi.org/10.1080/1745039X.2016.1184368.

Bayrami, A., Shirdel, A., Pouran, S. R., Mahmoudi, F., Habibi-Yangjeh, A., Singh, R., & Raman, A. A. A. (2020). Co-regulative effects of chitosan-fennel seed extract system on the hormonal and biochemical factors involved in the polycystic ovarian syndrome. *Materials Science and Engineering: C*, *117*, 111351.

Bento-Silva, A., Koistinen, V. M., Mena, P., Bronze, M. R., Hanhineva, K., Sahlstrøm, S., ... Aura, A. M. (2020). Factors affecting intake, metabolism and health benefits of phenolic acids: Do we understand individual variability? *European Journal of Nutrition*, *59*(4), 1275–1293.

Bhat, R. R., Rehman, M. U., Shabir, A., Mir, M. U. R., Ahmad, A., Khan, R., ... Ganaie, M. A. (2019). Chemical Composition and Biological Uses of *Artemisia absinthium* (Wormwood). *Plant and Human Health*, *3*. Available from https://doi.org/10.1007/978-3-030-04408-4_3.

Bhattacharjee, B., Singha, A. K., Banik, R., Dinda, B., & Maiti, D. (2012). Medicinal properties of stem bark extract of *Artocarpus chaplasa* (Moraceae). *Journal of Applied Bioscience*, *38*(2), 192–196.

Bursal, E., & Gülçin, İ. (2011). Polyphenol contents and in vitro antioxidant activities of lyophilised aqueous extract of kiwifruit (Actinidia deliciosa). *Food Research International*, *44*(5), 1482–1489.

Chander, V., Aswal, J. S., Dobhal, R., & Uniyal, D. P. (2017). A review on Pharmacological potential of Berberine; an active component of Himalayan Berberis aristata. *Journal of Phytopharmacology*, *6*(1), 53–58.

Chandrasekara, A. (2016). *Roots and tuber crops as functional foods. Reference Series in Phytochemistry*, 1–29. Available from https://doi.org/10.1007/978-3-319-54528-8_37-1.

Chandrasekara, A. (2018). *Roots and tubers as functional foods* (pp. 1–29). Cham: Springer. Available from https://doi.org/10.1007/978-3-319-54528-8_37-1.

Channabasavaraj, K. P., Badami, S., & Bhojraj, S. (2008). Hepatoprotective and antioxidant activity of methanol extract of *Ficus glomerata*. *Journal of Natural Medicine*, *62*, 379–383.

Cheynier, V. (2012). Phenolic compounds: From plants to foods. *Phytochemistry Reviews*, *11*(2), 153–177.

Cocato, M. L., Lobo, A. R., Azevedo-Martins, A. K., Filho, J. M., de Sá, L. R. M., & Colli, C. (2019). Effects of a moderate iron overload and its interaction with yacon flour, and/or phytate, in the diet on liver antioxidant enzymes and hepatocyte apoptosis in rats. *Food Chemistry*, *285*, 171–179. Available from https://doi.org/10.1016/j.foodchem.2019.01.142.

Cooper, R., Morré, D. J., & Morré, D. M. (2005). Medicinal benefits of green tea: Part I. Review of noncancer health benefits. *The Journal of Alternative and Complementary Medicine*, *11*(3), 521–528. Available from https://doi.org/10.1089/acm.2005.11.521.

D'Eliseo, D., Pannucci, E., Bernini, R., Campo, M., Romani, A., Santi, L., & Velotti, F. (2019). In vitro studies on anti-inflammatory activities of kiwifruit peel extract in human THP-1 monocytes. *Journal of Ethnopharmacology*, *233*, 41–46.

Da Silva, J. K., Cazarin, C. B. B., Colomeu, T. C., Batista, Â. G., Meletti, L. M. M., Paschoal, J. A. R., ... de Lima Zollner, R. (2013). Antioxidant activity of aqueous extract of passion fruit (*Passiflora edulis*) leaves: In vitro and in vivo study. *Food Research International*, *53*(2), 882–890. Available from https://doi.org/10.1016/j.foodres.2012.12.043.

Dabe, N. E., & Kefale, A. T. (2017). Antidiabetic effects of artemisia species. A systematic review. *Ancient Science of Life*, *36*(Issue 4).

Dahham, S. S., Agha, M. T., Tabana, Y. M., & Majid, A. M. S. A. (2015). Antioxidant activities and anticancer screening of extracts from banana fruit (*Musa sapientum*). *Academic Journal of Cancer Research*, *8*, 28–34.

Daily, J. W., Yang, M., & Park, S. (2016). Efficacy of turmeric extracts and curcumin for alleviating the symptoms of joint arthritis: A systematic review and *meta*-analysis of randomized clinical trials. *Journal of Medicinal Food*, *19*(8), 717–729.

Dar, N. J., Hamid, A., & Ahmad, M. (2015). Pharmacologic overview of Withania somnifera, the Indian Ginseng. *Cellular and Molecular Life Sciences*, *72*(23).

Das, S., Mishra, B., Gill, K., Ashraf, M. S., Singh, A. K., Sinha, M., ... Singh, T. P. (2011). Isolation and characterization of novel protein with anti-fungal and anti-inflammatory properties from *Aloe Vera* leaf gel. *International Journal of Biological Macromolecules*, *48*, 38–43.

de Pascual-Teresa, S., & Sanchez-Ballesta, M. T. (2008). Anthocyanins: from plant to health. *Phytochemistry Reviews*, *7*(2), 281–299.

Dehkordi, F. R., & Kamkhah, A. F. (2008). Antihypertensive effect of Nigella sativa seed extract in patients with mild hypertension. *Fundamental & Clinical Pharmacology*, *22*(4), 447.

Dekebo, A. (2019). *Introductory chapter: Plant extracts*. Plant Extracts. <https://doi.org/10.5772/intechopen.85493>.

Deotale, S. M., Dutta, S., Moses, J. A., & Anandharamakrishnan, C. (2019). Coffee oil as a natural surfactant. *Food Chemistry*, *295*, 180–188.

Dhasthakeer, A. B. G., Kavitha, S., & Vishnupriya, V. (2020). Evaluation of in vitro antidiabetic potential of lavender oil. *Drug Invention Today*, *14*(7).

Dhawefi, N., Jedidi, S., Rtibi, K., Jridi, M., Sammeri, H., Abidi, C., ... Sebai, H. (2021). Antidiarrheal, antimicrobial, and antioxidant properties of the aqueous extract of Tunisian persimmon (*Diospyros kaki* Thunb.) fruits. *Journal of Medicinal Food*.

Divakaran, D., Lakkakula, J. R., Thakur, M., Kumawat, M. K., & Srivastava, R. (2019). Dragon fruit extract capped gold nanoparticles: Synthesis and their differential cytotoxicity effect on breast cancer cells. *Materials Letters*, *236*, 498–502.

Doostan, F., Vafafar, R., Zakeri-Milani, P., Pouri, A., Afshar, R. A., & Abbasi, M. M. (2017). Effects of pomegranate (*Punica granatum* L.) seed and peel methanolic extracts on oxidative stress and lipid profile changes induced by methotrexate in rats. *Advanced Pharmaceutical Bulletin*, *7*(2), 269.

Duansak, D., Somboonwong, J., & Patumraj, S. (2003). Effects of Aloe Vera on leukocyte adhesion and TNF-α and IL-6 levels in burn wounded rats. *Clinical Hemorheology and Microcirculation*, *29*, 239–246.

Durga, R., Kumar, P. S., Hameed, S. A. S., Dheeba, B., & Saravanan, R. (2018). Evaluation of in-vitro anticancer activity of Hibiscus rosa sinensis against hela cell line. *Journal of Global Pharma Technology*, *10* (1), 1–10.

Eberhardt, M., Lee, C., & Liu, R. H. (2000). Antioxidant activity of fresh apples. *Nature*, *405*, 903–904.

Eddouks, M., Maghrani, M., Lemhadri, A., Ouahidi, M. L., & Jouad, H. (2002). Ethnopharmacological survey of medicinal plants used for the treatment of diabetes mellitus, hypertension and cardiac diseases in the south-east region of Morocco (Tafilalet). *Journal of Ethnopharmacology*, *82*, 97–103.

El Moussaoui, A., Nijs, M., Paul, C., Wintjens, R., Vincentelli, J., Azarkan, M., et al. (2008). Revisiting the enzymes stored in the laticifers of Carica papaya in the context of their possible participation in the plant defence mechanism. *Cellular and Molecular Life Sciences*, *58*, 556–570.

Fadeyi, S. A., Fadeyi, O. O., Adejumo, A. A., Okoro, C., & Myles, E. L. (2013). In vitro anticancer screening of 24 locally used Nigerian medicinal plants. *BMC Complementary Medicine and Therapies*, *13*, 79.

Fallah Hossini, H., FakhrZade, H., Larijani, B., & Sheikh Samani, A. (2005). A review of medicinal plants used in diabetics. *Diabetes Special Issue*, *5*, 1–8.

Farhadi Chitgar, M., Aalami, M., Maghsoudlou, Y., & Milani, E. (2017). Comparative study on the effect of heat treatment and sonication on the quality of barberry (*Berberis Vulgaris*) juice. *Journal of Food Processing and Preservation*, *41*, 12956.

Fasano, E., Serini, S., Mondella, N., Trombino, S., Celleno, L., Lanza, P., ... Calviello, G. (2014). Antioxidant and anti-inflammatory effects of selected two human immortalized keratinocyte lines. *BioMed Research International*, *2014*, 327452.

Franzen, F. L., & Bolini, H. M. A. (2019). The medicinal and nutritional importance of plant extracts and the consumption of healthy foods—A review. *Acta Scientific nutritional Health Care*, *7*, 131–136.

Goud, B. J., & Swamy, B. C. (2015). A review on history, controversy, traditional use, ethnobotany, phytochemistry and pharmacology of *Artemisia absinthium* Linn. *International Journal of Advanced Research in Engineering and Applied Sciences*, *4*(5), 77–107.

Grauso, L., Emrick, S., de Falco, B., Lanzotti, V., & Bonanomi, G. (2019). Common dandelion: A review of its botanical, phytochemical and pharmacological profiles. *Phytochemistry*.

Gulcin, I., Kaya, R., Goren, A. C., Akincioglu, H., Topal, M., Bingol, Z., ... Alwasel, S. (2019). Anticholinergic, antidiabetic and antioxidant activities of cinnamon (Cinnamomum verum) bark extracts: Polyphenol contents analysis by LC-MS/MS. *International Journal of Food Properties*, *22*(1), 1511–1526.

Guler, M. C., Tanyeli, A., Eraslan, E., Bozhuyuk, M. R., Akdemir, F. N. E., Toktay, E., ... Ozkan, G. (2021). Persimmon (diospyros kaki) alleviates ethanol-induced gastric ulcer in rats/persimmon (Diospyros kaki L.) sicanlarda etanol ile induklenen mide ulserini hafifletir. *Southern Clinics of Istanbul Eurasia (SCIE)*, *32* (1), 1–8.

Guo, Q, Wang, N, & Liu, H (2020). The bioactive compounds and biological functions of Asparagus officinalis L. – A review. *J Funct Foods*, *65*, 103727. Available from https://doi.org/10.1016/J.JFF.2019.103727.

Guo, Q., Wang, N., Liu, H., Li, Z., Lu, L., & Wang, C. (2019). The bioactive compounds and biological functions of Asparagus officinalis L.—A review. *Journal of Functional Foods*, 103727. Available from https://doi.org/10.1016/j.jff.2019.103727.

Gutheil, W., Reed, G., Ray, A., Anant, S., & Dhar, A. (2012). Crocetin: An agent derived from saffron for prevention and therapy for cancer. *Current Pharmaceutical Biotechnology*, *13*(1), 173–179.

Hadipour, E., Taleghani, A., Tayarani-Najaran, N., & Tayarani-Najaran, Z. (2020). Biological effects of red beetroot and betalains: A review. *Phytotherapy Research*. Available from https://doi.org/10.1002/ptr.6653.

Hameed, I. H., Cotos, M. R. C., & Hadi, M. Y. (2017). Antimicrobial, antioxidant, hemolytic, anti-anxiety, and antihypertensive activity of passiflora species. *Research Journal of Pharmacy and Technology*, *10*(11), 4079–4084.

Hamza, A. A. (2010). Ameliorative effects of *Moringa oleifera* Lam seed extract on liver fibrosis in rats. *Food and Chemical Toxicology*, *48*, 345–355. Available from https://doi.org/10.1016/j.fct.2009.10.022.

Hsu, C. K., Yeh, J. Y., & Wei, J. H. (2011). Protective effects of the crude extracts from yam (Dioscorea alata) peel on tert-butylhydroperoxide-induced oxidative stress in mouse liver cells. *Food Chemistry*, *126*, 429–434. Available from https://doi.org/10.1016/j.foodchem.2010.11.004.

Hsu, L. S., Ho, H. H., Lin, M. C., Chyau, C. C., Peng, J. S., & Wang, C. J. (2012). Mulberry water extracts (MWEs) ameliorated carbon tetrachloride-induced liver damages in rat. *Food and Chemical Toxicology*, *50*(9), 3086–3093.

Hussein, J., Abo-elmatty, D., El-Khayat, Z., Abdel Latif, Y., Saleh, S., Farrag, A. R., & Abd-El-Ghany, W. (2015). Kiwifruit extract attenuates DNA damage and vitamins reduction in indomethacin-induced experimental gastric ulcer. *Jokull J*, *65*, 2–16.

Imenshahidi, M., & Hosseinzadeh, H. (2016). Berberis vulgaris and berberine: An update review. *Phytotherapy Research*, *30*, 1745–1764.

Jaramillo, S., Muriana, F. J. G., Guillen, R., Jimenez-Araujo, A., Rodriguez-Arcos, R., & Lopez, S. (2016). Saponins from edible spears of wild asparagus inhibit AKT, p70S6K, and ERK signalling, and induce apoptosis through G0/G1 cell cycle arrest in human colon cancer HCT-116 cells. *Journal of Functional Foods*, *26*, 1–10.

Jeong, R. J., Ji, S. L., Chu, H. L., Jong, Y. K., Soon, D. K., & Doo, H. N. (2006). Effect of ethanol extract of dried Chinese yam (Dioscorea batatas) flour containing dioscin on gastrointestinal function in rat model. *Archives of Pharmacal Research*, *29*, 348–353. Available from https://doi.org/10.1007/BF02968583.

Jung, K. A., Song, T. C., Han, D., Kim, I. H., Kim, Y. E., & Lee, C. H. (2005). Cardiovascular protective properties of kiwifruit extracts in vitro. *Biological and Pharmaceutical Bulletin*, *28*(9), 1782–1785.

Kalappurayil, T. M., & Joseph, B. P. (2017). A review of pharmacognostical studies on Moringa oleifera lam. flowers. *Pharmacognosy Journal*, *9*(1).

Kapadia, G., Azuine, M. A., Subba Rao, G., Arai, T., Iida, A., & Tokuda, H. (2011). Cytotoxic effect of the red beetroot (Beta vulgaris L.) extract compared to doxorubicin (Adriamycin) in the human prostate (PC-3) and breast (MCF-7) cancer cell lines. *Anti-Cancer Agents in Medicinal Chemistry (Formerly Current Medicinal Chemistry-Anti-Cancer Agents)*, *11*(3), 280–284.

Kiokias, S., Proestos, C., & Oreopoulou, V. (2018). Effect of natural food antioxidants against LDL and DNA oxidative changes. *Antioxidants*, *7*, 133.

Konda, P. Y., Egi, J. Y., Dasari, S., Katepogu, R., Jaiswal, K. K., & Nagarajan, P. (2020). Ameliorative effects of Mentha aquatica on diabetic and nephroprotective potential activities in STZ-induced renal injury. *Comparative Clinical Pathology*, *29*(1), 189–199.

Koul, B., Taak, P., Kumar, A., Khatri, T. A., & Sanyal, I. (2017). The artemisia genus: A review on traditional uses, phytochemical constituents, pharmacological properties and germplasm conservation. *Journal of Glycomics & Lipidomics*, *7*, 1. Available from https://doi.org/10.4172/2153-0637.1000142.

Kowalska, K., Olejnik, A., Zielińska-Wasielica, J., & Olkowicz, M. (2019). Raspberry (Rubus idaeus L.) fruit extract decreases oxidation markers, improves lipid metabolism and reduces adipose tissue inflammation in hypertrophied 3T3-L1 adipocytes. *Journal of Functional Foods*, *62*, 103568.

Kumar, P. K., Annapurna, A., Ramya, G., Sheba, D., Krishna, G., & Sudeepthi, L. (2014). Gastroprotective effect of flower extracts of Hibiscus rosa sinensis against acute gastric lesion models in rodents. *Journal of Pharmacognosy and Phytochemistry*, *3*(3), 137–145.

Lattanzio, V. (2013). Phenolic compounds: Introduction 50. *Natural Products*, 1543–1580.

Lee, S. G., Lee, Y. J., Jang, M. H., Kwon, T. R., & Nam, J. O. (2017). Panax ginseng leaf extracts exert anti-obesity effects in high-fat diet-induced obese rats. *Nutrients*, *9*(9), 999.

Li, Y., Yang, Z., & Jia, S. (2016). Yuan K. Protective effect and mechanism of action of mulberry marc anthocyanins on carbon tetrachloride-induced liver fibrosis in rats. *Journal of Functional Foods*, *24*, 595–601. Available from https://doi.org/10.1016/j.jff.2016.05.001.

Li, R. W., Myers, S. P., Leach, D. N., Lin, G. D., & Leach, G. (2003). A cross-cultural study: Anti-inflammatory activity of Australian and Chinese plants. *Journal of Ethnopharmacology*, *85*, 25–32.

Li, S., Al-Misned, F. A., El-Serehy, H. A., & Yang, L. (2021). Green synthesis of gold nanoparticles using aqueous extract of Mentha Longifolia leaf and investigation of its anti-human breast carcinoma properties in the in vitro condition. *Arabian Journal of Chemistry*, *14*(2), 102931.

Li, Y., Yang, P., Luo, Y., Gao, B., Sun, J., Lu, W., ... Yu, L. L. (2019). Chemical compositions of chrysanthemum teas and their anti-inflammatory and antioxidant properties. *Food Chemistry*, *286*, 8–16.

Li, Y., Zheng, M., Zhai, X., Huang, Y., Khalid, A., Malik, A., ... Hou, X. O. (2015). Effect of *Gymnema sylvestre, Citrullus colocynthis and Artemisia absinthium* on blood glucose and lipid profile a Diabetic Human. *Acta Poloniae Pharmaceutical - Drug Research*, *72*(5), 981–985.

Li, Y.-Y., Feng, J., Zhang, X.-L., & Cui, Y.-Y. (2015). Pine bark extracts: Nutraceutical, pharmacological, and toxicological evaluation. *Journal of Pharmacology and Experimental Therapeutics*, *353*(1), 9–16. Available from https://doi.org/10.1124/jpet.114.220277.

Li, Z. J., Yang, F. J., Yang, L., & Zu, Y. G. (2018). Comparison of the antioxidant effects of carnosic acid and synthetic antioxidants on tara seed oil. *Chemistry Central Journal*, *12*(1), 1–6.

Lins, P. G., Pugine, S. M. P., Scatolini, A. M., & de Melo, M. P. (2018). In vitro antioxidant activity of olive leaf extract (Olea europaea L.) and its protective effect on oxidative damage in human erythrocytes. *Heliyon*, *4*(9), e00805.

Liu, L., Tang, D., Zhao, H., Xin, X., & Aisa, H. A. (2017). Hypoglycemic effect of the polyphenols rich extract from Rose rugosa Thunb on high fat diet and STZ induced diabetic rats. *Journal of Ethnopharmacology*, *200*, 174–181.

Liu, Y., Carver, J. A., Calabrese, A. N., & Pukala, T. L. (2014). Gallic acid interacts with α-synuclein to prevent the structural collapse necessary for its aggregation. *Biochimica et Biophysica Acta (BBA)-Proteins and Proteomics*, *1844*(9), 1481−1485.

Lorenzo, J. M., & Munekata, P. E. S. (2016). Phenolic compounds of green tea: Health benefits and technological application in food. *Asian Pacific Journal of Tropical Biomedicine*, *6*(8), 709−719. Available from https://doi.org/10.1016/j.apjtb.2016.06.010.

Loussouarn, M., Krieger-Liszkay, A., Svilar, L., Bily, A., Birtić, S., & Havaux, M. (2017). Carnosic acid and carnosol, two major antioxidants of rosemary, act through different mechanisms. *Plant Physiology*, *175*(3), 1381−1394.

Macias-Ceja, D. C., Cosín-Roger, J., Ortiz-Masiá, D., Salvador, P., Hernández, C., Calatayud, S., ... Barrachina, M. D. (2016). The flesh ethanolic extract of Hylocereus polyrhizus exerts anti-inflammatory effects and prevents murine colitis. *Clinical Nutrition*, *35*(6), 1333−1339.

Malairajan, P., Gopalakrishnan, G., Narasimhan, S., & Kavimani, S. (2007). Antiulcer activity of *Ficus glomerata. Pharmaceutical Bioogyl*, *45*, 674−677.

Manayi, A., Abdollahi, M., Raman, T., Nabavi, S. F., Habtemariam, S., Daglia, M., & Nabavi, S. M. (2016). Lutein and cataract: From bench to bedside. *Critical Reviews in Biotechnology*, *36*(5), 829−839.

Masaldan, S., & Iyer, V. V. (2014). Exploration of effects of emodin in selected cancer cell lines: Enhanced growth inhibition by ascorbic acid and regulation of LRP1 and AR under hypoxia-like conditions. *Journal of Applied Toxicology*, *34*, 95−104.

Mir, S. A., Shah, M. A., Ganai, S. A., Ahmad, T., & Gani, M. (2019). Understanding the role of active components from plant sources in obesity management. *Journal of Saudi Society of Agricultural Sciences*, *18*, 168−176.

Missoum, A. (2018). An update review on Hibiscus rosa sinensis phytochemistry and medicinal uses. *Journal of Ayurvedic and Herbal Medicine*, *4*, 135−146.

Mohammed, A., Wudil, A. M., Alhassan, A. J., Imam, A. A., Muhammad, I. U., & Idi, A. (2017). Hypoglycemic activity of Curcuma longa Linn root extracts on alloxan induced diabetic rats. *Saudi Journal of Life Science*, *2*, 43−49.

Mojica, L., Meyer, A., Berhow, M. A., & de Mejía, E. G. (2015). Bean cultivars (Phaseolus vulgaris L.) have similar high antioxidant capacity, in vitro inhibition of α-amylase and α-glucosidase while diverse phenolic composition and concentration. *Food Research International*, *69*, 38−48.

Montagut, G., Blade, C., Blay, M., Fernandez-Larrea, J., Pujadas, G., Salvado, M. J., ... Ardevol, A. (2010). Effects of a grapeseed procyanidin extract (GSPE) on insulin resistance. *The Journal of Nutritional Biochemistry*, *21*, 961−967. Available from https://doi.org/10.1016/j.jnutbio.2009.08.001.

Motohashi, N., Shirataki, Y., Kawase, M., Tani, S., Sakagami, H., Satoh, K., ... Molnár, J. (2002). Cancer prevention and therapy with kiwifruit in Chinese folklore medicine: A study of kiwifruit extracts. *Journal of Ethnopharmacology*, *81*(3), 357−364.

Mourtzinos, I., Prodromidis, P., Grigorakis, S., Makris, D. P., Biliaderis, C. G., & Moschakis, T. (2018). Natural food colorants derived from onion wastes: Application in a yoghurt product. *Electrophorosis*, *39* (15), 1975−1983. Available from https://doi.org/10.1002/elps.201800073.

Nayak, B. S., Pereira, L. P., & Maharaj, D. (2007). Wound healing activity of Carica papaya L. in experimentally induced diabetic rats. *Indian Journal of Experimental Biology*, *45*, 739−743.

Nigam, M., Atanassova, M., Mishra, A. P., Pezzani, R., Devkota, H. P., Plygun, S., ... Sharifi-Rad, J. (2019). Bioactive compounds and health benefits of *Artemisia* species. *Natural Product Communications*, *14*(7). Available from https://doi.org/10.1177/1934578x19850354.

Nurzyńska-Wierdak, R., & Zawiślak, G. (2016). Chemical composition and antioxidant activity of lavender (Lavandula angustifolia Mill.) aboveground parts. *Acta Scientiarum Polonorum Hortorum Cultus*, *15*(5), 225−241.

Oreopoulou, A., Papavassilopoulou, E., Bardouki, H., Vamvakias, M., Bimpilas, A., & Oreopoulou, V. (2018). Antioxidant recovery from hydrodistillation residues of selected Lamiaceae species by alkaline extraction. *Journal of Applied Research on Medicinal and Aromatic Plants*, *8*, 83–89.

Panda, V., Sonkamble, M., Sanjeev Panda, V., Kundnani, P.K.M., 2012. *Anti-ulcer activity of Ipomoea batatas tubers (sweet potato)*. Functional Foods in Health and Disease; Volume 2. <ffhdj.com>.

Pandurangan, A. K., Mohebali, N., Norhaizan, M. E., & Looi, C. Y. (2015). Gallic acid attenuates dextran sulfate sodium-induced experimental colitis in BALB/c mice. *Drug Design, Development and Therapy*, *9*, 3923.

Pérez-García, F., Adzet, T., & Cañigueral, S. (2000). Activity of artichoke leaf extract on reactive oxygen species in human leukocytes. *Free Radical Research*, *33*(5), 661–665. Available from https://doi.org/10.1080/10715760000301171.

Pethe, M., Yelwatkar, S., Gujar, V., Varma, S., & Manchalwar, S. (2017). Antidiabetic, hypolipidimic and antioxidant activities of Hibiscus rosa sinensis flower extract in alloxan induced diabetes in rabbits. *International Journal of Biomedical and Advance Research*, *8*(4), 138–143.

Poma, A., Fontecchio, G., Carlucci, G., & Chichiricco, G. (2012). Anti-inflammatory properties of drugs from saffron crocus. *Anti-Inflammatory & Anti-Allergy Agents in Medicinal Chemistry (Formerly Current Medicinal Chemistry-Anti-Inflammatory and Anti-Allergy Agents)*, *11*(1), 37–51.

Poormoosavi, S. M., Najafzadehvarzi, H., Behmanesh, M. A., & Amirgholami, R. (2018). Protective effects of *Asparagus officinalis* extract against Bisphenol A-induced toxicity in Wistar rats. *Toxicology Reports*, *5*, 427–433. Available from https://doi.org/10.1016/j.toxrep.2018.02.010.

Potestà, M., Minutolo, A., Gismondi, A., Canuti, L., Kenzo, M., Roglia, V., ... Montesano, C. (2019). Cytotoxic and apoptotic effects of different extracts of Moringa oleifera Lam on lymphoid and monocytoid cells. *Experimental and therapeutic medicine*, *18*(1), 5–17.

Qabaha, K., Al-Rimawi, F., Qasem, A., & Naser, S. A. (2018). Oleuropein is responsible for the major anti-inflammatory effects of olive leaf extract. *Journal of Medicinal Food*, *21*(3), 302–305.

Radha, M. H., & Laxmipriya, N. P. (2015). Evaluation of biological properties and clinical effectiveness of Aloe Vera: A systematic review. *Journal of Traditional and Complementary Medicine*, *5*(1), 21–26. Available from https://doi.org/10.1016/j.jtcme.2014.10.006.

Raes, K., Doolaege, E. H., Deman, S., Vossen, E., & De Smet, S. (2015). Effect of carnosic acid, quercetin and α-tocopherol on lipid and protein oxidation in an in vitro simulated gastric digestion model. *International Journal of Food Sciences and Nutrition*, *66*(2), 216–221.

Rahimi-Madiseh, M., Lorigoini, Z., Zamani-gharaghoshi, H., & Rafieian-Kopaei, M. (2017). Berberis vulgaris: Specifications and traditional uses. *Iranian Journal of Basic Medical Sciences*, *20*, 569–587. Available from https://doi.org/10.22038/IJBMS.2017.8690.

Rahmani, M., Hamel, L., Toumi-Benali, F., Dif, M. M., Moumen, F., & Rahmani, H. (2018). Determination of antioxidant activity, phenolic quantification of four varieties of fenugreek *Trigonella foenum graecum* L. seed extract cultured in west Algeria. *Journal of Materials and Environmental Science*, *9*(6), 1656–1661.

Razak, S. I. A., Anwar Hamzah, M. S., Yee, F. C., Kadir, M. R. A., & Nayan, N. H. M. (2017). A review on medicinal properties of saffron toward major diseases. *Journal of Herbs, Spices & Medicinal Plants*, *23*(2), 98–116.

Rebecca, O. P. S., Boyce, A. N., & Chandran, S. (2010). Pigment identification and antioxidant properties of red dragon fruit (*Hylocereus polyrhizus*). *African Journal of Biotechnology*, *9*(10), 1450–1454.

Regnier, T., Combrinck, S., & Du Plooy, W. (2012). Essential Oils and Other Plant Extracts as Food Preservatives. *Progress in Food Preservation*, 539–579. Available from https://doi.org/10.1002/9781119962045.ch26.

Saleh, M. S. M., Jalil, J., Zainalabidin, S., Asmadi, A. Y., Mustafa, N. H., & Kamisah, Y. (2021). Genus parkia: Phytochemical, medicinal uses, and pharmacological properties. *International Journal of Molecular Sciences*, *22*, 618. Available from https://doi.org/10.3390/ijms22020618.

Salehi, B., Albayrak, S., Antolak, H., Kręgiel, D., Pawlikowska, E., Sharifi-Rad, M., . . . Sharifi-Rad, J. (2018). Aloe genus plants: From farm to food applications and phytopharmacotherapy. *International Journal of Molecular Sciences*, *19*(9), 2843. Available from https://doi.org/10.3390/ijms19092843.

Salehi, B., Selamoglu, Z., Sener, B., Kilic, M., Kumar Jugran, A., de Tommasi, N., & Cho, W. C. (2019). Berberis plants—Drifting from farm to food applications, phytotherapy, and phytopharmacology. *Foods*, *8* (10), 522. Available from https://doi.org/10.3390/foods8100522.

Salem, M. B., Affes, H., Ksouda, K., Dhouibi, R., Sahnoun, Z., Hammami, S., & Zeghal, K. M. (2015). Pharmacological studies of artichoke leaf extract and their health benefits. *Plant Foods for Human Nutrition*, *70*(4), 441–453. Available from https://doi.org/10.1007/s11130-015-0503-8.

Samadi-Noshahr, Z., Hadjzadeh, M. A. R., Moradi-Marjaneh, R., & Khajavi-Rad, A. (2021). The hepatoprotective effects of fennel seeds extract and trans-Anethole in streptozotocin-induced liver injury in rats. *Food Science & Nutrition*, *9*(2), 1121–1131.

Sarraf, M., Beig-babaei, A., & Naji-Tabasi, S. (2019). Investigating functional properties of barberry species: An overview. *Journal of the Science of Food and Agriculture*. Available from https://doi.org/10.1002/jsfa.9804.

Shad, F. S., & Haghighi, M. J. (2018). Study of the effect of the essential oil (extract) of rhubarb stem (shoot) on glycosylated hemoglobin and fasting blood glucose levels in patients with type II diabetes. *Biomedicine (Taipei)*, *8*(4), 24.

Shao, Y., Chin, C. K., Ho, C. T., Ma, W., Garrison, S. A., & Huang, M. T. (1996). Anti-tumor activity of the crude saponins obtained from asparagus. *Cancer Letters*, *104*, 31–36. Available from https://doi.org/10.1016/0304-3835(96)04233-4.

Shete, S. V., Mundada, S. J., & Dhande, S. (2017). Comparative effect of crude extract of Parkia biglandulosa and Its isolate on regenerative angiogenesis In adult Zebrafish. *Indian Drug*, *54*, 51–57.

Shin, S., Lee, J. A., Son, D., Park, D., & Jung, E. (2017). Anti-skin-aging activity of a standardized extract from Panax ginseng leaves in vitro and in human volunteer. *Cosmetics*, *4*(2), 18.

Siriamornpun, S., & Kaewseejan, N. (2017). Quality, bioactive compounds and antioxidant capacity of selected climacteric fruits with relation to their maturity. *Scientia Horticulturae*, *221*, 33–42.

Siriwardhana, S. S. K. W., & Shahidi, F. (2002). Antiradical activity of extracts of almond and its by-products. *Journal of the American Oil Chemists' Society*, *79*(9), 903–908.

Souza, M. C., Santos, M. P., Sumere, B. R., Silva, L. C., Cunha, D. T., Martinez, J., . . . Rostagno, M. A. (2020). Isolation of gallic acid, caffeine and flavonols from black tea by on-line coupling of pressurized liquid extraction with an adsorbent for the production of functional bakery products. *LWT*, *117*, 108661.

Srivastava, P., Singh, M., Devi, G., & Chaturvedi, R. (2014). Herbal medicine and biotechnology for the benefit of human health. *Animal Biotechnology*, 563–575. Available from https://doi.org/10.1016/b978-0-12-416002-6.00030-4.

Srivastava, S., Srivastava, M., Misra, A., Pandey, G., & Rawat, A. (2015). A review on biological and chemical diversity in Berberis (Berberidaceae). *EXCLI Journal*, *14*, 247–267.

Stagos, D. (2019). Antioxidant activity of polyphenolic plant extracts. *Antioxidants*, *9*(1), 19.

Surjushe, A., Vasani, R., & Saple, D. G. (2008). Aloe Vera: A short review. *Indian Journal of Dermatology*, *53*, 163–166.

Taïwe, G.S., & Kuete, V. (2017). *Passiflora edulis. Medicinal spices and vegetables from Africa* (pp. 513–526). Available from https://doi.org/10.1016/b978-0-12-809286-6.00024-8.

Takagaki, A., & Nanjo, F. (2015). Effects of metabolites produced from (-)-epigallocatechin gallate by rat intestinal bacteria on angiotensin I-converting enzyme activity and blood pressure in spontaneously hypertensive rats. *Journal of Agricultural and Food Chemistry*, *63*, 8262–8266.

Takahashi, J. A., Rezende, F. A. G. G., Moura, M. A. F., Dominguete, L. C. B., & Sande, D. (2020). Edible flowers: Bioactive profile and its potential to be used in food development. *Food Research International*, *129*, 108868.

Tatyasaheb, P., Snehal, P., Anuprita, P., & Shreedevi, P. (2014). Carica papaya leaf extracts—An ethnomedicinal boon. *International Journal of Pharmacognosy and Phytochemical Research*, *6*(2), 260–265.

Terra, X., Montagut, G., Bustos, M., Llopiz, N., Ardevol, A., Blade, C., ... Arola, L. (2009). Grape-seed procyanidins prevent low-grade inflammation by modulating cytokine expression in rats fed a high-fat diet. *The Journal of Nutritional Biochemistry*, *20*, 210–218.

Thitipramote, N., Maisakun, T., Chomchuen, C., Pradmeeteekul, P., Nimkamnerd, J., Vongnititorn, P., ... Pintathong, P. (2019). Bioactive compounds and antioxidant activities from pomegranate peel and seed extracts. *Food and Applied Bioscience Journal*, *7*(3), 152–161.

Thomas, J. V., Rao, J., John, F., Begum, S., Maliakel, B., Krishnakumar, I. M., & Khanna, A. (2020). Phytoestrogenic effect of fenugreek seed extract helps in ameliorating the leg pain and vasomotor symptoms in postmenopausal women: A randomized, double-blinded, placebo-controlled study. *PharmaNutrition*, *14*, 100209.

Tijani, A. Y., Okhale, S. E., Salawu, T. A., Onigbanjo, H. O., Obianodo, L. A., Akingbasote, J. A., ... Emeje, M. (2009). Antidiarrhoeal and Antibacterial properties of crude aqueous stem bark extract and fractions of Parkia biglobosa (Jacq.) R. Br. Ex G. Don. *African Journal of Pharmacy and Pharmacology*, *3*(7).

Trisha, S. (2018). Role of hesperdin, luteolin and coumaric acid in arthritis management: A Review. *International Journal of Physiology, Nutrition and Physical Education*, *3*, 1183–1186.

Tsimogiannis, D., Choulitoudi, E., Bimpilas, A., Mitropoulou, G., Kourkoutas, Y., & Oreopoulou, V. (2017). Exploitation of the biological potential of Satureja thymbra essential oil and distillation by-products. *Journal of Applied Research on Medicinal and Aromatic Plants*, *4*, 12–20.

Van Wyk, A., & Prinsloo, G. (2018). Medicinal plant harvesting, sustainability and cultivation in South Africa. *Biological Conservation*, *227*, 335–342.

Vasudevan, K., Sophia, D., Balakrishanan, S., & Manoharan, S. (2007). Antihyperglycemic and antilipidperoxidative effects of *Ficus racemosa* (Linn.) bark extracts in alloxan induced diabetic rats. *Journal of Medical Sciences (Taipei, Taiwan)*, *7*, 330–338.

Veiga, M., Costa, E. M., Silva, S., & Pintado, M. (2018). Impact of plant extracts upon human health: A review. *Critical Reviews in Food Science and Nutrition*, 1–14. Available from https://doi.org/10.1080/10408398.2018.1540969.

Vijayakumar, K., Prasanna, B., Rengarajan, R. L., Rathinam, A., Velayuthaprabhu, S., & Vijaya Anand, A. (2020). Anti-diabetic and hypolipidemic effects of Cinnamon cassia bark extracts: An in vitro, in vivo, and in silico approach. *Archives of Physiology and Biochemistry*, 1–11.

Virtamo, J., Taylor, P. R., Kontto, J., Männistö, S., Utriainen, M., Weinstein, S. J., ... Albanes, D. (2014). Effects of α-tocopherol and β-carotene supplementation on cancer incidence and mortality: 18-Year postintervention follow-up of the Alpha-Tocopherol, Beta-Carotene Cancer Prevention Study. *International Journal of Cancer*, *135*(1), 178–185.

Vishnu, V.R., Renjith, R.S., Mukherjee, A., Anil, S.R., Sreekumar, J., Jyothi, A.N., 2019. Comparative Study on the Chemical Structure and In Vitro Antiproliferative Activity of Anthocyanins in Purple Root Tubers and Leaves of Sweet Potato (Ipomoea batatas). <https://doi.org/10.1021/acs.jafc.8b05473>.

Wei, C., Yuting, L., Tao, B., & Vemana, G. (2017). Mulberry fruit extract affords protection against ethyl carbamate-induced cytotoxicity and oxidative stress. *Oxidative Medicine and Cellular Longevity*. Available from https://doi.org/10.1155/2017/1594963.

Wu, H. C., Horng, C. T., Tsai, S. C., Lee, Y. L., Hsu, S. C., Tsai, Y. J., ... Yang, J. S. (2018). Relaxant and vasoprotective effects of ginger extracts on porcine coronary arteries. *International Journal of Molecular Medicine*, *41*(4), 2420–2428.

Xu, Y. B., Chen, G. L., & Guo, M. Q. (2019). Antioxidant and anti-inflammatory activities of the crude extracts of Moringa oleifera from Kenya and their correlations with flavonoids. *Antioxidants*, *8*(8), 296.

Yamamoto, J., Tadaishi, M., Yamane, T., Oishi, Y., Shimizu, M., & Kobayashi-Hattori, K. (2015). Hot water extracts of edible Chrysanthemum morifolium Ramat. exert antidiabetic effects in obese diabetic KK-Ay mice. *Bioscience, Biotechnology, and Biochemistry*, *79*(7), 1147–1154.

Yang, H., & Shin, Y. (2017). Antioxidant compounds and activities of edible roses (Rosa hybrida spp.) from different cultivars grown in Korea. *Applied Biological Chemistry*, *60*(2), 129–136.

Yang, P.-F., Yang, Y.-N., Feng, Z.-M., Jiang, J.-S., & Zhang, P.-C. (2019). Six new compounds from the flowers of Chrysanthemum morifolium and their biological activities. *Bioorganic Chemistry*, *82*, 139–144.

Yashin, A., Yashin, Y., Xia, X., & Nemzer, B. (2017). Antioxidant activity of spices and their impact on human health: A review. *Antioxidants*, *6*(3), 70.

Yi, Q. Y., Li, H. B., Qi, J., Yu, X. J., Huo, C. J., Li, X., ... Kang, Y. M. (2016). Chronic infusion of epigallocatechin-3-O-gallate into the hypothalamic paraventricular nucleus attenuates hypertension and sympathoexcitation by restoring neurotransmitters and cytokines. *Toxicology Letters*, *262*, 105–113.

Zarei, A., Ashtiyani, S.C., Taheri, S., & Ramezani, M. (2015). A quick overview on some aspects of endocrinological and therapeutic effects of *Berberis vulgaris L.* Avicenna Journal of Phytomedicine, 5(6), 485–497.

Zhang, H., Ma, Z. F., Luo, X., & Li, X. (2018). Effects of mulberry fruit (Morus alba L.) consumption on health outcomes: A mini-review. *Antioxidants*, *7*(5), 69.

Zhang, Y., Duan, W., Owusu, L., Wu, D., & Xin, Y. (2015). Epigallocatechin 3 gallate induces the apoptosis of hepatocellular carcinoma LM6 cells but not non cancerous liver cells. *International Journal of Molecular Medicine*, *35*(1), 117–124.

Zhang, Z. Y. (2003). An experimental study on the immune function of asparagus-extracted crude polysaccharides. *Journal of Zhengzhou College of Animal Husbandry & Engineering*, *23*(2), 83–84, in Chinese.

Zheng, J., Lu, B., & Xu, B. (2021). An update on the health benefits promoted by edible flowers and involved mechanisms. *Food Chemistry*, *340*, 127940.

Zhu, X., Bian, H., & Gao, X. (2016). The potential mechanisms of berberine in the treatment of non-alcoholic fatty liver disease. *Molecules (Basel, Switzerland)*, *21*, 1336.

Zhao Q, Xie B, Yan J, et al (2012) In vitro antioxidant and antitumor activities of polysaccharides extracted from Asparagus officinalis. Carbohydr Polym 87:392–396. https://doi.org/10.1016/J.CARBPOL.2011.07.068

CHAPTER

Opportunities and challenges of plant extracts in food industry 13

V. Geetha Balasubramaniam[1,*], Sudha Rani Ramakrishnan[1,*] and Usha Antony[1,2]

[1]Department of Biotechnology, Centre for Food Technology, Anna University, Chennai, India [2]College of Fish Nutrition and Food Technology, Dr. J. Jayalalithaa Fisheries University, Chennai, India

13.1 Introduction

Plant extracts derived from different parts of the plant are used for various food applications because of their preservative, aromatic, antimicrobial, and medicinal properties. Plant extracts are obtained from different sources and parts of the plant such as leaf, fruit, seed, bark, peel, stem, root, and flower. The crude extracts from plants are still in use in folk medicine and traditional systems of therapy, as well as food supplements. Modern techniques of isolation, purification, and incorporation enhance the use of plant extracts in the food industry.

Research over the last two decades points to reduced risk of several chronic diseases such as diabetes, cancer, and cardiovascular diseases due to the regular consumption of many plant extracts (Liu, 2004; Mir, Shah, Ganai, Ahmad, & Gani, 2019). The demand for phytomedicines and herbal extracts is increasing significantly in various applications, including functional foods, nutraceuticals, and health care, due to the growing awareness of their health benefits among consumers. Many studies corroborate this and indicate a high growth in this sector both in developed and developing nations. Given this scenario, the opportunities of plant extracts to be used in foods applications are increasing rapidly. However, there are several challenges, especially with respect to their stability, efficacy, and formulation during production. Furthermore, the interaction of plant extracts with other components of the food matrix and bioavailability along with required standards and regulations are also the challenging factors.

13.2 Opportunities

13.2.1 Prebiotics

Prebiotics are nondigestible and selectively fermentable ingredients that allow specific changes in the composition and/or activity of the gastrointestinal microbiota to confer benefits upon host health and well being (Gibson, Probert, Rastall, & Roberfroid, 2004). They are largely

*Both authors have contributed equally.

Plant Extracts: *Applications in the Food Industry*. DOI: https://doi.org/10.1016/B978-0-12-822475-5.00002-8

plant-derived components, which are widely used in bakery products, sports drinks, sugar-free confectionery, fermented milks and yogurts, baby foods, and chewing gum (Azmi, Mustafa, Hashim, & Manap, 2012). Prebiotics comprise of short-chain carbohydrates, mainly oligosaccharides, for example, fructooligosaccharides (FOS), xylooligosaccharides, galactooligosaccharides (GOS), human milk oligosaccharides (HMO) and polysaccharides like inulin (Panitantum, 2004). Prebiotics occur naturally in fruits and vegetables and small amounts are found in the form of free sugars or glycoconjugates in human milk and animal colostrums (Bucke & Rastall, 1990). However, compounds with prebiotic properties are not limited to pure carbohydrates and fibers but are also expanded to polyphenols.

Polyphenols are secondary plant metabolites that represent a large group of 8000 different compounds. They are structurally made up of one or more aromatic rings attached to one or more hydroxyl groups (González-Centeno et al., 2013). Polyphenols are abundantly found in fruits, vegetables, cereals, nuts, spices and herbs; the common among them are flavonoids and phenolic acids (Araújo, Gonçalves, & Martel, 2011). Dietary polyphenols are mostly found in glycosylated forms with sugar residues conjugated to a hydroxyl group or the aromatic ring. While their absorption is low, they are metabolized in the colon by the colonic microflora (Mocanu, Nagy, & Szöllosi, 2015).

The crude extracts obtained from several plant materials with rich source of phenolic compounds have potential application as preservatives and are also used in the development of several functional foods and nutraceuticals. Prebiotics can sustain high processing temperatures and are stable under low pH conditions making their use in foods highly feasible. Prebiotic fibers, oligosaccharides and polyphenols are incorporated in bakery foods, cereals, beverages, dairy products, and food supplements.

Among the prebiotics, the functional oligosaccharides like FOS, GOS, XOS, IMO, and HMO produced as purified compounds from their respective carbohydrate substrates have multiple food applications such as sugar replacers, fat replacers, and prebiotic and weaning foods (Singla & Chakkaravarthi, 2017). With respect to polyphenols as prebiotic agents, research has been carried out largely with crude extracts from olive pomace, grape seed extract, walnut husk, pomegranate peel, red chicory by products and various other sources. Novel prebiotics and their effects on rheological, textural, sensorial and nutritional profiling of nutraceuticals need to be studied for their usage in the food industry. Prebiotics may find interesting applications as energy bars and meal replacement shakes to increase the health component associated with them (Spacova et al., 2020).

Single compounds such as catechin, epigallocatechins, resveratrol, rosamarinic acid, and anthocyanins have also been studied in various foods for their physiochemical benefits (Xie, VanAlstyne, Uhlir, & Yang, 2017). There is evidence that crude plant extracts often have greater pronounced in vitro or/and in vivo prebiotic activity due to synergistic effects than the isolated constituents at an equivalent dose (Rasoanaivo, Wright, Willcox, & Gilbert, 2011). The activity depends on a number of factors: the polarity of the solvent used for extract and solubility of the targeted bioactive (Chemat et al., 2020).

13.2.2 Herbs

The demand for phytomedicines and herbal extracts are gaining tremendous popularity among the consumers due to their immense health benefits and the preference for the use of natural products rather than the synthetic ones. They are widely used in functional foods, health care, nutraceuticals, skincare, and cosmetics (Oreopoulou, Tsimogiannis & Oreopoulou, 2019). The leaves of many plants, including aloe vera, rosemary, basil, thyme, clove leaves are rich in antioxidants and also offer antimicrobial properties; which has led to their extensive application (Oreopoulou, Tsimogiannis & Oreopoulou, 2019).

13.2.2.1 Herbal polyphenols

Aromatic and medicinal herbs are rich sources of polyphenols; rosemary (*Rosmarinus officinalis*), sage (*Salvia officinalis*), winter savory (*Satureja thymbra*) oregano (*Origanum vulgare ssp. hirtum*), and marjoram (*Majorana syriaca*) are among the most promising sources of polyphenols (Oreopoulou, Tsimogiannis & Oreopoulou, 2019). All of these herbs belong to the *Lamiaceae* family and are used as spices or for the recovery of essential oils through the process of hydrodistillation. The use of spices and essential oils in active packaging of processed food has seen a rising trend in food technology (Fernández-López & Martos, 2018). Green tea has been considered as a good source of polyphenolic compounds; catechins (also known as flavanols), including catechin gallate, epicatechin, epigallocatechin, epicatechin-gallate, epigallocatechin-gallate, gallocatechin gallate, and gallocatechin are the dominant phenolic compounds that have varied food applications (Musial, Kuban-Jankowska, & Gorska-Ponikowska, 2020). Basil extracts containing high contents of Vitamin A, Vitamin K, Vitamin C, iron, magnesium, potassium, and calcium are used extensively in food and beverage applications (Fernández-López & Martos, 2018).

The dried herbs or the residues after essential oil recovery, currently disposed as waste, could be utilized to obtain natural extracts rich in phenolic compounds and with high antioxidant activity. Even though essential oils are promising alternatives to chemical preservatives, they have special limitations, such as low water solubility, high volatility, and strong odor, which must be addressed before application in food systems. (Fernández-López & Martos, 2018).

13.2.2.2 Herbal alkaloids

Alkaloids are a vast group of naturally occurring secondary metabolites synthesized by plants and animals. They are derived from amino acids and contain a nitrogen atom or atoms (amino or amido in some cases) in their structures, the reason behind the alkalinity of these compounds (Aniszewski, 2015). They are generally the main bioactive component of the plant that is isolated naturally or sometimes synthesized for their applications in the food and pharma industries. Many alkaloids are found as elements of the human diet such as coffee seeds (caffeine), tea leaves (theophylline, caffeine), cacao seeds (theobromine and caffeine), tomatoes (tomatine), and potatoes (solanine) (Aniszewski, 2015). Caffeine is the most common alkaloid having application as an ingredient of soft drinks like Coca-Cola to enhance their taste and in sport drinks. Some alkaloids are components of modern food and spice mixes. The black, green, white pepper (*Piper nigrum* L.), and long peppers (*Piper longum L.*) contains piperine that is widely used in food. Capsicum peppers such as Peruvian pepper (*Capsicum baccatum L.*), chili or red pepper (*Capsicum annuum L.*), bird pepper or tabasco (*Capsicum frutescens L.*), ají pepper (*Capsicum chinese Jacq.*), and rocoto pepper (*Capsicum pubescens Ruiz et Pav.*) have a wide range of food applications in food packaging and preservation (Aniszewski, 2015). Quinine is a well-known alkaloid with no color and a bitter taste; whose bio-activity is used in food as additional supplement to the refreshment drinks with water, sugar, carbon dioxide, and protectors E 320 and E 211 (Aniszewski, 2015). More common alkaloids are pyrrolizidine, quinolizidine, beta-carboline, ergot, and steroid alkaloids. Some alkaloids such as piperine, nicotine, theobromine, and tropane have low or negligible risk in the normal food chain. However, their use outside the normal food chain, in pure doses is not safe and reasonable only in clinical practice (Aniszewski, 2015). Scientists still keep trying to design and synthetize more and more semisynthetic and synthetic alkaloids derived from natural sources of alkaloids.

There are many more herbs used by local communities and ethnic groups in South American, Middle Eastern, Asian, African, and Oceania countries, which are used at domestic or community level, but not studied in depth. The scope for leveraging such plant sources to benefit producers and consumers is still large.

13.2.3 Spices

Spices are parts of plants used as colorants, preservatives, or medicine, due to their functional properties and many have a long history of use. Spices are known for culinary application as seasoning ingredients in various cultures, for example, garlic, onion, cinnamon, anise, clove, and red pepper are preferred seasoning agents of Chinese culture while coriander and black pepper are likely consumed in the East Indian region (Opara & Chohan, 2014).

The wider classification of spices come from *Monocotyledoneae* plants, such as garlic, ginger, turmeric and vanilla or from *Dicotyledoneae* plants, such as paprika, pepper, nutmeg, cinnamon, and clove (Chhetri et al., 2018). Spices are derived from bark, fruit, seeds, or leaves of plants and often contain spice-specific phytochemicals. Polyphenols are the major chemical compounds that are present in herbs and spices, which, especially in their dried forms, generally contain relatively high levels of polyphenols compared to other polyphenol rich foods including broccoli, dark chocolate, red, blue and purple berries, grape and onion (Pérez-Jiménez, Neveu, Vos, & Scalbert, 2010). Different spices such as clove bud, turmeric, celery, parsley, mint, rosemary, thyme, sage, dill, curry and ginger contain high levels of polyphenols (Mocanu et al., 2015).

13.2.3.1 Spice-based polyphenols and oleoresins

The application of spices as natural colorants is highly recommended against the chemical or synthetic forms. The spices tint in different colors from yellow and orange to different variations of red (except chlorophyll from herbs). Commonly used spices for coloring are red pepper, paprika, ginger, mustard, parsley and turmeric. The coloring property of spices is due to the lutein and neoxanthin and carotenoids, such as β carotene (Bartley & Scolnik, 1995). Other compounds that provide these coloring properties are flavonoids with yellow colors, curcumin with orange and chlorophyll with green (Peter & Shylaja, 2012). There are reports in literature on processing of spice spent residues into value-added products, viz. preparation of bakery products, snacks, and bio-films. Essential oils and spice extracts have been widely explored for shelf stability of shallow and deep-fried meat, raw and processed chicken, dried cured meat, fermented meat and meat sausages (Lu, Kuhnle, & Cheng, 2018). The application of some spices as preservatives in food has been evaluated to determine their efficiency, since spices are natural sources and offer an opportunity to replace synthetic preservatives in food, such as nitrates, which have been claimed to possess negative effects on human health (Anand & Sati, 2013). Some of the chief chemical compounds in spices are eugenol in clove, cinammaldehyde in cinnamon, cuminaldehyde in cumin, and curcumin in turmeric which have been proven to prevent food spoilage and inhibit the growth of pathogenic microorganisms. They are best known for their strong antioxidant properties that exceed the levels in most foods.

Spice oil is a spice derivative, a secondary metabolite of spices with spice flavor and fragrance properties that is extracted generally by steam distillation process (Singhal & Kulkarni, 2003). The various food applications of spice oils include beverages, cosmetics, perfumery and toiletries. Spice oleoresins are the concentrated liquid form obtained from spices with same character and properties of spices. Oleoresins are commonly used as food flavorings, liquid seasonings for improving the flavor, aroma and taste profile in foods by processing industries (Singhal & Kulkarni, 2003). They have many applications as a coloring agent in butter, cheese, meats, snack foods, and cereals; in frozen foods, soups, desserts, meat sauces, fish preserved in oil, or any prepared food where a more vibrant color is desired; in jellies, jams, and preparation of gelatin. Oleoresins are typically dispersed in a dry neutral carrier or liquid such as vegetable oil to the desired strength. About 80%–90% of spices left over as spent residue in the spice oil and oleoresin industries are not commercially exploited for food application and provide ample scope for investigation to identify bioactive that can find suitable applications in foods (Singhal & Kulkarni, 2003).

13.2.4 Whole extracts versus purified components

Pure or isolated drugs that are industrially produced may be chosen for their high activity against human disease; however, they have their own disadvantages. They hardly have the same degree of activity as the crude extract at comparable concentrations or dose of the active component (Wagner & Ulrich-Merzenich, 2009). This could be attributed to the absence of interacting co substances present in the extract.

The crude compounds showing increased activity relates to the evidence of synergy that several mechanisms may be operating in parallel, although the exact mechanisms have not yet been delineated. In pharmacodynamic synergy, various substances act at different receptor targets involved in the disease to improve the overall therapeutic effect. The substances with diminutive or no activity on the causative agent, support the main active principle to reach the target by improving its bioavailability, or by decreasing the metabolism and excretion.

13.2.4.1 Herbs and spices

Synergy between the different constituents of the crude extract has been reported for many pharmacological activities (Houghton, 2009). Curcumin, the bioactive compound from turmeric, that is, *Curcuma longa* root and the whole turmeric itself is reported to have traditional remedies for malaria and fever (Mishra, Dash, Swain, & Dey, 2009). When used in combination with artemisinin (derived from *Qinghao*, a plant called sweet wormwood), curcumin prevents the recrudescence of malarial parasites and death in animal models (Mishra et al., 2009).

However, when used in combination with *Andrographis paniculata* and *Hedyotis corymbosa* extracts, curcumin exhibited a clear synergistic effect in vitro *and* in vivo in rodent malaria models (Mishra et al., 2009). Curcumin alone has shown to have poor oral bioavailability due to glucuronidation in the small intestine, but piperine from black pepper (*P. nigrum* seeds) enhanced the bioavailability of curcumin by 2000% in humans, by inhibiting glucuronidation and slowing the gastrointestinal transit (Shoba et al., 1998). Piperine has also shown to improve the bioavailability of epigallocatechin-gallate (EGCG) which might improve its activity as a multidrug resistance inhibitor in vivo (Rasoanaivo et al., 2011).

There are several reports of synergy between the extracts of different plants, which are traditionally combined, but the mechanisms of the synergy have not yet been clarified (Rasoanaivo et al., 2011). More clinical trials of combinations of pure compounds (such as piperine + artemisinin + curcumin) and of combinations of crude extracts (such as *C. longa* root + *Artemisia annua* leaves + *Piper nigum* seeds) is fundamental to understand the interaction between plant constituents. Such research focus can enhance the activity of existing pharmaceutical preparations as well as improve the effectiveness of the existing herbal remedies for use in food and modern drug applications (Rasoanaivo et al., 2011).

Till date, the exploitation of spices for the recovery of polyphenols and their applications as antioxidants in food systems, food supplements, or cosmetics has been very scarce. Considering spices, the decoction or extracted crude compounds are mostly used. However, few isolated compounds like rosmarinic acid, several essential oils from thyme, savory, marjoram, cinnamon, oregano and garlic are used as additives in biodegradable films and coatings for active food packaging applications (Fernández-López & Martos, 2018).

13.2.4.2 Prebiotics—nondigestible oligosaccharides (fructooligosaccharides, galactooligosaccharides, and xylooligosaccharides)

Fractionation and purification procedures are mandatory for oligosaccharide, as the elimination of mono- and disaccharides fractions is required to evaluate and enhance their functional properties (e.g., in vitro prebiotic activity). Purification also helps in gaining enriched bioactive fractions for their use as food ingredients in specialized products (for individuals with disorders like diabetes, lactose intolerance, etc.), and in low calorie foods with a reduction of mono- and disaccharides. The most commonly used purification steps are filtration, centrifugation and precipitation. The purification of functional oligosaccharides based on adsorption techniques using various adsorbents such as: aluminum hydroxide or oxide, activated carbon, bentonite, silica, titanium and porous synthetic materials have been stated in literature (Qing, Li, Kumar, & Wyman, 2013).

However, despite its promise for industrial-scale purification and concentration of oligosaccharide mixtures, fractionation of oligosaccharide mixtures is still a challenge. Selective fermentation (bioconversion) can be a plausible technological alternative for purification of GOS using yeast strains from the genera *Saccharomyces* and *Kluyveromyces* (Guerrero et al., 2014). The basis of this strategy is the selective removal of the metabolizable sugars (monosaccharides plus lactose, or monosaccharides only) from raw GOS by yeast fermentation (bioconversion).

The above-mentioned techniques present advantages and disadvantages that must be taken into consideration in the design during the purification at an industrial scale. Although, effective removal of monosaccharides is a necessary step, still it will require an enzymatic pre-hydrolysis step with the consequent reduction in productivity and increase in cost. A financial evaluation of these strategies in commercial production will throw light as to their real industrial potential.

13.2.4.3 Polyphenols—quercitin, catechin, epigallocatechin, curcumin, capcisin, and allicin

To obtain a high-quality extract for its suitable use in the food, cosmetic, and pharmaceutical industries, the extract must be purified to remove all inert and undesirable components. Purification improves the functional properties of plant extracts and minimizes any taste, odor, and color (Peschel et al., 2006). It has been observed that the purified secondary metabolites are usually more active than the crude extracts, on condition that, there is no synergism within the mixture, so that it gets suppressed when molecular entities are separated and purified (Peschel et al., 2006).

To obtain a single or pure fraction of polyphenolic compounds, it is necessary that an effective purification step should be carried out after extraction to remove the impurities. The purification of active component depends on the structure, stability, and quantity of the compound. There are different approaches used to purify polyphenolic crude extract after extraction. The common technique is organic solvent extraction, followed by evaporation and reconstitution with distilled water. Solid phase extraction on columns such as Sephadex LH-20, Diaion HP-20, or C-18 are used extensively for purification (Suwal & Marciniak, 2018). Polyphenols can also be recovered with methanol or aqueous acetonitrile when using C-18 cartridges (Suwal & Marciniak, 2018).

The method of chromatographic fingerprinting using reverse-phase high-performance liquid chromatography helps in metabolite profiling of crude extracts. The highly complex vegetable matrix requires a high-resolution metabolite profiling as well as rapid fingerprinting of crude plant extracts which can be achieved by ultra-high-pressure liquid chromatography (UHPLC). UHPLC allows higher separation efficiency and resolution, increases the speed of analysis, lower solvent consumption and high sensitivity. With the use of HPLC micro-fractionation, the constituents can be fractionated and collected into 96 well microplates for further biological screening. The action observed in the microplate wells can be directly corresponded to the component in the chromatogram, which allows rapid localization and further scale-up purification (Grosso, Jäger, & Staerk, 2013).

A successful and simple method such as semipreparative high-performance liquid chromatography was carried out to purify tea catechins: catechin (C), epicatechin (EC), gallocatechin (GC), epigallocatechin (EGC), epigallocatechin-3-O-gallate (EGCG), epicatechin-3-O-gallate (ECG), and epigallocatechin-3-O-(3-O-methyl)-gallate (EGCG3"Me) (Gong et al., 2017). Literature have also shown that a high recovery efficiency of EGCG can be obtained by using cyclic ion exchange chromatography (IEC) techniques. The process uses minimal solvents (30% ethanol) as compared to conventional techniques (Acikara, 2013). Polyphenols such as epigallocatechin-gallate (EGCG), epicatechin-gallate (ECG), and epigallocatechin (EGC) were separated by using weakly acidic cation exchange gels (dextran based) from crude tea extracts without using any solvent (Banerjee & Chatterjee, 2015). Thin layer chromatography was observed as a simple, economical and efficient technique to isolate and purify spices such as curcuminoids, allicin, capsascin, and cinnamldehydes (Pawar, Gavasane, & Choudhary, 2018).

In this fast-growing field of bioactives from plant extracts, the opportunities are wide, with scope for utilization, once the scientific understanding is fairly comprehensive and due consideration is given to challenges arising.

13.3 Challenges

13.3.1 Food versus supplements

Functional foods (also referred to as foods for special dietary use/health supplements/nutraceuticals/similar such foods) should be distinguished from conventional foods, and therefore such products may be formulated as capsules, granules, jellies, liquids, powders, tablets, and other dosage forms, which are meant for oral administration (FSSAI – Food Safety & Standard Act, 2006).

According to different studies, the use of food supplements is mainly associated with being female, high education level, high household income, older age, and use of over-the-counter drugs (Timbo, Ross, McCarthy, & Lin, 2006). Despitethis, the consumers are now generally aware and careful about their health status and the growing use of botanicals has been associated with a possible risk of adverse effects, taking also into consideration the thousands of plant food supplements (PFS) present in the market. Although the incidence of mild to severe adverse effects is generally low (about 4% according to European and US data from surveys among consumers) and some plants have been traditionally used from thousands of years, many of them have not been tested in scientific studies and their efficacy and safety not established sufficiently (Lüde et al., 2016).

Functionality through randomized controlled trials (RCTs) in normal healthy and individuals with disease throws light on the use of plant extracts for specific purposes. Studies by Onakpoya, Davies, Posadzki, and Ernst (2013) and Lee, Chung, Fu, Choi, and Lee (2020) focused on (1) healthy individuals (less than 20% had cardiovascular disease/diabetes/people who were obese); (2) oral intake of IGOB131, a proprietary patented form of *Irvingia gabonensis* seed extract, or any other preparation of *I. gabonensis* seed extract; (3) outcomes related to weight (body weight, body fat, and waist circumference), cardiovascular biomarkers (high-density lipoproteins, low-density lipoproteins, total cholesterol, triglycerides, and blood pressure); and (4) parallel or crossover RCTs. RCTs of *I. gabonensis* seed extract supplementation on anthropometric measures and cardiovascular biomarkers were identified from 4 databases by Lee et al. (2020). Among 5 RCTs, 4 were rated with high risk of bias (ROB) assessment, and only 1 with low ROB. Random-effect *meta*-analysis of the 5 RCTs showed decrease in body weight, body fat, and waist circumference to *I. gabonensis* seed extract supplementation. However, RCT with low ROB did not have significantly different outcomes. Overall efficacy of *I. gabonensis* seed extract supplementation on weight loss seemed positive but was limited due to poor methodological quality and insufficient reporting of clinical trials. Further, high-quality RCTs are needed to determine the effectiveness of *I. gabonensis* seed extract supplement on weight-related health outcomes. A 2013 systematic review identified only 3 RCTs and concluded that *I. gabonensis* cannot be recommended for weight loss (Onakpoya et al., 2013).

Critchley, Zhang, Suthisisang, Chan, and Tomlinson (2000) suggested that the controls should be carefully chosen such that they closely match with the intervention group as well as standardized for factors such as color, odor, duration, and frequency of intake. Nevertheless, in certain natural products such as ginger, which has a peculiar odor, choosing a matching control is an uphill task. The few randomized, double-blind tests that attracted attention in mainstream publications were often questioned based on methodological grounds or interpretation in terms of reproducible results (Pittler, Abbot, Harkness, & Ernst, 2000). Jiménez, Delgado, and Muguerza (2004) reported that intake of Funciona (330 mL/day), a blend of antioxidant-rich fruit extracts enriched with antioxidant vitamins reduced oxidative stress in elderly people. Antioxidant supplementation showed optimum level of defense against the free radical generation associated with aging and exercise in 400 healthy aged (58–86 years) subjects on moderately intense exercise (50 min sessions, 3 sessions/week).

Camellia sinensis (green tea) has been associated with hepatotoxic effects mainly due to the intake of high amounts of epigallocatechin-3-gallate (EGCG). This effect is also mediated by the type of extraction used: hydroalcoholic extracts were more involved in adverse effects (Di Lorenzo et al., 2015).

13.3.2 Stability

Given the current trend of health promotion through diet, understanding the processing effects is critical for conserving active phytochemicals. Phytochemicals present in most foods are lost by heat processing such as dehydration, pasteurization, and sterilization. Thermal processing caused marked losses in total anthocyanins in blackberries (Hager, Howard, Liyanage, Lay, & Prior, 2008) and blueberries (Brownmiller, Howard, & Prior, 2008). It also decreased the biological activity of drumstick leaves extract (Arabshahi, Devi, & Urooj, 2007). However, no difference in activity of carrot tuber extracts was observed before and after heat treatment, while in some cases, processing induced the formation of novel compounds that either maintained or even increased the potential of various fruit and vegetable extracts (Nicoli, Anese, & Parpinel, 1999).

During thermal processing of food, total phenolic contents decreased by 20.21% on heating at 70°C for 30 min. Catechin was stable at room temperature, but on brewing at 98°C for 7 h, it degraded by 20% (Chen, Zhu, Tsang, & Huang, 2001). The degradation of gallic acid was 15% at 80°C after 4 h of exposure (Volf, Ignat, Neamtu, & Popa, 2014). Drying temperatures of 100 and 140°C reduced the total polyphenol content of red grape pomace peels by 18.6% and 32.6%, respectively (Larrauri, Rupérez, & Saura-Calixto, 1997). Storage conditions directly affected the polyphenol content due to hydrolysis, oxidation, and complexation (Zafrilla et al., 2003). At 18°C, 28°C, and 38°C, hydrocinnamic acid of orange juice decreased by 13%, 22%, and 32%, respectively after 6 months of storage (Klimczak, Małecka, Szlachta, & Gliszczyńska-Świgło, 2007). At 4°C, polyphenols were stable for longer periods, which was attributed to the inhibition of phenol oxidases at low temperature (Wei & Ying-tuan, 2008).

Significant changes in individual isoflavone levels were observed during the storage of ultra-high temperature processed chocolate flavored high protein beverage containing soy protein isolates depending on storage temperatures (4°C, 23°C, and 38°C), but total isoflavones remained the same irrespective of storage temperature or duration (Hayes, Unklesbay, & Grun, 2004). Hexane extract from *Garcinia* was more suitable in biscuit preparation than turmeric powder, as it retained high antioxidant activity after baking, followed by 2 months of storage (Nandita et al., 2009). Flower extracts of *Peltophorum ferrugineum* (Nanditha, Jena & Prabhasankar, 2009); and marjoram, spearmint, peppermint, and basil powders or their purified extracts (Bassiouny, Hassanien, Ali, & El-Kayati, 1990) were also found to retain their antioxidant activity during baking process. Biscuits with extracts of raisins, amla, and drumstick leaves were stable during 6 weeks of storage (Reddy, Urooj, & Kumar, 2005).

Ingredient formulators use spray-drying method to convert liquids into easy-to-handle powders and release of polyphenols without affecting their functional properties (Sansone et al., 2011). Maltodextrins are the commonly used coating material, especially for anthocyanins from sources such as apple pomace. However, inlet air temperature above 160°C–180°C can cause loss of anthocyanins (Tonon, Brabet, & Hubinger, 2010). Microwave-assisted extraction (MAE) is an advanced, efficient, and rapid method for extracting phenolic compounds. Extraction rate of flavonoids in pomegranate slag increased with increase in extraction time, which was highest at 6 min (Cai & Zhang, 2012). MAE can prevent the degradation of polyphenols due to the shorter extraction time.

In general, when pH value is lower, the stability of polyphenols will be greater. According to Chethan and Malleshi (2007), the phenolic contents in the millet seed coat extract remained constant at highly acidic to nearly neutral pH (6.5) but decreased as the alkalinity increased to pH 10. Tea catechins in aqueous solutions were stable at $pH < 4$ and unstable at $pH > 6$ (Ananingsih, Sharma, & Zhou, 2013), while *Galla chinensis* extracts with substantial amounts of tannins were unstable at neutral and alkaline conditions (Huang et al., 2012). In acidic aqueous solutions, anthocyanins exist in equilibrium as four main species namely, flavylium cation, quinonoidal base, pseudobase, and chalcone. At pH 1, flavylium cation (red color) is predominant. As pH increases, due to loss of proton, the quinonoidal base is formed. Between pH 5 and 6, pseudobase and chalcone are observed. At $pH > 6$, anthocyanins are degraded (Castañeda-Ovando, de Lourdes Pacheco-Hernández, Páez-Hernández, Rodríguez, & Galán-Vidal, 2009).

Duration and magnitude of high-pressure processing (HPP) have strong influence on polyphenols' stability. The polyphenols content did not change in pressure-treated strawberry puree although the total anthocyanins decreased (Marszałek, Mitek, & Skąpska, 2015). Cyanidin-3-O-glucoside (Cy3gl) subjected to HPP at 600 MPa and temperature of 70°C with longer holding time (6 h) reduced by 35% due to condensation of (Cy3gl) with pyruvic acid by formation of a new pyran ring by cycloaddition (Corrales, Butz, & Tauscher, 2008). Although HPP has a negative effect on stability, content, and antioxidant activity of polyphenols, it was slighter compared to thermal processing.

13.3.3 Interactions

A comprehensive understanding of the interaction between proteins and phenolic compounds as well as their characteristics is essential to develop the novel conjugates with improved functional and bioactive properties for better applications in food or related systems.

Basically, proteins and polyphenols can interact together via either noncovalent (hydrophobic, ionic, and hydrogen bonding) or covalent bonds (You, Luo, & Wu, 2014). However, the conjugates formed by covalent bonds are more preferably used in food applications owing to their stronger and more permanent interactions with high stability (Liu, Ma, Gao & McClements, 2017). Noncovalent protein–polyphenol interactions generally result from a combination of different interactions. Even though the bonds formed are potentially reversible and have low energy, the noncovalent protein–polyphenol interactions may play an important role in food industries for improvement of functional and quality of food products (Liu et al., 2017).

There are two main factors affecting the interaction between proteins and polyphenols. Those include extrinsic (pH and temperature) and intrinsic (structure and type of polyphenols and proteins) parameters (Czubinski & Dwiecki, 2017). These factors determine the protein–polyphenol conjugates formed by noncovalent or covalent interactions. Interaction of these components generally affects the functional attributes of food products and eating quality. The interaction between proteins and polyphenols plays an important role in quality improvement of certain food products. For example, protein–epigallocatechin-gallate (EGCG) conjugates have been reported to exhibit better antioxidant activity than unmodified proteins (Gu et al., 2017). Changes in hydrophilic/

hydrophobic balance of the protein might affect the solubility along with other important functional properties including emulsifying, foaming, and gelation properties of the protein–polyphenol conjugates (Rawel, Czajka, Rohn, & Kroll, 2002).

Thermal stability and mechanical properties of gelatine gel could be enhanced via the interaction between proteins and polyphenols (Maqsood, Benjakul, & Shahidi, 2013). An irreversible interaction of quinone with the sulfhydryl and amino groups of proteins and further condensation reactions of quinones will result in the formation of high molecular weight brown pigments. Reaction of quinone with amino group is also known to affect the digestibility and bioavailability of amino acids such as lysine and cysteine (Damodaran, 2008).

Larger phenolic compounds such as the arubigin and the aflavin from black tea showed higher preference for reaction toward milk proteins than flavanol (catechin) monomer because of higher binding sites available for interaction in case of large phenolic compounds (Dubeau, Samson, & Tajmir-Riahi, 2010). Most polyphenols have a distinctive astringency and bitterness that result from their interactions, particularly of procyanidins with glycoproteins of saliva (Dai & Mumper, 2010).

Addition of antioxidant extracts of rosemary in food products can lead to unacceptable flavor and aroma due to the presence of residual volatile compounds such as camphor, verbenone, and borneol (Carrillo & Tena, 2006). Arts et al. (2002) reported the outcome of interaction between flavonoids and proteins on the total antioxidant capacity. It was observed that addition of catechin to β-casein increased the antioxidant capacity of the β-casein solution.

Addition of silver nanoparticles (AgNPs) + guava leaves extract (GLE) solution to corn starch (CS) film caused structural changes between CS matrix through interactions among CS chains and components of the AgNPs + GLE solution (Fortunati et al., 2014).

The application of essential oils (EOs) in food may be limited due to changes in organoleptic and textural quality of food or interactions of EOs with food components (Devlieghere, Vermeulen, & Debevere, 2004). Accordingly, a challenge for practical application of EOs is to develop optimized low dose combinations to maintain product safety and shelf-life, thereby minimizing the undesirable flavor and sensory changes associated with the addition of high concentrations of EOs.

13.3.4 Toxicity

Although plant extracts are used in food applications, toxicological information such as acceptable dietary intake (ADI) and no-observed-adverse-effect level (NOAEL) are unavailable. Owing to problems in standardization of extracts and owing to batch-wise compositional variability, it becomes difficult to assign ADI or NOAEL. The marker compounds in extracts are affected by the variety of plant, geographical origin, plant part used, age, growth conditions, methods of extraction or drying, preparation, packaging, and storage. If the botanical ingredient existed in the market before October 1994, it is then exempted from the food additives category, and the generally recognized as safe (GRAS) status is not mandatory, according to the Dietary Supplement Health and Education Act (DSHEA – Dietary Supplement Health & Education Act, 1994). Some botanicals may not have a history of use as food ingredients but may be derived from sources that have been used in herbal medicinal products in various parts of the world. Examples include *Ginkgo biloba*, Ginseng extract, *Hypericum perforatum* (St. John's Wort). Further, materials with no history of human use such as, phytostanols derived as a by-product from wood, shikimic acid isolated from water-soluble extract of pine needles, may be considered for use in foods, if a checklist of tests to establish safety of phytochemicals added to foods is available (Negi, 2012).

The plant materials for use in food must be consistent with respect to quality and quantity of the active ingredient, and the method of preparation must meet good manufacturing practices. Risk assessment of natural products may require adequate specification of identity and composition as it may be the whole plant, extracts thereof, or purified components. Variability among plant sources, and the process used to obtain the constituents will be a limiting factor in adopting a generic approach to their risk assessment. The nature of the compound, prior knowledge of human consumption, likely exposure, and nutritional impact will determine the approach for toxicological testing of such compounds. Generally, for herbs or complex extracts, it is not possible to make a risk assessment based on a single active component, as more than one component may be of toxicological significance, and the food matrix may affect their bioavailability. A decision tree has been suggested as an aid to the safety evaluation process for plant material intended for food use (Walker, 2004), and general framework for safety assessment of botanicals has been described (Van den Berg, Serra-Majem, Coppens, & Rietjens, 2011).

An antioxidant-rich extract of *Phyllostachys nigra* (Lodd.) Munro leaves containing a high level of polyphenols has been reported to be nontoxic. Acute oral toxicity tests showed that the maximum tolerated dose was greater than 10 g/kg body weight in both rats and mice, without mutagenic effects. A sub-chronic administration for up to 90 days resulted in NOAEL at a dose of 4.30 g/kg per day (Lu, Wu, Tie, Zhang, & Zhang, 2005). When pregnant rats were treated with this extract at NOAEL dose, they did not show significant changes in fertility and gestation index, and there were no effects on embryo-fetal number, viability, sex ratio, and development observed (Lu, Wu, Shi, Dong, & Zhang, 2006).

Triterpenoid-rich extract from stem of *P. nigra* var. *henonis* (Mitford) Rendle showed no toxic effect. The oral maximum tolerated dose was over 10 g/kg body weight in both rats and mice. No mutagenicity was found by Ames, mouse bone marrow cell micronucleus, or mouse sperm abnormality tests. No abnormal symptoms, clinical signs or deaths were observed in rats during a 30-day sub-chronic feeding study using doses up to 830 mg/kg body weight per day. No abnormalities in organ development and hematological parameters were associated with feeding of this product (Zhang, Wu, Ren, Fu, & Zhang, 2004).

The *in silico* toxicophorical analyses were performed to screen the potential hazards of major compounds found in green coffee fruit extracts by Faria et al. (2020). Prediction results did not reveal any toxicological potential for most parameters assessed, with no detected risk of Ames toxicity, carcinogenicity, hERG inhibition, binding to the estrogen receptor, and skin sensitization. In contrast, only a hepatotoxicity risk was predicted for caffeine. Although no toxicity was predicted in the Ames test, which is a short-term bacterial reverse mutation assay, it is known that mutagenic compounds can be formed from trigonelline when green coffee is roasted (Wu, Skog, & Jägerstad, 1997). The compounds were also predicted as noninhibitors of the hERG channel, whose inhibition may lead to ventricular arrhythmia (Priest, Bell, & Garcia, 2008). A nondetected risk was predicted for skin sensitization, a potential adverse effect for dermally applied products (Pires, Blundell, & Ascher, 2015). No binding of compounds to the estrogen receptor was predicted, which may cause disruption of the endocrine system and reproductive toxicity (Shanle & Xu, 2011).

Plants producing pyrrolizidine alkaloids must be mentioned for their harmful effects on human health with hepatotoxic and carcinogenic effects. The risk of accidental contamination of other plants used for animal feed and human nutrition is concrete, as documented by several studies (Li, Xia, Ruan, Fu & Lin, 2011).

13.3.5 Regulations

As procurement of raw materials is the initial step and plays a crucial role in defining the quality of the final product, biological (bacteria, viruses, molds, and related toxins) and chemical (pesticides and heavy metals) contaminants must be stringently controlled. To reduce the risks for consumers, farmers should follow the rules of good agricultural practice for the use of pesticides or fertilizers or comply with the regulations for organic agriculture, depending on the agronomic strategy used (WHO – World Health Organization, 2004). Other factors related to raw materials and potentially contributing to adverse effects are the misidentification of botanicals used in PFS. The main problem is the use of common names instead of the binomial Latin names, as suggested by the Good Production Practices (Tankeu, Vermaak, Chen, Sandasi, & Viljoen, 2016).

Hundreds of plant derivatives are ingredients commonly used in food industry, and they are regulated by the food law as (1) functional food, (2) novel food, (3) traditional food from third countries, and (4) food/dietary supplements. Although there is no international agreement, according to the consensus document on "Scientific Concepts of Functional Foods in Europe" of the European Commission Concerted Action on Functional Food Science in Europe (FUFOSE), functional food is usually considered a product to which one or more ingredients have been added (or more rarely subtracted) with a positive consequence on the functionality of human organs or systems (EFSA – European Food Safety Authority & EFSA Scientific Committee, 2009). It is, therefore, a food that has not only the function of providing calories and nutrients but intends to carry out a favorable action on the consumer's health. This effect must be reached with the quantity of food normally consumed, it must be in "traditional" form (and not in pharmaceutical form) and must guarantee the safety of the subjects taking the product.

Colombo, Restani, Biella, and Di Lorenzo (2020) reviewed that there are numerous food supplements obtained from plants, and most of them are derived from the tradition of use, that is the preparation of infusions or decoctions. Plants have been used by the industries to prepare extracts, to enrich the products with active molecules and enhance the expected positive properties. Supplements may contain a single plant ingredient or their mixture. When greater number of botanicals are present in the finished product, complex problems on correct usage by the consumer and quality control issues arise.

Another critical aspect derives from the fact that there is no harmonization at the international level on the lists of plants allowed in food supplements. Many countries have published positive and/or negative lists of botanical ingredients, which are allowed or prohibited in food/dietary supplements. Unfortunately, there are few correlations between them. The inconsistent situation of the products with botanical ingredients leads to several consequences, including the difficulty in discriminating "healthy" or "therapeutic" information for the same plant used in different product classes. Researchers know that the dose makes the difference, but discriminating the dose with physiological effects from the one that is suitable for clinical applications is a complex objective even for plants of more ancient tradition. These difficulties are reflected in very small number of claims approved by European Food Safety Authority (EFSA) in the field of botanicals. To obtain authorization to associate a nutrition/health claim with a botanical in European Union (EU), the manufacturer must submit a dossier to EFSA, which evaluates the scientific evidence of the studies provided and publishes the relative opinion. The guidelines provided by EFSA – European Food Safety Authority, and EFSA Scientific Committee (2020) describes in detail the studies required for the approval of nutritional/physiological claims.

In the guidelines published by EFSA,

1. "For dietary supplements, food industries cannot claim any therapeutic effect, but only physiological ones"—As a consequence, apart from few exceptions, it is difficult to prove a healthy activity that reduces disease risk factors in the long term.
2. "To obtain statistical significance, it is necessary to recruit a very large "healthy" population that is willing to take a certain product for very long periods"—This is economically unsustainable, and the results could still be affected by the dietary habits of the subjects considered.

It seems unreasonable that the tradition of use has been accepted for traditional medicines and not for food supplements (Colombo et al., 2020).

13.3.6 Economic and ecological costs

Economic analyses performed by McNulty et al. (2020) indicates that the unit production cost of antimicrobial proteins (AMP) in plants at commercial scale is $3.00–6.88/g. The base case manufacturing facility scenario produces 500 kg of AMP per year at 92% purity including a 42% loss in extraction, downstream processing, and formulation. The base case manufacturing facility requires $50.1 million capital expenditure (CAPEX) and $3.44 million per year operating expenditure (OPEX). Upstream and downstream processes represent 58% and 42% of OPEX, respectively. Of the $2.01 million/year upstream OPEX, the seeding operation represents the majority (79%) of the cost. Chromatography (38%) and ultrafiltration (UF) operations (35%) represent most of the downstream processing OPEX of $1.43 million/year. The downstream CAPEX accounts for 62% of the overall CAPEX with the clarification and UF units representing the largest portion (49%) of the downstream capital investment costs. Downstream processing is the main contributor to cost of goods sold (COGS) at low production levels, while upstream processing contributes to COGS at high production levels. At 100 kg/year, downstream processing represents 64% ($8.51/g) of the COGS, while at 1000 kg/year the contribution is reduced to 35% ($2.15/g) of the COGS.

Economies of scale shows that unit material price decreases as yearly production increases. This becomes a more important consideration when evaluating COGS over a wide yearly production range. In addition, development and regulatory approval of any product to be added to food is a complex, lengthy, and usually costly process.

The utilization of plant extracts is a green and environmentally friendly approach because most of the phytochemicals are water-soluble metabolites like organic acids, quinones, phenolic compounds, flavonoids, alkaloids, catechins, terpenoids, coenzymes etc., including amino acids, plants derived proteins, polysaccharides, and vitamins. Because of their renewable nature, reliable, versatile, biodegradable, biocompatible, and ease of application, several plant extracts have been used earlier (Costa et al., 2015).

However, aspects such as toxicity and bioaccumulation need to be considered prior to their use. Moreover, the effect of solvents selected for extract preparation on the surrounding environment must be considered, as traditional extraction procedures require the use of highly harmful organic solvents. Supercritical fluids represent a new class of alternative solvents for the preparation of plant extracts which permit selective separation of phytochemicals from the extracts at moderate temperatures and optimum processing time (Mari, Bautista-Baños, & Sivakumar, 2016).

13.4 Conclusion

Plants of various species are sources of nourishment and therapeutic formulations for human populations since early times. Evidence indicates that extracts of edible and nonedible parts of plants, herbs, and spices such as leaves, stem, roots, flowers, fruits, and seeds have been in vogue. Aqueous extracts both cold and hot, as well as alcohol based are known, as are lipid based or ash based. A large number have been and continue to be used in traditional systems of medicine. While some are recorded, scientific validation in the current modern paradigm is relatively low. In recent times, the role of bioactive components from such sources, including micronutrients, phytochemicals, botanicals, and prebiotics have become an area of immense interest to health professionals and researchers, contributing to cumulative evidence of their role in health and disease, driving their application in foods and nutraceutical formulations/supplements. The opportunities are multifarious: (1) identification of plant varieties with high levels of the bioactives, (2) development of clean and sustainable extraction methodologies, (3) blending and formulation for optimal beneficial activities, stability, (4) optimization to enhance food quality and safety, (5) function as replacement of synthetic food additives, and (6) add value to existing foods or novel ones.

Leveraging these opportunities will demand appropriate means to address the several challenges. There is an urgent need to develop and harmonize methodologies for the evaluation of these components for their physiological and systemic benefits, while giving due credit to specific scientific issues. These include the level of bioactive component/nutrient; the matrix/food used for the formulation; selection of a representative sample of the study population; their exposure to the food/component; double blinding of both the subjects and the investigators. Continued and concerted efforts by all stake holders are the key to moving forward and deriving the benefits that can be transmitted to consumers.

References

Acikara, O. B. (2013). Ion-Exchange Chromatography and Its Applications, in, Martin D (Ed), Column Chromatography, IntechOpen, Avaliable from: https://doi.org/10.5772/47823, ISBN: 978-953-51-1074-3, eBook (PDF) ISBN: 978-953-51-6329-9. https://www.intechopen.com/books/3487

Anand, S. P., & Sati, N. (2013). Artificial preservatives and their harmful effects looking toward nature for safer alternatives. *International Journal of Pharmaceutical Sciences and Research*, *4*, 2496–2501.

Ananingsih, V. K., Sharma, A., & Zhou, W. (2013). Green tea catechins during food processing and storage: A review on stability and detection. *Food Research International*, *50*(2), 469–479.

Aniszewski, T. (2015). *Applied potential and current applications of alkaloids*. 10.1016/B978-0-444-59433-4.00006-7.

Arabshahi, D. S., Devi, D. V., & Urooj, A. (2007). Evaluation of antioxidant activity of some plant extracts and their heat, pH and storage stability. *Food Chemistry*, *100*, 1100–1105.

Araújo, J. R., Gonçalves, P., & Martel, F. (2011). Chemopreventive effect of dietary polyphenols in colorectal cancer cell lines. *Nutrition Research (New York, N.Y.)*, *31*, 77–87.

Arts, M. J., Haenen, G. R., Wilms, L. C., Beetstra, S. A., Heijnen, C. G., Voss, H. P., & Bast, A. (2002). Interactions between flavonoids and proteins: Effect on the total antioxidant capacity. *Journal of Agricultural and Food Chemistry*, *50*(5), 1184–1187.

Azmi, A. F. M. N., Mustafa, S., Hashim, D. M., & Manap, Y. A. (2012). Prebiotic activity of polysaccharides extracted from *Gigantochloa levis* (Buluh beting) shoots. *Molecules (Basel, Switzerland)*, *17*(2), 1635–1651.

Banerjee, S., & Chatterjee, J. (2015). Efficient extraction strategies of tea (*Camellia sinensis*) biomolecules. *Journal of Food Science and Technology*, *52*(6), 3158–3168.

Bartley, G. E., & Scolnik, P. A. (1995). Plant carotenoids: Pigments for photoprotection, visual attraction, and human health. *The Plant Cell*, *7*, 1027–1038.

Bassiouny, S. S., Hassanien, F. R., Ali, F. A. E. R., & El-Kayati, S. M. (1990). Efficiency of antioxidants from natural sources in bakery products. *Food Chemistry*, *37*(4), 297–305.

Brownmiller, C., Howard, L. R., & Prior, R. L. (2008). Processing and storage effects on monomeric anthocyanins, percent polymeric color, and antioxidant capacity of processed blueberry products. *Journal of Food Science*, *73*(5), H72–H79.

Bucke, C., & Rastall, R. A. (1990). Synthesising sugars by enzymes in reverse. *Chern Britain*, *26*, 675–678.

Cai, L., & Zhang, R. G. (2012). Microwave-assisted extraction of flavonoids from pomegranate pomace. *Academic Periodical of Farm Products Processing*, *11*, 53–56.

Carrillo, J. D., & Tena, M. T. (2006). Determination of volatile compounds in antioxidant rosemary extracts by multiple headspace solid-phase microextraction and gas chromatography. *Flavour and fragrance journal*, *21*(4), 626–633.

Castañeda-Ovando, A., de Lourdes Pacheco-Hernández, M., Páez-Hernández, M. E., Rodríguez, J. A., & Galán-Vidal, C. A. (2009). Chemical studies of anthocyanins: A review. *Food Chemistry*, *113*(4), 859–871.

Chemat, F., Abert, V. M., Fabiano-Tixier, A. S., Nutrizio, M., Režek, J. A., Munekata, P. E. S., Lorenzo, J. M., Barba, F. J., Arianna, B., & Giancarlo, C. (2020). A review of sustainable and intensified techniques for extraction of food and natural products. *Green Chemistry*, *22*(8), 2325–2353.

Chen, Z. Y., Zhu, Q. Y., Tsang, D., & Huang, Y. (2001). Degradation of green tea catechins in tea drinks. *Journal of Agricultural and Food Chemistry*, *49*(1), 477–482.

Chethan, S., & Malleshi, N. G. (2007). Finger millet polyphenols: Optimization of extraction and the effect of pH on their stability. *Food Chemistry*, *105*(2), 862–870.

Chhetri, P., Vijayan, A., Bhat, S., Gudade, B., & Bora, S. (2018). Projects: Spices and Plantation Crops. *An Overview of Grouping of Spices*.

Colombo, F., Restani, P., Biella, S., & Di Lorenzo, C. (2020). Botanicals in functional foods and food supplements: Tradition, efficacy and regulatory aspects. *Applied Sciences*, *10*(7), 2387.

Corrales, M., Butz, P., & Tauscher, B. (2008). Anthocyanin condensation reactions under high hydrostatic pressure. *Food Chemistry*, *110*(3), 627–635.

Costa, D. C., Costa, H. S., Albuquerque, T. G., Ramos, F., Castilho, M. C., & Sanches-Silva, A. (2015). Advances in phenolic compounds analysis of aromatic plants and their potential applications. *Trends in Food Science & Technology*, *45*(2), 336–354.

Critchley, J. A. J. H., Zhang, Y., Suthisisang, C. C., Chan, T. Y. K., & Tomlinson, B. (2000). Alternative therapies and medical science: Designing clinical trials of alternative/complementary medicines—Is evidence-based traditional Chinese medicine attainable? *The Journal of Clinical Pharmacology*, *40*(5), 462–467.

Czubinski, J., & Dwiecki, K. (2017). A review of methods used for investigation of protein–phenolic compound interactions. *International Journal of Food Science & Technology*, *52*(3), 573–585.

Dai, J., & Mumper, R. J. (2010). Plant phenolics: Extraction, analysis and their antioxidant and anticancer properties. *Molecules (Basel, Switzerland)*, *15*(10), 7313–7352.

Damodaran, S. (2008). *Amino acids, peptides and proteins* (Vol. 4, pp. 217–329). Boca Raton, FL: CRC Press.

Devlieghere, F., Vermeulen, A., & Debevere, J. (2004). Chitosan: Antimicrobial activity, interactions with food components and applicability as a coating on fruit and vegetables. *Food Microbiology*, *21*(6), 703–714.

Di Lorenzo, C., Ceschi, A., Kupferschmidt, H., Lüde, S., De Souza Nascimento, E., Dos Santos, A., ... Restani, P. (2015). Adverse effects of plant food supplements and botanical preparations: A systematic review with critical evaluation of causality. *British Journal of Clinical Pharmacology*, *79*(4), 578–592.

DSHEA – Dietary Supplement Health and Education Act. (1994). *United States Department of Health and Human Services*. Public Law, 103–417.

Dubeau, S., Samson, G., & Tajmir-Riahi, H. A. (2010). Dual effect of milk on the antioxidant capacity of green, Darjeeling, and English breakfast teas. *Food Chemistry*, *122*(3), 539–545.

EFSA – European Food Safety Authority, & EFSA Scientific Committee. (2009). Guidance on Safety assessment of botanicals and botanical preparations intended for use as ingredients in food supplements. *EFSA Journal*, *7*(9), 1249.

EFSA – European Food Safety Authority, & EFSA Scientific Committee. (2020). Available from: https://www.efsa.europa.eu/en/topics/topic/food-supplements (accessed 26.08 20).

Faria, W. C. S., Oliveira, M. G., da Conceição, E. C., Silva, V. B., Veggi, N., Converti, A., ... Bragagnolo, N. (2020). Antioxidant efficacy and in silico toxicity prediction of free and spray-dried extracts of green Arabica and Robusta coffee fruits and their application in edible oil. *Food Hydrocolloids*, 106004.

Fernández-López, J., & Martos Viuda, M. (2018). Introduction to the Special Issue: Application of Essential Oils in Food Systems. *Foods (Basel, Switzerland)*, *7*(4), 56. Available from https://doi.org/10.3390/foods7040056.

Fortunati, E., Rinaldi, S., Peltzer, M., Bloise, N., Visai, L., Armentano, I., ... Kenny, J. M. (2014). Nano-biocomposite films with modified cellulose nanocrystals and synthesized silver nanoparticles. *Carbohydrate Polymers*, *101*, 1122–1133.

FSSAI – Food Safety and Standard Act. (2006). Available from: https://archive.fssai.gov.in/home/fss-legislation/food-safetyand-standards-act.html (accessed 26.08.20).

Gibson, G. R., Probert, H. M., Rastall, R. A., & Roberfroid, M. B. (2004). Dietary modulation of the human colonic microbiota: Updating the concept of prebiotics. *Nutrition Research Reviews*, *17*, 259–275.

Gong, Z., Chen, S., Gao, J., Li, M., Wang, X., Lin, J., & Yu, X. (2017). Isolation and purification of seven catechin compounds from fresh tea leaves by semi-preparative liquid chromatography. *Se pu = Chinese Journal of Chromatography*, *35*(11), 1192–1197.

González-Centeno, M. R., Jourdes, M., Femenia, A., Simal, S., Rosselló, C., & Teissedre, P. L. (2013). Characterization of polyphenols and antioxidant potential of white grape pomace by products (*Vitis vinifera* L.). *Journal of Agricultural Food Chemistry*, *61*(47), 11579–11587.

Grosso, C., Jäger, A. K., & Staerk, D. (2013). Coupling of a high-resolution monoamine oxidase-A inhibitor assay and HPLC–SPE–NMR for advanced bioactivity profiling of plant extracts. *Phytochemical Analysis*, *24*(2), 141–147.

Gu, L., Peng, N., Chang, C., McClements, D. J., Su, Y., & Yang, Y. (2017). Fabrication of surface-active antioxidant food biopolymers: Conjugation of catechin polymers to egg white proteins. *Food Biophysics*, *12*(2), 198–210.

Guerrero, C., Vera, C., Novoa, C., Dumont, J., Acevedo, F., & Illanes, A. (2014). Purification of highly concentrated galacto-oligosaccharide preparations by selective fermentation with yeasts. *International Dairy Journal*, *39*(1), 78–88.

Hager, T. J., Howard, L. R., Liyanage, R., Lay, J. O., & Prior, R. L. (2008). Ellagitannin composition of blackberry as determined by HPLC-ESI-MS and MALDI-TOF-MS. *Journal of Agricultural and Food Chemistry*, *56*(3), 661–669.

Hayes, S. A., Unklesbay, N., & Grun, I. U. (2004). Isoflavone stability in chocolate beverages. *In Nutraceutical Beverages*, *871*, 189–199.

Houghton, P. J. (2009). Synergy and polyvalence: Paradigms to explain the activity of herbal products. *Evaluation of Herbal Medicinal Products*, *85*, 94.

Huang, X., Cheng, L., Exterkate, R. A. M., Liu, M., Zhou, X., Li, J., & Ten Cate, J. M. (2012). Effect of pH on Galla chinensis extract's stability and anti-caries properties in vitro. *Archives of Oral Biology*, *57*(8), 1093–1099.

Jiménez, R., Delgado, M. A., & Muguerza, B. (2004). Physical activity and antioxidant vitamins for the elderly: Beneficial effects of a long-term dietary treatment with antioxidant vitamins on the prooxidant effects of physical activity in a large-scale study in aged healthy people. *Free Radical Biology & Medicine*, *36*(Suppl 1), 16.

Klimczak, I., Małecka, M., Szlachta, M., & Gliszczyńska-Świgło, A. (2007). Effect of storage on the content of polyphenols, vitamin C and the antioxidant activity of orange juices. *Journal of Food Composition and Analysis*, *20*(3–4), 313–322.

Larrauri, J. A., Rupérez, P., & Saura-Calixto, F. (1997). Effect of drying temperature on the stability of polyphenols and antioxidant activity of red grape pomace peels. *Journal of Agricultural and Food Chemistry*, *45*(4), 1390–1393.

Lüde, S., Vecchio, S., Sinno-Tellier, S., Dopter, A., Mustonen, H., Vucinic, S., … Ceschi, A. (2016). Adverse effects of plant food supplements and plants consumed as food: Results from the poisons centres-based PlantLIBRA study. *Phytotherapy research*, *30*(6), 988–996.

Lee, J., Chung, M., Fu, Z., Choi, J., & Lee, H. J. (2020). The effects of *Irvingia gabonensis* seed extract supplementation on anthropometric and cardiovascular outcomes: A systematic review and *meta*-analysis. *Journal of the American College of Nutrition*, *39*(5), 388–396.

Liu, R. H. (2004). Potential synergy of phytochemicals in cancer prevention: Mechanism of action. *Journal of Nutrition*, *134*(12), 3479S–3485S.

Liu, F., Ma, C., Gao, Y., & McClements, D. J. (2017). Food-grade covalent complexes and their application as nutraceutical delivery systems: A review. *Comprehensive Reviews in Food Science and Food Safety*, *16*(1), 76–95.

Li, N., Xia, Q., Ruan, J., Fu, P. P., & Lin, G. (2011). Hepatotoxicity and tumorigenicity induced by metabolic activation of pyrrolizidine alkaloids in herbs. *Current Drug Metabolism*, *12*(9), 823–834.

Lu, B., Wu, X., Tie, X., Zhang, Y., & Zhang, Y. (2005). Toxicology and safety of anti-oxidant of bamboo leaves. Part 1: Acute and subchronic toxicity studies on anti-oxidant of bamboo leaves. *Food and Chemical Toxicology*, *43*(5), 783–792.

Lu, B., Wu, X., Shi, J., Dong, Y., & Zhang, Y. (2006). Toxicology and safety of antioxidant of bamboo leaves. Part 2: Developmental toxicity test in rats with antioxidant of bamboo leaves. *Food and Chemical Toxicology*, *44*(10), 1739–1743.

Lu, F., Kuhnle, G. K., & Cheng, Q. (2018). The effect of common spices and meat type on the formation of heterocyclic amines and polycyclic aromatic hydrocarbons in deep-fried meatballs. *Food Control*, *92*, 399–411.

Maqsood, S., Benjakul, S., & Shahidi, F. (2013). Emerging role of phenolic compounds as natural food additives in fish and fish products. *Critical Reviews in Food Science and Nutrition*, *53*(2), 162–179.

Mari, M., Bautista-Baños, S., & Sivakumar, D. (2016). Decay control in the postharvest system: Role of microbial and plant volatile organic compounds. *Postharvest Biology and Technology*, *122*, 70–81.

Marszałek, K., Mitek, M., & Skąpska, S. (2015). The effect of thermal pasteurization and highpressure processing at cold and mild temperatures on the chemical composition, microbial and enzyme activity in strawberry purée. *Innovative Food Science & Emerging Technologies*, *27*, 48–56.

McNulty, M. J., Gleba, Y., Tusé, D., Hahn-Löbmann, S., Giritch, A., Nandi, S., & McDonald, K. A. (2020). Techno-economic analysis of a plant-based platform for manufacturing antimicrobial proteins for food safety. *Biotechnology Progress*, *36*(1), e2896.

Mir, S. A., Shah, M. A., Ganai, S. A., Ahmad, T., & Gani, M. (2019). Understanding the role of active components from plant sources in obesity management. *Journal of the Saudi Society of Agricultural Sciences*, *18* (2), 168–176.

Mishra, K., Dash, A. P., Swain, B. K., & Dey, N. (2009). Anti-malarial activities of *Andrographis paniculata* and *Hedyotis corymbosa* extracts and their combination with curcumin. *Malaria Journal*, *8*, 26. Available from https://doi.org/10.1186/1475-2875-8-26.

Mocanu, M.-M., Nagy, P., & Szöllosi, J. (2015). Chemoprevention of breast cancer by dietary polyphenols. *Molecules (Basel, Switzerland)*, *20*, 22578–22620.

Musial, C., Kuban-Jankowska, A., & Gorska-Ponikowska, M. (2020). Beneficial properties of green tea catechins. *International journal of molecular sciences*, *21*(5), 1744. Available from https://doi.org/10.3390/ijms21051744.

Nanditha, B. R., Jena, B. S., & Prabhasankar, P. (2009). Influence of natural antioxidants and their carry-through property in biscuit processing. *Journal of the Science of Food and Agriculture*, *89*(2), 288–298.

Negi, P. S. (2012). Plant extracts for the control of bacterial growth: Efficacy, stability and safety issues for food application. *International Journal of Food Microbiology*, *156*(1), 7–17.

Nicoli, M. C., Anese, M., & Parpinel, M. (1999). Influence of processing on the antioxidant properties of fruit and vegetables. *Trends in Food Science & Technology*, *10*(3), 94–100.

Onakpoya, I., Davies, L., Posadzki, P., & Ernst, E. (2013). The efficacy of *Irvingia gabonensis* supplementation in the management of overweight and obesity: A systematic review of randomized controlled trials. *Journal of Dietary Supplements*, *10*(1), 29–38.

Opara, E. I., & Chohan, M. (2014). Culinary herbs and spices: Their bioactive properties, the contribution of polyphenols and the challenges in deducing their true health benefits. *International journal of molecular sciences*, *15*(10), 19183–19202.

Oreopoulou, A., Tsimogiannis, D., & Oreopoulou, V. (2019). *Extraction of Polyphenols from Aromatic and Medicinal Plants: An Overview of the Methods and the Effect of Extraction Parameters. in: Watson, Ronald Ross. Polyphenols in Plants, Isolation, Purification and Extract Preparation, 2nd edition*, pp. 243–259. UK (Academic Press).

Panitantum, V. (2004, January 26). *The story of probiotics, prebiotics and synbiotics*. A seminar presentation at Kasetsart University, Bangkok, under the auspices of BIOTEC. Bangkok, Thailand: National Science and Technology Development Agency, pp. 157–161.

Pawar, H., Gavasane, A., & Choudhary, P. (2018). A novel and simple approach for extraction and isolation of curcuminoids from turmeric rhizomes. *Advances in Recycling & Waste Management*, 06. Available from https://doi.org/10.4172/2475-7675.1000300.

Peschel, W., Sánchez-Rabaneda, F., Diekmann, W., Plesche, A., Gartzía, A., Jiménez, D., ... Codina, C. (2006). An industrial approach in the search of natural antioxidants from vegetable and fruit wastes. *Food Chemistry*, *97*(1), 137–150.

Peter, K. V., & Shylaja, M. (2012). *Introduction to herbs and spices: Definitions, trade and applications*. 10.1533/9780857095671.1.

Pires, D. E., Blundell, T. L., & Ascher, D. B. (2015). pkCSM: Predicting small-molecule pharmacokinetic and toxicity properties using graph-based signatures. *Journal of Medicinal Chemistry*, *58*(9), 4066–4072.

Pittler, M. H., Abbot, N. C., Harkness, E. F., & Ernst, E. (2000). Location bias in controlled clinical trials of complementary/alternative therapies. *Journal of Clinical Epidemiology*, *53*(5), 485–489.

Pérez-Jiménez, J., Neveu, V., Vos, F., & Scalbert, A. (2010). A systematic analysis of the content of 502 polyphenols in 452 foods and beverages—An application of the phenol-explorer database. *Journal of Agricultural and Food Chemistry*, *58*, 4959–4969.

Priest, B., Bell, I. M., & Garcia, M. (2008). Role of hERG potassium channel assays in drug development. *Channels*, *2*(2), 87–93.

Qing, Q., Li, H., Kumar, R., & Wyman, C. E. (2013). Xylooligosaccharides production, quantification, and characterization in context of lignocellulosic biomass pretreatment. In: *Aqueous pretreatment of plant biomass for biological and chemical conversion to fuels and chemicals*, pp. 391–415.

Rasoanaivo, P., Wright, C. W., Willcox, M. L., & Gilbert, B. (2011). Whole plant extracts vs single compounds for the treatment of malaria: Synergy and positive interactions. *Malaria Journal*, *10*(Suppl. 1), S4. Available from https://doi.org/10.1186/1475-2875-10-S1-S4.

Rawel, H. M., Czajka, D., Rohn, S., & Kroll, J. (2002). Interactions of different phenolic acids and flavonoids with soy proteins. *International Journal of Biological Macromolecules*, *30*(3-4), 137–150.

Reddy, V., Urooj, A., & Kumar, A. (2005). Evaluation of antioxidant activity of some plant extracts and their application in biscuits. *Food Chemistry*, *90*(1-2), 317–321.

Sansone, F., Mencherini, T., Picerno, P., d'Amore, M., Aquino, R. P., & Lauro, M. R. (2011). Maltodextrin/pectin microparticles by spray drying as carrier for nutraceutical extracts. *Journal of Food Engineering*, *105*(3), 468–476.

Shanle, E. K., & Xu, W. (2011). Endocrine disrupting chemicals targeting estrogen receptor signalling: Identification and mechanisms of action. *Chemical Research in Toxicology*, *24*(1), 6–19.

Shoba, G., Joy, D., Joseph, T., Majeed, M., Rajendran, R., & Srinivas, P. S. (1998). Influence of piperine on the pharmacokinetics of curcumin in animals and human volunteers. *Planta Medica*, *64*(4), 353–356.

Singhal, R. S., & Kulkarni, P. R. (2003). 18 - Herbs and spices. In M. Lees (Ed.), *Woodhead publishing series in food science, technology and nutrition, food authenticity and traceability* (pp. 386–414). Woodhead Publishing.

Singla, V., & Chakkaravarthi, S. (2017). Applications of prebiotics in food industry: A review. *Food Science and Technology International*, *23*(8), 649–667.

Spacova, I., Dodiya, H. B., Happel, A. U., Strain, C., Vandenheuvel, D., Wang, X., & Reid, G. (2020). Future of probiotics and prebiotics and the implications for early career researchers. *Frontiers in Microbiology*, *11*.

Suwal, S., & Marciniak, A. (2018). Technologies for the extraction, separation and purification of polyphenols—A review. *Nepal Journal of Biotechnology*, *6*(1), 74–91.

Tankeu, S., Vermaak, I., Chen, W., Sandasi, M., & Viljoen, A. (2016). Differentiation between two "fang ji" herbal medicines, *Stephania tetrandra* and the nephrotoxic *Aristolochia fangchi*, using hyperspectral imaging. *Phytochemistry*, *122*, 213–222.

Timbo, B. B., Ross, M. P., McCarthy, P. V., & Lin, C. T. J. (2006). Dietary supplements in a national survey: Prevalence of use and reports of adverse events. *Journal of the American Dietetic Association*, *106*(12), 1966–1974.

Tonon, R. V., Brabet, C., & Hubinger, M. D. (2010). Anthocyanin stability and antioxidant activity of spray-dried açai (*Euterpe oleracea* Mart.) juice produced with different carrier agents. *Food Research International*, *43*(3), 907–914.

Van den Berg, S. J., Serra-Majem, L., Coppens, P., & Rietjens, I. M. (2011). Safety assessment of plant food supplements (PFS). *Food & Function*, *2*(12), 760–768.

Volf, I., Ignat, I., Neamtu, M., & Popa, V. I. (2014). Thermal stability, antioxidant activity, and photo-oxidation of natural polyphenols. *Chemical Papers*, *68*, 121–129.

Wagner, H., & Ulrich-Merzenich, G. (2009). Synergy research: Approaching a new generation of phytopharmaceuticals. *Phytomedicine: International Journal of Phytotherapy and Phytopharmacology*, *16*(2–3), 97–110.

Walker, R. (2004). Criteria for risk assessment of botanical food supplements. *Toxicology Letters*, *149*(1-3), 187–195.

Wei, L., & Ying-tuan, Z. (2008). The antioxidation effect of *Pyracantha fortuneana*polyphenol in vitro. *Science and Technology of Food Industry* (9), 33.

WHO – World Health Organization. (2004). *WHO guidelines on safety monitoring of herbal medicines in pharmacovigilance systems*. Available from: https://apps.who.int/iris/handle/10665/43034 (accessed 26.08.20).

Wu, X., Skog, K., & Jägerstad, M. (1997). Trigonelline, a naturally occurring constituent of green coffee beans behind the mutagenic activity of roasted coffee? *Mutation Research/Genetic Toxicology and Environmental Mutagenesis*, *391*(3), 171–177.

Xie, J., VanAlstyne, P., Uhlir, A., & Yang, X. (2017). A review on rosemary as a natural antioxidation solution. *European Journal of Lipid Science and Technology*, *119*(6), 1600439.

You, J., Luo, Y., & Wu, J. (2014). Conjugation of ovotransferrin with catechin shows improved antioxidant activity. *Journal of Agricultural and Food Chemistry*, *62*(12), 2581–2587.

Zafrilla, P., Morillas, J., Mulero, J., Cayuela, J. M., Martínez-Cachá, A., Pardo, F., & &López Nicolás, J. M. (2003). Changes during storage in conventional and ecological wine: Phenolic content and antioxidant activity. *Journal of Agricultural and Food Chemistry*, *51*(16), 4694–4700.

Zhang, Y., Wu, X., Ren, Y., Fu, J., & Zhang, Y. (2004). Safety evaluation of a triterpenoid-rich extract from bamboo shavings. *Food and Chemical Toxicology*, *42*(11), 1867–1875.

Index

Note: Page numbers followed by "*f*" and "*t*" refer to figures and tables, respectively.

Printed in the United States
by Baker & Taylor Publisher Services